BIOMECHANICS
VIII-B

International Series on Biomechanics,
Volume 4B

BIOMECHANICS
VIII-B

**Proceedings of the Eighth International
Congress of Biomechanics
Nagoya, Japan**

EDITORS

Hideji Matsui, D.M.S.
Research Center of Health,
  Physical Fitness and Sports
Nagoya University
Nagoya, Japan

Kando Kobayashi, Ph.D.
Research Center of Health,
  Physical Fitness and Sports
Nagoya University
Nagoya, Japan

HUMAN KINETICS PUBLISHERS
Champaign, Illinois

Publications Director
Richard D. Howell

Production Director
Margery Brandfon

Editorial Staff
John Sauget, Copyeditor
Dana Finney, Proofreader

Typesetters
Sandra Meier
Carol McCarty

Text Layout
Denise Peters
Lezli Harris

Library of Congress Catalog Card Number: 82-84703

ISSN: 0360-344X
ISBN: 0-931250-42-0 (Two-vol. set)
ISBN: 0-931250-44-7 (Vol. B)

Human Kinetics Publishers, Inc.
Box 5076, Champaign, Illinois 61820

CONTENTS

CONTRIBUTORS

Adrian, Marlene J. (869, 883, 903), Department of Physical Education for Women, Biomechanics Laboratory, Washington State University, Pullman, Washington 99164, U.S.A.

Ae, Michiyoshi (648, 737, 762), Institute of Health and Sport Science, The University of Tsukuba, Sakuramura, Niihari-gun, Ibaraki 305, Japan

Amano, Yoshihiro (498, 663), Aichi University of Education, 1, Hirosawa, Igaya-cho, Kariya-shi, Aichi 448, Japan

Amidror, Itzhak (1207), Information and Computer Sciences, Toyohashi University of Technology, 1-1, Hibarigaoka, Tempaku-cho, Toyohashi, Aichi 440, Japan

Andersson, G.B.J. (386, 543), Departments of Clinical Neurophysiology and Orthopedic Surgery, Sahlgren Hospital, S-413 45 Göteborg, Sweden

Andrews, James G. (923, 939), Institute for Developmental Research, Aichi Prefectural Colony, Kamiyacho, Kasugai, Aichi 480-03, Japan

Antonsson, E. (1104), Department of Mechanical Engineering, Massachusetts Institute of Technology, Cambridge, Massachusetts 02139, U.S.A.

Aoki, Hisashi (223, 413), Department of Orthopaedic Surgery, Akita University, School of Medicine, 1-1-1, Hondo, Akita 010, Japan

Arai, Michio (380), Department of Orthopaedic Surgery, Akita University School of Medicine, Akita, Japan

Asami, Toshio (695), University of Tokyo, 3-8-1, Komaba, Meguro-ku, Tokyo 153, Japan

Azuma, Akira (35), Institute of Interdisciplinary Research, Faculty of Engineering, University of Tokyo, 4-6-1, Komaba, Meguro-ku, Tokyo 153, Japan

Balsevich, Vadim Konstantinovich (1032), Department of Biomechanics, Institute of Physical Culture, Maslennikova 144, 644063 Omsk, U.S.S.R.

Bates, Barry T. (251, 574, 635), Biomechanics/Sports Medicine Laboratory, Department of Physical Education, University of Oregon, Eugene, Oregon 97403, U.S.A.

Bauer, Wilhelm L. (801), Sensomotorik-Labor, Universität Bremen, Badgasteiner Strasse, 2800 Bremen 33, B.R.D.

Baumann, Wolfgang (722), Institut für Biomechanik, Deutsche Sporthochschule Köln, Carl-Diem-Wag, D-5000 Köln, B.R.D.

Bechtold, Joan E. (403), Department of Orthopaedics, Mayo Clinic, Rochester, Minnesota 55901, U.S.A.

Bengtzelius, Ulf (861), Chalmer's Institute of Technology, Göteborg, Sweden

Björk, Roland (386), Departments of
Clinical Neurophysiology and Ortho-
pedic Surgery, Sahlgren Hospital,
S-413 45 Göteborg, Sweden

Bober, Tadeusz (244, 1144), Academy
of Physical Education, Biomechanics
Laboratory, Al. Olimpijska 35,
51-612 Wrocław, Poland

Bobet, J. (1239), Department of Kinesiology, University of Waterloo,
Waterloo, Ontario, Canada N2L 3G1

Boon, K.L. (363), Department of
Anatomy and Biomechanics, Vrije
Universiteit, 1007 MC Amsterdam,
Pustbue 7161, The Netherlands

Bouisset, Simon (615), Université de
Paris-Sud, Laboratoire de Physiologie
de Mouvement, 91405 Orsay, France

Boukes, R.J. (363), Department of
Anatomy and Biomechanics, Vrije
Universiteit, 1007 MC Amsterdam,
Pustbue 7161, The Netherlands

Bourassa, Paul (582), Department of
Mechanical Engineering, Faculty of
Applied Science, University of Sher-
brooke, Sherbrooke, Province of
Quebec, Canada J1K 2R1

Bourgeois, Marc (978), Unité de
Recherche de Biomécanique du
Mouvement, Université Libre de
Bruxelles ISEPK-Laboratoire de l'Ef-
fort, C.P, 168, 28 av. Paul Heger
1050 Bruxelles, Belgium

Broeck, M. Van Den (951), Instituut
voor Morfologie, Experimental Anatomy, Faculty of Medicine, Vrije Universiteit, Brussels, Belgium

Brouwer, Hendrik Rogier (171),
Faculty of Medicine-Faculty of Ap-
plied Science, Vrije Universiteit
Brussel, Pleinlaan 2, B-1050 Brussels,
Belgium

Brüggemann, Peter (793), Institut für
Sport und Sportwissenschaften,

Ginnheimer Landstr. 39, 6 Frankfurt-
Main 90, B.R.D.

Bunch, Richard P. (1089), Biomechanics Laboratory, The Pennsylvania State University, University Park,
Pennsylvania 16802, U.S.A.

Burstein, A.H. (1075), Biomechanics
Department, Hospital for Special
Surgery, Cornell University, Medical
College, 535 East 70 Street, New
York 10021, U.S.A.

Cahill, Byron P. (403), Department
of Orthopedics, Mayo Clinic,
Rochester, Minnesota 55901, U.S.A.

Cardon, A. (171), Faculty of
Medicine-Faculty of Applied Science,
Vrije Universiteit Brussel, Pleinlaan
2, B-1050 Brussels, Belgium

Cappozzo, Aurelio (669, 1067),
Laboratorio di Biomeccanica, Instituto di Fisiologia, Umana Università degli Studi, Roma, Italy

Cavanagh, Peter R. (641, 928, 1081,
1089), Biomechanics Laboratory,
Pennsylvania State University, University Park, Pennsylvania 16802,
U.S.A.

Cerretelli, Paolo (703), Department
de Physiologie, Université de Genève,
20, rue de l'Ecole de Médecine, 1211
Genève 4, Switzerland

Chao, Edmund Y.S. (403, 490),
Department of Orthopedics, Mayo
Clinic, Rochester, Minnesota 55901,
U.S.A.

Chiba, G. (467, 485), Department of
Orthopaedic Surgery, Nagasaki Uni-
versity, School of Medicine, 7-1,
Sakamoto-machi, Nagasaki-shi,
Nagasaki 852, Japan

Chikama, H. (97, 110), Department
of Orthopaedic Surgery, Faculty of
Medicine, Kyushu University, 3-1-1,
Maidashi, Higashi-ku, Fukuoka, 812,
Japan

Cho, K. (648), Institute of Health and Sport Science, The University of Tsukuba, Sakura-mura, Niihari-gun, Ibaraki 305, Japan

Clarys, Jan P. (951), Instituut voor Morfologie—Experimental Anatomy —Faculty of Medicine, Vrije Universiteit, Laarbeeklaan 103, B-1090 Brussels, Belgium

Conati, E. (1104), Department of Mechanical Engineering, Massachusetts Institute of Technology, Cambridge, Massachusetts 02139, U.S.A.

Cornelis, Jan (986), Department of Electronics, Vrije Universiteit Brussel, Pleinlaan 2, 1050 Brussels, Belgium

Cotton, C.E. (553), Department of Kinanthropology, University of Ottawa, Ottawa, Ontario, Canada K1N 6N5

Crowninshield, R.D. (1023), Biomechanics Laboratory, Department of Orthopaedic Surgery, University of Iowa Hospitals, Iowa City, Iowa, U.S.A.

Dainty, David A. (553), Department of Kinanthropology, University of Ottawa, Ottawa, Ontario, Canada K1N 6N5

Dalrymple, G. (1104), Department of Mechanical Engineering, Massachusetts Institute of Technology, Cambridge, Massachusetts 02139, U.S.A.

Davis, Ken (915), School of Education, Deakin University, Victoria 3217, Australia

Docter G.L. (363), Department of Anatomy and Biomechanics, Vrije Universiteit, 1007 MC Amsterdam, Pustbue 7161, The Netherlands

Doi, T. (105), Department of Orthopaedic Surgery, Faculty of Medicine, University of Tokyo, 7-3-1, Hongo, Bunkyo-ku, Tokyo 113, Japan

Ebashi, Hiroshi (895), Physical Fitness Research Institute, Meiji Foundation of Health and Welfare, 1-1-18, Shiroganedai, Minato-ku, Tokyo 108, Japan

Ekblom, Berit (567), Department of Human Movement Studies, University of Queensland, St. Lucia Queensland 4067, Australia

Ekström, Hans (861), Department of Mechanical Engineering, Linköping Institute of Technology, Linköping University, S-581 83 Linköping, Sweden

Engin, Ali Erkan (125), Department of Engineering Mechanics, The Ohio State University, Boyd Laboratory, 155 West Woodruff Avenue, Columbus, Ohio 43210, U.S.A.

Enoka, R.M. (301), Department of Kinesiology, University of Washington, Seattle, Washington 98195, U.S.A.

Ensink, J. (363), Department of Anatomy and Biomechanics, Vrije Universiteit, 1007 MC Amsterdam, Pustbue 7161, The Netherlands

Fabian, D.F. (1075), Biomechanics Department, Hospital for Special Surgery, Cornell University, Medical College, 535 East 70 Street, New York 10021, USA

Felici, F. (669), Laboratorio di Biomeccanica, Instituto di Fisiologia Umana, Università degli Studi, Città Universitaria - 00100 Roma, Italy

Fidelus, Kazimirez (1175), Institute of Sport, Ceglowska Str., 68-70, 01-809 Warsaw, Poland

Figura, F. (669), Laboratorio di Biomeccanica, Institute di Fisiologia Umana, Università degli Studi, Città Universitaria - 00100 Roma, Italy

Fucci, S. (851), Institute of Sports

Medicine of C.O.N.I., Via dei Campi Sportivi 46, 00197 Roma, Italy

Fuchimoto, T. (754), Laboratory of Exercise Physiology, Osaka College of Physical Education, 1-1, Gakuen-cho, Ibaraki-shi, Osaka 567, Japan

Fujimaki, Etsuo (141, 162), Department of Orthopaedic Surgery, School of Medicine, Showa University, 1-5-8, Hatanodai, Shinagawa-ku, Tokyo 142, Japan

Fujimatsuo, H. (604), Laboratory for Exercise Physiology and Biomechanics, School of Physical Education, Chukyo University, 101, Tokodate, Kaizu-cho, Toyota, Aichi 470-03, Japan

Fujita, K. (604), Laboratory for Exercise Physiology and Biomechanics, School of Physical Education, Chukyo University, 101, Tokodate, Kaizu-cho, Toyota, Aichi 470-03, Japan

Fujita, Masaaki (467, 485), Department of Orthopaedic Surgery, Nagasaki University, School of Medicine, 7-1, Sakamoto-machi, Nagasaki-shi, Nagasaki 852, Japan

Fujiwara, Katsuo (209), Institute of Health and Sport Science, The University of Tsukuba, Sakura-mura, Niihari-gun, Ibaraki 305, Japan

Fukashiro, Senshi (258), Laboratory for Exercise Physiology, Faculty of Education, University of Tokyo, 7-3-1, Hongo, Bunkyo-ku, Tokyo 113, Japan

Fukubayashi, Toru (105), Department of Orthopaedic Surgery, Faculty of Medicine, University of Tokyo, 7-3-1, Hongo, Bunkyo-ku, Tokyo 113, Japan

Fukunaga, Tetsuo (676, 959), Department of Exercise Physiology and Sports Science, College of General Education, University of Tokyo, 3-8-1, Komaba, Meguro-ku, Tokyo 153, Japan

Funk, Sandy (869), Department of Physical Education for Women, Biomechanics Laboratory, Washington State University, Pullman, Washington 99164, U.S.A.

Furukawa, Ryozoh (503), Department of Orthopedic Surgery, Aichi Medical University, Nagakute-cho, Aichi-gun, Aichi 480-11, Japan

Furuya, K. (1198), Department of Orthopaedics Surgery, Tokyo Medical and Dental University, 1-5-45, Yushima, Bunkyo-ku, Tokyo 113, Japan

Gheluwe, B. Van (876, 986), Vrije Universiteit Brussel, H.I.L.O.K. Pleinlaan 2, 1050 Brussels, Belgium

Gollnick, Philip D. (9), Department of Physical Education for Men, Washington State University, Pullman, Washington 99164, U.S.A.

Goto, H. (419), Osaka City University, Osaka 558, Japan

Goto, Yukihiro (1097), Department of Physical Education, Osaka City University, Sumiyoshi, Osaka 558, Japan

Goya, Toshiaki (683), Aichi University of Education, 1, Hirosawa, Igaya-cho, Kariya-shi, Aichi 448, Japan

Grainger, J. (1239), Department of Kinesiology, University of Waterloo, Waterloo, Ontario, Canada N2L 3G1

Grieve, D.W. (527), Royal Free Hospital, School of Medicine, University of London, Biomechanics Laboratory, Department of Anatomy, Clinical Sciences Building, The Royal Free Hospital, Pond Street, London NW3 2QG, England

Hamill, J. (251, 635), Biomechanics/

Sports Medicine Laboratory, Department of Physical Education, University of Oregon, Eugene, Oregon 97403, U.S.A.

Hang, Yi-Shiong (70), National Taiwan University Hospital, Chang-te Street, Taipei, Taiwan, R.O.C.

Hasegawa, Tatsuhiko (89), Century Research Center Corporation, 4-68, Kitakyutaro, Higashi-ku, Osaka 541, Japan

Hashihara, T. (737), Institute of Health and Sport Science, The University of Tsukuba, Sakura-mura, Niihari-gun, Ibaraki 305, Japan

Hashimoto, Fujio (157, 440), College of General Education, Osaka Electro-Communication University, 18-8, Hatsu-machi, Neyagawa-shi, Osaka 572, Japan

Hatano, Izumi (157), Department of Orthopaedic Surgery, Kansai Medical University, 1, Fumizono-cho, Moriguchi-shi, Osaka 570, Japan

Hattori, Tomokazu (503), Department of Orthopedic Surgery, Aichi Medical University, Nagakute-cho, Aichi-gun, Aichi 480-11, Japan

Hay, James G. (923, 939), Department of Physical Education, University of Iowa, Iowa City, Iowa 52242, U.S.A.

Hayashi, Ryoichi (597, 971), Institute of Equilibrium Research, Gifu University School of Medicine, Tsukasa-machi 40, Gifu 500, Japan

Hayashi, S. (1198), Department of Orthopedic Surgery, Tokyo Medical and Dental University, 1-5-45, Yushima, Bunkyo-ku, Tokyo 113, Japan

Hayashi, T. (467, 485), Department of Orthopaedic Surgery, Nagasaki University, School of Medicine, 7-1, Sakamoto-machi, Nagasaki-shi, Nagasaki 852, Japan

Hennig, Ewald M. (1081, 1089), Biomechanics Laboratory, The Pennsylvania State University, University Park, Pennsylvania 16802, U.S.A.

Hermann, H. (1053), Deutsche Hochschule für Körperkultur, 7010 Leipzig, F.-L.-Jahn-Allee 59, D.D.R.

Himeno, Shinkichi (132), Department of Orthopedic Surgery, Fukuoka Children's Hospital, 2-5-1, Tojin-machi, Chuo-ku, Fukuoka 810, Japan

Hinrichs, Richard N. (641), Department of Physical Education, University of South Carolina, Columbia, South Carolina 29208, U.S.A.

Homma, Saburo (189, 444), Department of Physiology, School of Medicine, Chiba University, 1-8-1, Inohana, Chiba 280, Japan

Honjoh, Hiroshi (503), Department of Orthopedic Surgery, Aichi Medical University, Nagakute-cho, Aichi-gun, Aichi 480-11, Japan

Hori, Masami (503), Department of Orthopedic Surgery, Aichi Medical University, Nagakute-cho, Aichi-gun, Aichi 480-11, Japan

Hoshikawa, Tamotsu (498, 663, 683), Aichi Prefectural University, 3-28, Takada-cho, Mizuho-ku, Nagoya 467, Japan

Howard, A. (1223), Department of Human Movement Studies, University of Queensland, St. Lucia Queensland 4067, Australia

Hutton, R.S. (301), Department of Kinesiology, University of Washington, Seattle, Washington 98195, U.S.A.

Hyodo, K. (959), Department of Exercise Physiology and Sports Science, College of General Education, University of Tokyo, 3-8-1, Komaba, Meguro-ku, Tokyo 153, Japan

Igarashi, Hisato (787), Department of Health, Physical Education and Recreation, University of Oklahoma, 151 West Brooks, Norman, Oklahoma 73019, U.S.A.

Ikegami, Yasuo (963), Research Center of Health, Physical Fitness and Sports, Nagoya University, Furo-cho, Chikusa-ku, Nagoya 464, Japan

Inoue, Yoshimitsu (1097), Kobe University, School of Medicine, 7-12, Kusunoki-cho, Ikuta-ku, Kobe 650, Japan

Insall, J.N. (1075), Biomechanics Department, Hospital for Special Surgery, Cornell University, Medical College, 535 East 70 Street, New York 10021, U.S.A.

Ishii, Kihachi (239, 773), Laboratory of Physiological and Kinesiological Performance, Nippon College of Health and Physical Education, 7-1-1, Fukazawa, Setagaya-ku, Tokyo 158, Japan

Ishii, Nobuko (773), Laboratory of Physiological and Kinesiological Performance, Nippon College of Health and Physical Education, 7-1-1, Fukazawa, Setagaya-ku, Tokyo 158, Japan

Ishiko, Toshihiro (816), Juntendo University, 5-4-54, Fujisaki, Narashino-shi, Chiba 275, Japan

Ito, Akira (754), Laboratory of Exercise Physiology, Osaka College of Physical Education, 1-1, Gakuen-cho, Ibaraki-shi, Osaka 567, Japan

Ito, Masami (1129), Automatic Control Laboratory, School of Engineering, Nagoya University, Furo-cho, Chikusa-ku, Nagoya 464, Japan

Iwai, Akira (380), Miyagi Education

College, Aoba, Aramaki, Sendai-shi, Miyagi 980, Japan

Iwai, Takeshi (773), Laboratory of Physiological and Kinesiological Performance, Nippon College of Health and Physical Education, 7-1-1, Fukazawa, Setagaya-ku, Tokyo 158, Japan

Iwata, Kazuaki (1160), Department of Production Engineering, Faculty of Engineering, Kobe University, Rokko, Nada-ku, Kobe 657, Japan

Jack, Martha L. (889), P.O. Box 776, Richland, Washington 99352, U.S.A.

Jonsson, Bengt (561), Work Physiology Division, National Board of Occupational Safety and Health, Box 6104, S-900 06 Umeå, Sweden

Kakeno, Hidetatsu (591), Toyota Technical College, 2-1, Eisei-cho, Toyota-shi, Aichi 471, Japan

Kameyama, Osamu (157), Department of Orthopaedic Surgery, Kansai Medical University, 1, Fumizono-cho, Moriguchi-shi, Osaka 570, Japan

Kanehisa, Hiroaki (258), Laboratory for Exercise and Biomechanics, Faculty of Education, University of Tokyo, 7-3-1, Hongo, Bunkyo-ku, Tokyo 158, Japan

Kaneko, Masahiro (754), Laboratory of Exercise Physiology, Osaka College of Physical Education, 1-1, Gakuen-cho, Ibaraki-shi, Osaka 567, Japan

Karpeev, A.G. (1032), Department of Biomechanics, Institute of Physical Culture, Maslennikova 144, 644063 Omsk, U.S.S.R.

Kashiwase, Toshio (1011), Department of Electrical Engineering, Chiba University, 1-33, Yayoi-cho, Chiba 260, Japan

Katamoto, Shizuo (816), Juntendo

University, 5-4-54, Fujisaki, Narashino-shi, Chiba 275, Japan

Kato, Hisao (1167), Nagoya Municipal Industrial Research Institute, 3-24, Rokuban-cho, Atsuta-ku, Nagoya 456, Japan

Katoh, Yoshihisa (490), Department of Orthopedics, Mayo Clinic, Rochester, Minnesota 55901, U.S.A.

Kawabe, Shoko (231), Laboratory of Human Movements, Faculty of Letters, Nara Women's University, Nara 630, Japan

Kawachi, S. (1198), Department of Orthopedics Surgery, Tokyo Medical and Dental University, 1-5-45, Yushima, Bunkyo-ku, Tokyo 113, Japan

Kawahats, Kiyonori (289), Human Performance Laboratory, College of Liberal Arts, Kyoto University, Yoshida-Nihonmatsu-cho, Sakyo-ku, Kyoto 606, Japan

Kawai, Tadahiko (132), Institute of Industrial Science, Tokyo University, 7-22-1, Roppongi, Minato-ku, Tokyo 106, Japan

Kawano, Tsuneo (1160), Department of Production Engineering, Faculty of Engineering, Kobe University, Rokko, Nada-ku, Kobe 657, Japan

Kazei, Nobuyuki (419), Bukkyo University, 96, Kitahananobo-cho, Murasakino, Kita-ku, Kyoto 603, Japan

Kijima, Akira (239), Department of Literature, Laboratory of Physical Education, University of Kanto Gakuin, 4834, Mutsuura-cho, Kanazawa-ku, Yokohama-shi 236, Japan

Kinoshita, Hiroshi (574), Biomechanics/Sports Medicine Laboratory, Department of Physical Education, University of Oregon, Eugene, Oregon 97403, U.S.A.

Kira, H. (485), Department of Orthopaedic Surgery, Nagasaki University School of Medicine, 7-1, Sakamoto-nachi, Nagasaki 852, Japan

Kito, Nobukazu (498), Aichi University of Education, 1, Hirosawa, Igaya-cho, Kariya-shi, Aichi 448, Japan

Klinger, Anne K. (882), Clatsop Community College, 16th and Jerome Streets, Astoria, Oregon 97103, U.S.A.

Kojima, Takeji (321), Department of Physical Education, University of Tokyo, Komaba, Meguro-ku, Tokyo 153, Japan

Kondo, M. (959), Department of Exercise Physiology and Sports Science, College of General Education, University of Tokyo, 3-8-1, Komaba, Meguro-ku, Tokyo 153, Japan

Kornecki, Stefan (244), Academy of Physical Education, Biomechanics Laboratory, Al. Olimpijska 35, 51-612 Wrocław, Poland

Kulig, Kornelia (1144), Academy of Physical Education, Biomechanics Laboratory, Al. Olimpijska 35, 51-612 Wrocław, Poland

Kumamoto, Minoyori (157, 419, 440, 809, 828), College of Liberal Arts, Kyoto University, Yoshida-Nihonmatsu-cho, Sakyo-ku, Kyoto 606, Japan

Kuriyama, Setsuro (141), Department of Orthopaedic Surgery, School of Medicine, Showa University, 1-5-8, Hatanodai, Shinagawa-ku, Tokyo, Japan

Kurokawa, Takao (294), Department of Biophysical Engineering, Faculty of Engineering Science, Osaka University, Machikaneyama-cho, 1-1, Toyonaka, Osaka 560, Japan

Kurosawa, H. (105), Department of Orthopaedic Surgery, Faculty of Medicine, University of Tokyo, 7-3-1 Hongo, Bunkyo-ku, Tokyo 113, Japan

LaFortune, Mario A. (928), Biomechanics Laboratory, The Pennsylvania State University, University Park, Pennsylvania 16802, U.S.A.

Landjerit, B. (455), Laboratoire de Biomecanique, École Nationale Supérieure des Arts et Metiers, Paris, France

Lanshammar, Håkan (397, 1123), Institute of Technology, Uppsala University, Box 256, S-751 21, Uppsala, Sweden

Laughlin, Cynthia K. (903), Department of Physical Education for Women, Biomechanics Laboratory, Washington State University, Pullman, Washington 99164, U.S.A.

Laughman, R. Keith (403, 490), Department of Orthopedics, Mayo Clinic, Rochester, Minnesota 55901, U.S.A.

Leemputte, M.F. Van (264, 997, 1138), Laboratory for Biomechanics, Institut voor Lichamelijke Opleiding Katholieke Universiteit te Leuven Tervuursevest 101 B 3030 Heverlee, Belgium

Leo, Tommaso (1067), Instituto di Automatica, Universita degli Studi, Ancona, Italy

Leonardo, Maria (851), Institute of Sports Medicine of C.O.N.I., Via dei Campi Sportivi 46, 00197 Roma, Italy

Lewillie, Léon (978), Unité de Recherche de Bioméchanique du Movement, Université Libre de Bruxelles ISEPK—Laboratoire de l'Effort, C.P. 168, 28 av. Paul Héger B-1050 Bruxelles, Belgium

Macellari, Velio (1067), Laboratorio di Technologie Biomediche, Instituto

Superiore di Sanità, Roma, Italy

Macmillan, Norman H. (1081, 1089), Materials Research Laboratory, Pennsylvania State University, University Park, Pennsylvania 16802, U.S.A.

Maeshima, T. (816), Senshu University, 3-8, Kanda Jinbo-cho, Chiyoda-ku, Tokyo 101, Japan

Mann, Robert W. (1104, 1181), Department of Mechanical Engineering, Massachusetts Institute of Technology, 77 Massachusetts Avenue, Rm. 3-144, Cambridge, Massachusetts 02139, U.S.A.

Mano, T. (281), Department of Physiology, Hamamatsu University, School of Medicine, Hamamatsu 431-31, Japan

Marchetti, Marco (669), Laboratorio di Biomeccanica, Instituto di Fisiologia Umana, Università degli Studi, Città Universitaria—00100 Roma, Italy

Marhold, Gert (1053), Deutsche Hochschule für Körperkultur, 7010 Leipzig, F.-L.-Jahn-Allee 59, D.D.R.

Martin, E.E. (1032), Department of Biomechanics, Institute of Physical Culture, Maslennikova 144, 644063 Omsk, U.S.S.R.

Maruyama, Hirotake (419), Seibo Junior College, Kyoto 612, Japan

Mason, Michael (553), Department of Kinanthropology, University of Ottawa, Ottawa, Ontario, K1N 6NS Canada

Massez, C. (951), Instituut voor Morfologie—Experimental Anatomy—Faculty of Medicine, Vrije Universiteit, Laarbeeklaan 103, B-1090 Brussels, Belgium

Masuda, Makoto (423), Department of Physiology, The Jikei University, School of Medicine, 3-25-8,

Nishishinbashi, Minato-ku, Tokyo 105, Japan

Matake, Tomokazu (1115), Department of Mechanical Engineering, Nagasaki University, 1-14, Bunkyo-cho, Nagasaki 852, Japan

Matoba, Hideki (217), Laboratory of Biomechanics and Physiology, College of General Education, Yamaguchi University, 1-1677, Yoshida, Yamaguchi 753, Japan

Maton, Bernard (455), Laboratoire de Physiologie du Travail, CHU Pitié-Salpetrière, 91 Bd. de l'Hôpital, Paris 75634 cedex 13, France

Matsui, Hideji (3, 498, 683), Research Center of Health, Physical Fitness and Sports, Nagoya University, Furo-cho, Chikusa-ku, Nagoya 464, Japan

Matsuo, Akifumi (676, 959), Department of Exercise Physiology and Sports Science, College of General Education, University of Tokyo, 3-8-1, Komaba Meguro-ku, Tokyo 153, Japan

Matsuo, T. (373), Shinko-En Children Hospital, Kaminofu, Shingu, Kasuya 811-01, Japan

Matsusaka, N. (467, 485), Department of Orthopaedic Surgery, Nagasaki University, School of Medicine, 7-1, Sakamoto-machi, Nagasaki 852, Japan

McGill, S. (553), Department of Kinanthropology, University of Ottawa, Ottawa, Ontario, Canada KlN 6N5

Miki, Shunichiroh (503), Department of Orthopedic Surgery, Aichi Medical University, Nagakute-cho, Aichi-gun, Aichi 480-11, Japan

Miller, Doris I. (822), Department of Kinesiology, University of Washington, Seattle, Washington 98195, U.S.A.

Mimatsu, K. (223, 413), Institute for Developmental Research, Aichi Prefectural Colony, 713-8, Kamiya-cho, Kasugai, Aichi 480-03, Japan

Misaki, Norimasa (1160), Department of Production Engineering, Faculty of Engineering, Kobe University, Rokko, Nada, Kobe 657, Japan

Mishima, Ken (294), Department of Biophysical Engineering, Faculty of Engineering Science, Osaka University, 1-1, Machikaneyama-cho, Toyonaka, Osaka 560, Japan

Mita, Katsumi (223, 413), Institute for Developmental Research, Aichi Prefectural Colony, 713-8, Kamiya-cho, Kasugai, Aichi 480-03, Japan

Mita, Tsutomu (1011), Department of Electrical Engineering, Chiba University, 1-33, Yayoi-cho, Chiba 260, Japan

Mitarai, Genyo (281), Department of Aerospace Physiology, Research Institute of Environmental Medicine, Nagoya University, Furo-cho, Chikusa-ku, Nagoya 464, Japan

Miyake, Akihide (597, 971), Institute of Equilibrium Research, Gifu University, School of Medicine, Tsukasa-machi 40, Gifu 500, Japan

Miyamura, Miharu (963), Research Center of Health, Physical Fitness and Sports, Nagoya University, Furo-cho, Chikusa-ku, Nagoya 464, Japan

Miyanaga, Yutaka (105), Department of Orthopaedic Surgery, Faculty of Medicine, University of Tokyo, 7-3-1, Hongo, Bunkyo-ku, Tokyo 113, Japan

Miyashita, Mitsumasa (180, 258, 480, 629, 842), Laboratory for Biomechanics and Exercise Physiology, Faculty of Education, University of Tokyo, 7-3-1, Hongo, Bunkyo-ku, Tokyo 113, Japan

Miyozaki, M. (444, 467), Department of Orthopaedic Surgery, Nagasaki University, School of Medicine, 7-1, Sakamoto-machi, Nagasaki 852, Japan

Mizuno, Yoshio (597), Institute of Equilibrium Research, Gifu University, School of Medicine, 40, Tsukasa-machi, Gifu 500, Japan

Mizutani, Shiro (663), Mie University, 1515, Uehama-cho, Tsu-shi 514, Japan

Mizutani, Y. (380), Department of Orthopaedic Surgery, Akita University, School of Medicine, 1-1-1, Hondo, Akita 010, Japan

Monte, A. Dal (851), Institute of Sports Medicine of C.O.N.I. Via dei Campi Sportivi 46, 00197 Roma, Italy

Morecki, Adam (341), International Federation for the Theory Machines and Mechanisms, Central Office, Al. Neipodleglosci, 222 R. 206, 00-663 Warszawa, Poland

Mori, Takemi (180), Department of Orthopaedic Surgery, Tokyo Welfare-Pension Hospital, 23, Tsukudo-cho, Shinjuku-ku, Tokyo 162, Japan

Morimoto, Shigeru (423), Department of Physiology, The Jikei University School of Medicine, 3-25-8, Nishishinbashi, Minato-ku, Tokyo 105, Japan

Moritani, Toshio (312, 432), Bio-dynamics Laboratory, Department of Physical Education, University of Texas at Arlington, Arlington, Texas 76019, U.S.A.

Moriwaki, Toshimichi (1160), Department of Production Engineering, Faculty of Engineering, Kobe University, Rokko, Nada, Kobe 657, Japan

Morrey, B.F. (490), Department of Orthopedics, Mayo Clinic, Rochester, Minnesota 55901, U.S.A.

Morrison, W. (553), Department of Kinanthropology, University of Ottawa, Ottawa, Ontario, Canada K1N 6N5

Munro, A.R. (306), Department of Human Movement and Recreation Studies, University of Western Australia, Nedlands, W.A. 6009, Australia

Murakami, Naotoshi (217), Department of Physiology, Yamaguchi University, School of Medicine, 1144, Oguchi, Ube-shi, Japan

Murakami, Teruo (97, 110), Department of Mechanical Engineering, Kyushu University, 6-10-1, Hakozaki, Higashi-ku, Fukuoka 812, Japan

Muraki, Yukito (762), Institute of Health and Sport Science, University of Tsukuba, Sakura-mura, Niihari-gun, Ibaraki 305, Japan

Murase, Ken-ichi (1023), Department of 2nd Physiology, Fukushima Medical College, 5-75, Sugizuma-cho, Fukushima 960, Japan

Muro, Masuo (312, 432), Tokyo College of Pharmacy, 1432-1, Horinouchi, Hachioji-shi, Tokyo 192-03, Japan

Mutoh, Yoshiteru (165, 180), Department of Orthopaedic Surgery, Tokyo Welfare-Pension Hospital, 23, Tsukudo-cho, Shinjuku-ku, Tokyo 162, Japan

Nachemson, A.L. (543), Departments of Clinical Neurophysiology and Orthopedic Surgery, Sahlgren Hospital, Göteborg, Sweden

Nagata, Akira (412, 432), Laboratory of Bio-dynamics, Faculty of Science, Tokyo Metropolitan University, 1-1-1, Yakumo, Meguro-ku, Tokyo 146, Japan

Nakagawa, Hiroshi (419), Osaka University of Economics, Higashiyodogawa-ku, Osaka 533, Japan

Nakagawa, N. (809), Osaka University of Economics, Osaka 533, Japan

Nakamura, Yoshio (157, 180), Laboratory for Exercise, Physiology and Biomechanics, Faculty of Education, University of Tokyo, 7-3-1, Hongo, Bunkyo-ku, Tokyo 113, Japan

Narikiyo, Tatsuo (1129), Automatic Control Laboratory, School of Engineering, Nagoya University, Furo-cho, Chikusa-ku, Nagoya 464, Japan

Nicol, Klaus (1231), Institut für Leibesübungen, Westfälische Wilhelms–Universität Münster, Horstmarer Landweg 62B, 4400 Münster, B.R.D.

Nigg, Benno M. (1041), Biomechanics Laboratory, Faculty of Physical Education, The University of Calgary, 2500 University Drive N.W., Calgary, Alberta, Canada T2N 1N4

Niinomi, Shigeru (1215), Labor Accident Prosthetic and Orthotics Center, 1-10-5, Komei, Minato-ku, Nagoya 455, Japan

Nishio, A. (97, 110), Department of Orthopaedic Surgery, Faculty of Medicine, Kyushu University, 3-1-1, Maidashi, Higashi-ku, Fukuoka 812, Japan

Nishizaki, Hiromi (97, 110), Department of Orthopaedic Surgery, Faculty of Medicine, Kyushu University, 3-1-1, Maidashi, Higashi-ku, Fukuoka 812, Japan

Nissinen, Mauno A. (781), Institut für Sport und Sportwissenschaften, Ginnheimer Landstr. 39, 6 Frankfurt/Main 90, B.R.D.

Niwa, Shigeo (503), Department of Orthopaedic Surgery, Aichi Medical University, 21, Nagakute-cho, Aichi-gun, Aichi 480-11, Japan

Nomura, Haruo (1160), Department of Health and Physical Education, Faculty of Liberal Arts, Kobe University, Rokko, Nada, Kobe 657, Japan

Nomura Takeo (842), Institute of Sports Science, The University of Tsukuba, Sakura-mura, Niihari-gun, Ibaraki 305, Japan

Norimatsu, Toshiharu (467, 485), Department of Orthopaedic Surgery, Nagasaki University, School of Medicine, 7-1, Sakamoto-machi, Nagasaki 852, Japan

Norman, Robert William (1239), Department of Kinesiology, University of Waterloo, Waterloo, Ontario, Canada N2L 3G1

Notte, Volker (695), Institut für Biomechanik, Deutsche Sporthochschule, Carl-Diem-Wag, D-5000, Köln, B.R.D.

Nowacki, Zbigniew (1144), Academy of Physical Education, Biomechanics Laboratory, Al. Olimpijska 35, 51-612, Wrocław, Poland

Nyssen, M. (986), Medical Informatics, Vrije Universiteit Brussel, Pleinlaan 2, 1050 Brussels, Belgium

Ohenheimer, D. (1104), Department of Mechanical Engineering, Massachusetts Institute of Technology, Cambridge, Massachusetts 02139, U.S.A.

Ohkuwa, Tetsuo (963), Nagoya Institute of Technology, Gokiso-cho, Showa-ku, Nagoya 460, Japan

Ohmichi, Hitoshi (258, 480), Laboratory for Biomechanics and Exercise Physiology, Faculty of Education, University of Tokyo, 7-3-1, Hongo, Bunkyo-ku, Tokyo 113, Japan

Ohtsuki, N. (97, 110), Department of Mechanical Engineering, Kyushu University, 6-10-1, Hakozaki, Higashi-ku, Fukuoka 812, Japan

Ohtsuki, Tatsuyuki (231), Laboratory

of Human Movements, Faculty of Letters, Nara Women's University, Nara 630, Japan

Oka, Hideo (157, 419, 809), Osaka Kyoiku University, High School, 1-5-1, Midorigaoka, Ikeda-shi, Osaka 563, Japan

Okada, Morihiko (209), Institute of Health and Sport Science, The University of Tsukuba, Sakura-mura, Niihari-gun, Ibaraki 305, Japan

Okamoto, Tsutomu (157, 419, 809, 829), Kansai Medical School, 18-89, Uyamahigashi-cho, Hirakata-shi, Osaka 573, Japan

Okawa, Yoshikuni (147), Faculty of Engineering, Gifu University, 3-1, Naka-monzen-cho, Kakamigahara, Gifu 504, Japan

Okumura, H. (485), Department of Radiation Biophysics, Atomic Disease Institute, Nagasaki University, School of Medicine, 7-1, Sakamoto-machi, Nagasaki 852, Japan

Okumura, Shinji (1198), Department of Orthopedic Surgery, Tokyo Medical and Dental University, 1-5-45, Yushima, Bunkyo-ku, Tokyo 113, Japan

Oonishi, Hironobu (89), Department of Orthopedic Surgery, Osaka-Minami National Hospital, 677-6, Kido-cho, Kawachinagano, Osaka, 586, Japan

Örtengren, Roland (386, 543), Departments of Clinical Neurophysiology and Orthopedic Surgery, Sahlgren Hospital, S-413 45 Göteborg, Sweden

Osternig, L.R. (251, 635), Biomechanics/Sports Medicine Laboratory, Department of Physical Education, University of Oregon, Eugene, Oregon 97403, U.S.A.

Otis, James C. (1075), Biomechanics Department, Hospital for Special

Surgery, Cornell University, Medical College, 535 East 70 Street, New York 10021, U.S.A.

Ottonsson, Stig (861), Department of Mechanical Engineering, Linköping Institute of Technology, Linköping University, S-581 83 Linköping, Sweden

Payne, Andrew H. (746), Physical Education Department, University of Birmingham, Birmingham, P.O. Box 363, Birmingham B15 2TT, England

Pedersen, D.R. (1023), Biomechanics Laboratory, Department of Orthopaedic Surgery, University of Iowa Hospitals, Iowa City, Iowa 52242, U.S.A.

Peeraer, L. (1138), Laboratory for Biomechanics, Instituut voor Lichamelijke Opleiding, Katholieke Universiteit te Leuven, Tervuursevest, 101, B-3030 Heverlee, Belgium

Peres, G. (455), Laboratoire de Physiologie du Travail du CNRS, CHU Pitie-Salpetriere 91, Boulevard de l'Hôpital, Paris 756 34 cedex 13, France

Persyn, U. (833), Institute of Physical Education, Katholieke Universiteit, Leuven, Tervuursevest 101, B-3030 Heverlee, Belgium

Philippe, C. (455), Laboratoire de Biomécanique, Ecole Nationale Superieure des Arts et Métiers, Paris, France

Piette, G. (951), Instituut voor Morfologie — Experimental Anatomy — Faculty of Medicine, Vrije Universiteit, Laarbeeklaan 103, B-1090 Brussels, Belgium

Prampero, Pietro E. di (703), Centro Studi di Fisiologia del Lavoro Muscolare del C.N.R., Milano, Italy

Purcell, Michael (869), Department of Physical Education for Women,

Biomechanics Laboratory, Washington State University, Pullman, Washington 99164, U.S.A.

Putnam, Carol A. (688), School of Physical Education, Dalhousie University, Halifax, Nova Scotia, Canada B3H 3J5

Pyke, F.S. (306), Department of Sports Studies, Camberra College of Advanced Education, University of Western Australia, Nedlands, Western Australia 6009, Australia

Rau, Günter (513), Helmholtz-institute for Biomedical Engineering, Aachen, B.R.D.

Robeaux, R. (951), Instituut voor Morfologie — Experimental Anatomy — Faculty of Medicine, Vrije Universiteit, Laarbeeklaan 103, B-1090 Brussels, Belgium

Rossignol, P.F. (306), Department of Human Movement and Recreation Studies, University of Western Australia, Nedlands, Western Australia 6009, Australia

Rowell, Derek (1104, 1181), Department of Mechanical Engineering, Massachusetts Institute of Technology, 77 Massachusetts Avenue, Rm. 3-144, Cambridge, Massachusetts 02139, U.S.A.

Ryushi, T. (959), Department of Exercise Physiology and Sports Science, College of General Education, University of Tokyo, 3-8-1, Komaba, Meguro-ku, Tokyo 153, Japan

Saeki, C. (373), Shinko-En Children's Hospital, Kaminofu, Shingu, Kasuya 811-01, Japan

Sagawa, Kazunori (239), Laboratory of Physiological and Kinesiological Performance, Nippon College of Health and Physical Education, 7-1-1, Fukazawa, Setagaya-ku, Tokyo 158, Japan

Saibene, Franco (703), Centro Studi di Fisiologia de Lavoro Muscolare del C.N.R., Milano, Italy

Saito, Mitsuru (963), Toyota Technological Institute, 12-2, Hisakata Tempaku-ku, Nagoya 468, Japan

Saito, Shinichi (648, 762), Institute of Health and Sport Science, University of Tsukuba, Sakura-mura, Niihari-gun, Ibaraki 305, Japan

Saito, Susumu (1023), Department of 2nd Physiology, Fukushima Medical College, 5-75, Sugizuma-cho, Fukushima 960, Japan

Sakamoto, T. (762), Institute of Health and Sport Science, University of Tsukuba, Sakura-mura, Niihari-gun, Ibaraki 305, Japan

Sakurai, Shinji (629), Laboratory for Exercise Physiology and Biomechanics, Faculty of Education, University of Tokyo, 7-3-1, Hongo, Bunkyo-ku, Tokyo 113, Japan

Sawai, Kazuhiko (503), Department of Orthopedic Surgery, Aichi Medical University, Nagakute-cho, Aichi-gun, Aichi 480-11, Japan

Sawhill, J.A. (251, 635), Biomechanics/Sports Medicine Laboratory, Department of Physical Education, University of Oregon, Eugene, Oregon 97403, U.S.A.

Schandevijl, H. Van (876), Vrije Universiteit Brussel H.I.L.O.K. Pleinlaan, 2, 1050 Brussels, Belgium

Schneider, Erich (403, 490), M.E.M. Institut für Biomechanik, der Univeristät Bern, Murtenstrasse 35, CHO-3010 Bern, Switzerland

Schultz, A.B. (543), Department of Materials Engineering, University of Illinois at Chicago Circle, Chicago, Illinois, U.S.A.

Sellars, I.E. (116), Biomedical Engineering, Medical School, University of Cape Town, Observatory, 7925, Cape Town, South Africa

Seluyanov, V.N. (1152), State Central Institute of Physical Education, Department of Biomechanics, Syrenevyi Blvd. 4, 105008 Moscow, U.S.S.R.

Shibayama, Hidetaro (895), Physical Fitness Research Institute, Meiji Life Foundation of Health and Welfare, 1-1-18, Shiroganedai, Mibato-ku, Tokyo 108, Japan

Shibukawa, Kanji (648, 737, 762), Institute of Health and Sport Science, University of Tsukuba, Sakura-mura, Niihari-gun, Ibaraki 305, Japan

Shimada, Masatoshi (1097), Department of Physics, Osaka Kyoiku University, Tennoji, Osaka 543, Japan

Shirasaki, Y. (105), Department of Orthopaedic Surgery, Faculty of Medicine, University of Tokyo, 7-3-1, Hongo, Bunkyo-ku, Tokyo 113, Japan

Shitama, S. (373), Department of Control Engineering, Kyushu Institute of Technology, 1-1, Sensui, Tobata-ku, Kitakyushu 804, Japan

Spaepen, A.J. (264, 997, 1138), Laboratory for Biomechanics, Instituut voor Lichamelijke Opleiding, Katholieke Universiteit te Leuven, Tervuursevest 101, B-3030 Heverlee, Belgium

Stijnen, V.V. (264, 997, 1138), Laboratory for Biomechanics, Instituut voor Lichamelijke Opleiding, Katholieke Universiteit te Leuven, Tervuursevest 101, B-3030 Heverlee, Belgium

Strandberg, Lennart (397, 1123), National Board of Occupational Safety and Health, Accident Research Section, Arbetarskyddsstyrelsen, S-171 84, Solna, Sweden

Sugiura, Yasuo (165), Division of Orthopaedic Surgery, Nishio Municipal Hospital, 2-1, Hananoki-cho, Nishio-shi, Aichi 445, Japan

Suzuki, Kenji (380), Department of Orthopaedic Surgery, Akita University, School of Medicine, 1-1-1, Hondo, Akita 010, Japan

Suzuki, Masayasu (239), Laboratory of Physiological and Kinesiology Performance, Nippon College of Health and Physical Education, 7-1-1, Fukazawa, Setagaya-ku, Tokyo 158, Japan

Suzuki, Ryohei (57, 467, 485), Department of Orthopaedic Surgery, Nagasaki University, School of Medicine, 7-1, Sakamoto-machi, Nagasaki 852, Japan

Suzuki, Shuji (301, 444), Department of Physiology, School of Medicine, Kyorin University, 6-20-2, Shinkawa, Mitaka-shi, Tokyo 181, Japan

Suzuki, Yoshitaka (1215), Labor Accidents Prosthetics and Orthotics Center, 1-10-5, Komei, Minato-ku, Nagoya 455, Japan

Tachi, Susumu (1181), Mechanical Engineering Laboratory, Ministry of International Trade and Industry, Tsukuba Scientific Research City, Ibaraki 305, Japan

Tada, Shigeru (648, 737), Institute of Health and Sport Science, The University of Tsukuba, Sakura-mura, Niihari-gun, Ibaraki 305, Japan

Tagawa, Yoshihiko (1005), Educational Center for Information Processing, Kurume Institute of Technology, 2228-66, Mukaino, Kurume 830, Japan

Takahama, M. (380), Department of Orthopaedic Surgery, Akita University, School of Medicine, 1-1-1, Hondo, Akita 010, Japan

Takahashi, Goro (842), Institute of Sports Science, The University of Tsukuba, Sakura-mura, Niihari-gun, Ibaraki 305, Japan

Takata, Kazuyuki (591), Toyota Technical College, 2-1, Eisei-cho, Toyota-shi, Aichi 471, Japan

Takeuchi, Norio (132), Department of Orthopedic Surgery, Fukuoka Children's Hospital, 2-5-1, Tojin-machi, Chuo-ku, Fukuoka 810, Japan

Takeuchi, Shinya (591), Aichi University of Education, 1, Hirosawa, Igaya-cho, Kariya-shi, Aichi 448, Japan

Takeuchi, T. (1198), Department of Orthopedic Surgery, Tokyo Medical and Dental University, 1-5-45, Yushima, Bunkyo-ku, Tokyo 113, Japan

Tamura, Hiroshi (294), Department of Biophysical Engineering, Faculty of Engineering Science, Osaka University, 1-1, Machikaneyama-cho, Toyonaka, Osaka 560, Japan

Taniguchi, Takao (373), Kyushu Institute of Technology, 1-1, Sensui, Tobata-ku, Kitakyushu 804, Japan

Tashiro, Yoshihisa (141), Department of Orthopaedic Surgery, School of Medicine, Showa University, 1-5-8, Hatanodai, Shinagawa-ku, Tokyo 142, Japan

Tatara, Yoichi (471), Department of Engineering, Shizuoka University, 3-5-1, Johoku, Hamamatsu-shi, Shizuoka 432, Japan

Tateishi, Tetsuya (105), Mechanical Engineering Laboratory, 1-2, Namiki, Sakura-mura, Niihari-gun, Ibaraki 305, Japan

Tetewsky, A.T. (1104), Department of Mechanical Engineering, Massachusetts Institute of Technology, Cambridge, Massachusetts 02139, U.S.A.

Tezuka, Masataka (869), Meiji University, 1-9-1, Eifuku, Suginami-ku, Tokyo 168, Japan

Therrien, R. (582), Department of Mechanical Engineering, Faculty of Applied Sciences, University of Sherbrooke, Sherbrooke, Province of Quebec, Canada J1K 2R1

Tillberg, Bengt (567), Department of Human Work Science, University of Luleå, S-951 87, Luleå, Sweden

Tokuhara, Yasuhiko (440, 809), College of General Education, Teikoku Women's University, 6-173, Fujita-cho, Moriguchi-shi, Osaka 570, Japan

Toyonaga, T. (97, 110), Department of Orthopaedic Surgery, Faculty of Medicine, Kyushu University, 3-1-1, Maidashi, Higashi-ku, Fukuoka 812, Japan

Toyooka, Jiro (754), Laboratory of Exercise Physiology, Osaka College of Physical Education, 1-1, Gakuen-cho, Ibaraki-shi, Osaka 567, Japan

Toyoshima, Shintaro (683), Aichi Prefectural University, 3-28, Takada-cho, Mizuho-ku, Nagoya 467, Japan

Trozzi, V. (851), Institute of Sports Medicine of C.O.N.I., Via dei Campi Sportivi 46, 00197 Roma, Italy

Tsuchiya, Kazuo (1215), Labor Accident Prosthetic and Orthotics Center, 1-10-5, Komei, Minato-ku, Nagoya 455, Japan

Tsujino, Akira (1097), Department of Physical Education, Osaka Kyoiku University, 3-1-1, Jonan, Ikeda, Osaka 563, Japan

Tsukahara, Susumu (1023), Department of 2nd Physiology, Fukushima Medical College, 5-75, Sugizuma-cho, Fukushima 960, Japan

Uemura, Shokichi (141), Department of Orthopaedic Surgery, School of Medicine, Showa University, 1-5-8, Hatanodai, Shinagawa-ku, Tokyo 142, Japan

Ueya, Kiyomi (654), Department of Physical Education, Yamanashi University 4-4-37, Takeda, Kofu-shi 400, Japan

Umazume, Yoshiki (423), Department of Physiology, The Jikei University, School of Medicine, 3-25-8, Nishishinbashi, Minato-ku, Tokyo 105, Japan

Uemoto, Kazumi (971), Institute of Equilibrium Research, Gifu University, School of Medicine, Tsukasa-machi 40, Gifu 500, Japan

Usui, Shiro (1207), Information and Computer Sciences, Toyohashi University of Technology, 1-1, Hibarigaoka, Tempaku-cho, Toyohashi, Aichi 440, Japan

Vaughan, C.L. (116, 923, 939), Biomedical Engineering, University of Cape Town, South Africa

Verhetsel, D. (833), Institute of Physical Education, Katholieke Universiteit, Leuven te Tervuursevest 101, B-3030 Heverlee, Belgium

Veraecke, H. (833), Institute of Physical Education, Katholieke Universiteit Leuven te Tervuursevest 101, 3030 Heverlee, Belgium

Viitasalo, Jukka T. (271), Department of Biology of Physical Activity, University of Jyväskylä, FIN-40100, Jyväskylä 10, Finland

Watanabe, Kazuhiko (856), Laboratory of Physiology and Sports Biomechanics, School of Education, Hiroshima University, 2-17, Midori, Fukuyama-shi, Hiroshima 720, Japan

Watanabe, Satoru (597, 971), Institute of Equilibrium Research, Gifu University, School of Medicine, Tsukasa-machi 40, Gifu 500, Japan

Watanabe, Shiroh (444), Department of Physiology, School of Medicine, Kyorin University, 6-20-2, Shinkawa, Mitaka-shi, Tokyo 181, Japan

Watanabe, Yosaku (591), Toyota Technical College, 2-1, Eisei-cho, Toyota-shi, Aichi 471, Japan

Welch, W. (171), Faculty of Medicine, Faculty of Applied Science, Free University of Brussels (V.U.B.), Pleinlaan 2, B-1050 Brussels, Belgium

Wielki, Czeslaw (1190), Université de Louvain, Fac. de Médicine, Inst. d'Educ. Phys. Lab. JECO, Place Pierre de Coubertain, B-1348 Louvain-La-Neuve, Belgium

Willems, E.J. (264, 997, 1138), Laboratory for Biomechanics, Instituut voor Lichamelijke Opleiding, Katholieke Universiteit te Leuven, Tervuursevest 101, B-3030 Heverlee, Belgium

Williams, Keith R. (641), Department of Physical Education, University of California at Davis, Davis, California 95616, U.S.A.

Wilson, Barry D. (1223), Department of Human Movement Studies, University of Queensland, St. Lucia, Queensland 4067, Australia

Winkel, Jörgen (567), Department of Human Work Science, University of Luleå, S-951 87, Luleå, Sweden

Winter, David A. (329, 1239), Faculty of Human Kinetics and Leisure Studies, Department of Kinesiology, University of Waterloo, Waterloo, Ontario, Canada N2L 3G1

Wit, Andrzej (1175), Institute of Sport, 01-809 Warsaw, Ceglowska Str. 68/70, Poland

Wood, G.A. (306), Department of

Human Movement and Recreation Studies, University of Western Australia, Nedlands, Western Australia 6009, Australia

Yabe, Kyonosuke (223, 413), Institute for Developmental Research, Aichi Prefectural Colony, Kamiya-cho, Kasugai, Aichi 480-03, Japan

Yamaguchi, K. (467, 485), Department of Orthopaedic Surgery, Nagasaki University, School of Medicine, 7-1, Sakamoto-machi, Nagasaki-shi, Nagasaki 852, Japan

Yamaguchi, Toru (1011), Department of Electrical Engineering, Chiba University, 1-33, Kamiya-cho, Kasugai, Aichi 480-03, Japan

Yamamoto, Takashi (604), Laboratory for Exercise Physiology and Biomechanics, School of Physical Education, Chukyo University, 101, Tokodate, Kaizu-cho, Toyota, Aichi 470-03, Japan

Yamashita, Noriyoshi (440, 809), College Liberal Arts, Kyoto University, Yoshida-Nihonmatsu-cho, Sakyo-ku, Kyoto 606, Japan

Yamashita, Tadashi (373, 1005), Department of Control Engineering, Kyushu Institute of Technology, 1-1, Sensui, Tobata-ku, Kitakyushu 804, Japan

Yamazaki, Setsumasa (503), Department of Orthopedic Surgery, Aichi Medical University, Nagakute-cho, Aichi-gun, Aichi-ken, 480-11, Japan

Yamazaki, Yoshihiko (281), Department of Health and Physical Education, Nagoya Institute of Technology, Gokiso, Showa-ku, Nagoya 466, Japan

Yanagida, Yasuyoshi (1160), Department of Health and Physical Education, Faculty of Liberal Arts, Kobe University, Rokko, Nada, Kobe 657, Japan

Yata, H. (959), Department of Exercise Physiology and Sports Science College of General Education, University of Tokyo, 3-8-1, Komaba, Meguro-ku, Tokyo 153, Japan

Yoshida, Akira (842), Institute of Sports Science, The University of Tsukuba, 1-1-1, Tennodai, Sakuramura, Niihari-gun, Ibaraki 305, Japan

Yoshizawa, Masatada (809, 828), College Education, Fukui University, 3-9-1, Bunkyo, Fukui 910, Japan

Zatsiorsky, Vladimir M. (1152), State Central Institute of Physical Education, Department of Biomechanics, Syrenevyi Blvd. 4, 105008 Moscow, U.S.S.R.

Zawadzki, Jerzy (244), Academy of Physical Education, Biomechanics Laboratory, Al. Olimpijska 35, 51-612 Wrocław, Poland

Zetterberg, Carl (386), Departments of Clinical Neurophysiology and Orthopedic Surgery, Sahlgren Hospital, S-413 45 Göteborg, Sweden

I.
FUNDAMENTAL
MOVEMENTS

Keynote Lecture

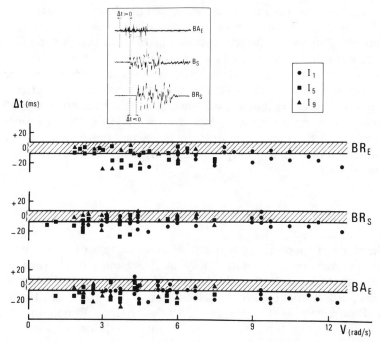

Figure 2—Chronology at the onset of activity in the main flexors. In the insert, a recording of flexion movement. The time interval (Δt), which is reported on the lower part of the figure, is measured in relation to the onset of biceps surface EMG (B_S). BR_E (and BR_S): intramuscular (and surface) EMG of brachio-radialis; BA_E : intramuscular activity of brachialis. Three conditions of inertia are considered: I_1 ($\bullet$) = 0.13 kgm²; I_5 ($\blacksquare$) = 0.39 kgm²; I_9 ($\blacktriangle$) = 0.64 kgm². (From Bouisset et al., 1977.)

A typical tracing of a normal movement is shown in Figure 1A. It is comprised of two parts: a displacement up to the target, followed in certain cases by small oscillations around the target. The latter, which corresponds to the subject's adjustments to the target, is not dealt with in this article.

Accelerometric curves are specially interesting since in this situation they equal a constant factor times the motor torque (Bouisset, 1973), and therefore provide a sensitive index of the organization of the movement. Study of accelerometric curves shows that their general form is biphasic: a deceleration phase follows an acceleration phase, except when the movement is stopped by impact with a buffer (see Figure 1B).

Besides, it was shown (Bouisset and Lestienne, 1974) that each of the form parameters (peak values, phases duration values, etc.) varies as a continuous function of a single variable which is the speed of the movement, irrespective of the additional inertial loads and of the direction of the movement (flexion or extension). Thus, the speed selected by the subject before starting the movement seems to be critical in determining the ultimate form of both the acceleration and deceleration phases.

The biphasic form of accelerometric curves does not result from mechanical necessity. Indeed, the only necessity is that, when a movement starts from and stops at a rest position—which is the case—the positive areas under the acceleration curve are equal to the negative ones. The biphasic form results from a nervous organization which selects this solution. This is the simplest solution among all the possible ones. The variations in motor torque, which are only due, in this experimental situation, to the forces developed by the elbow flexors and extensors, probably follow some principle of economy. This result seems to have a general scope.

Synergy between Agonists or between Antagonists

The study of coordination between different muscles of the same group was initiated in a qualitative way by Wachholder and Altenburger (1926a). More recently, the synergy between agonists has been characterized from the EMG of elbow flexors (Bouisset et al., 1977) and of elbow extensors (Lebozec et al., 1980) for elbow flexion and extension, respectively. It shows two main features: 1) the onset (as well as the cessation) of activity in the different muscles follows a stable mean chronology, i.e., biceps, brachioradialis and brachialis (see Figure 2) at the onset of flexion (and the reverse order at the cessation of activity) and, at the onset of extension, anconeus, lateral head of triceps, long head, and medial head almost simultaneously; 2) the excitation levels (integrated EMG) of the different muscles are directly proportional, regardless of the velocity and inertia (see Figure 3).

There is apparently a difference between elbow flexors and extensors: the time of the onset of each extensor activity (see Figure 4) depends on a velocity threshold (the higher the inertia, the lower the velocity threshold), contrary to flexors for which no velocity or inertia effect have been observed. This result could likely be explained by a wider range in muscle fiber types when extensor muscles are considered (Lebozec and Maton, personal communication).

The synergy among antagonist muscles has been characterized from the EMG of elbow flexors during extension (Lestienne, 1974) and of the elbow extensors during flexion (Lebozec et al., 1980). It shows the same two main common features and the same difference as in the case when elbow flexors and extensors act as agonist muscles. It can be noticed, however, that the excitation level of each muscle in the braking process is reduced to about $\frac{1}{3}$ of its value during the acceleration phase. Furthermore, there is a direct proportionality between the excitation levels of agonists and antagonists (see Figure 7). Thus, the synergy between the different muscles of the same group is characterized by a stable

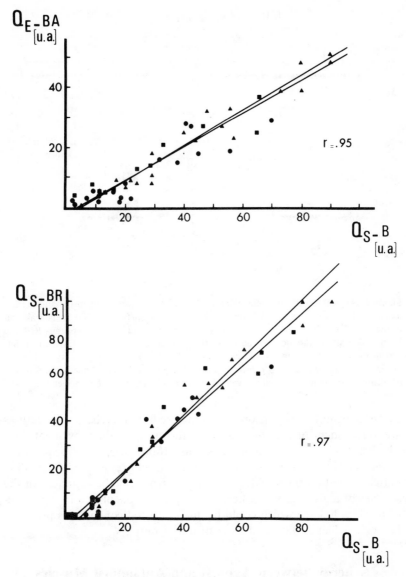

Figure 3—Correlation between integrated EMG activities of the three main elbow flexors. Q_E – BA: Integrated intramuscular EMG of brachialis; Q_S – BR: Integrated surface EMG of brachioradialis; Q_S – B: Integrated surface EMG of biceps brachii. Three conditions of inertia are shown: I_1 (•) = 0.13 kgm², I_5 (■) = 0.39 kgm² and I_9 (▲) = 0.64 kgm². The values of Q were measured during the positive phase of the acceleration curve and expressed in arbitrary units (u.a.). The two regression lines are shown. (From Bouisset et al., 1977.)

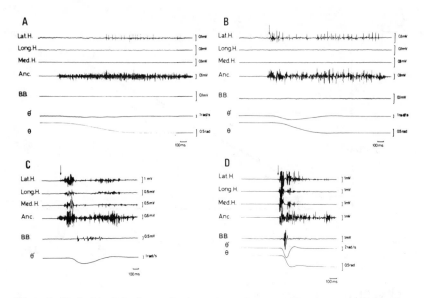

Figure 4—Typical records of extension movements of 40° performed against a same inertial load (Io = 0.021 kgm²) at four different velocities. Lat. H: surface EMG of the lateral head of the triceps muscle; Long. H: surface EMG of the long head of the triceps muscle; Med. H: surface EMG of the medial head of the triceps muscle; Anc.: surface EMG of the anconeus muscle; B.B.: surface EMG of the biceps brachii; θ′: angular velocity; θ: angular displacement; ↓: the arrow represents the onset of activity of the lateral head. The four sets of records correspond to increasing velocties from A to D. (From Lebozec et al., 1980.)

chronology and a proportionality in the excitation levels. These features are independent of the role played by the group in the movement, i.e., whether it acts as agonist or as antagonist. If one accepts the idea that the differences between flexors and extensors would result from individual muscle excitation thresholds, the same control could be assumed for the muscles of a given group, whatever the precise role played in the movement. This linkage which could be due to central connections probably favors a reduction in the number of motor outputs.

Synergy between Agonist and Antagonist Muscles

Since Wachholder and Altenburger's (1926b) pioneering study, it has been determined there exist different modalities of coordination between the agonist and the antagonist muscles, as the velocity increases.

More precisely, it was possible to define three patterns of EMG activity for elbow flexion (Lestienne and Bouisset, 1974) which were confirmed for elbow extension (Maton et al., 1980): 1) type L, characterized essentially by a continuous EMG activity of the agonist (to which may be associated slight bursts of antagonist activity) as seen in Figure 5; 2) type

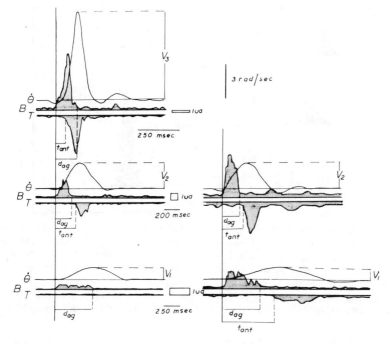

Figure 5—Agonist-antagonist coordination in elbow flexion. Each of the five sets of recordings represents the average of 20 movements performed at the same velocity and against the same inertial load. On the left: three sets of movements performed against a small inertial load (I = 0.190 kgm^2) and reaching three increasing peak velocities (V$_1$, V$_2$, V$_3$). On the right: two sets of movements performed against a higher inertial load (I = 0.580 kgm^2). The two peaks of velocity are V$_1$ and V$_2$, respectively. $\dot{\theta}$′: angular velocity; B and T: rectified and filtered surface activity of agonist (biceps) and antagonist (triceps), respectively; d$_{ag}$: duration of the burst of agonist activity; t$_{ant}$: time of the onset of antagonist activity. The shaded grey areas indicate the EMG activities of agonist and antagonist muscles. The vertical line indicates the onset of the burst of activity in the biceps. For each set of movements reaching the same peak of velocity (V), EMG activity is on the same scale. The surfaces delimited by squares or rectangles represent one arbitrary unit (ua) of the integrated EMG activity of biceps and triceps. (From Lestienne, Bouisset, and Fiori-Savary, to be published.)

S, defined by successive agonist and antagonist phasic activities, having a common electrical silence; and 3) type R, characterized by a partial overlap of the agonist and antagonist.

Furthermore, the type of coordination between agonist and antagonist does not only depend on the velocity of the movement (see Figure 6), but also on the inertia, that is to say, on the force developed by the agonist muscle group (Lestienne and Bouisset, 1974).

In the type L pattern, the braking process is only due to viscoelastic passive forces developed by the lengthening of the antagonist muscles. The existence of slight bursts of activity simultaneous to the continuous

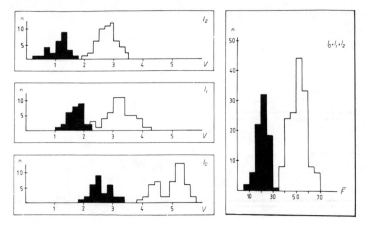

Figure 6—Distribution of the two extreme types of braking processes. This distribution regroups results from flexion movements performed at different peak velocities (V) and against three increasing inertial loads ($I_0 = 0.060$ kgm²; $I_1 = 0.120$ kgm²; $I_2 = 0.190$ kgm²). Two variables have been considered: on the left, peak velocity (V in rad/s) and on the right, mean value of the agonist force ($\bar{F}$ in N). Darkened area: movements characterized by a lack of antagonist activity (type L); white: movements characterized by a burst in EMG activity in the antagonists (type S and R); n: number of movements. (From Lestienne, 1979.)

activity of the agonist muscle indicates that the mechanisms of reciprocal inhibition have probably not come into play, or, if so, only very slightly. This is also true in type R pattern. Thus reciprocal inhibition cannot be considered as the rule during the course of a movement. Furthermore, for rapid movements, at least, mechanisms of peripheral origin might not have time to be activated and, then, the entire movement would be centrally programmed. Indeed, direct evidence of programming of neural patterns underlying pure kinetic movements has been established (Maton and Bouisset, 1975).

Moreover, taking an argument from close correlated relations (Figure 7, upper part) between the velocity and the timing of the two EMG bursts (duration of the agonist activity and onset of the antagonist activity), Lestienne (1979) stressed the idea that the movement would be under the control of centrally programmed agonist-antagonist pulses. However that may be, the organization of these elbow movements suggests that the number of agonist-antagonist combinations is reduced, which implies that the number of parameters requiring individual control is also reduced. In that sense, the elbow movements may be considered as an example of basic synergy, according to the views of Gelfand et al. (1971).

Finally, it should be emphasized that various patterns of muscle coordination correspond to the same acceleration, i.e., motor torque pattern. Passive as well as active muscular forces are cooperating precisely in order to fulfill the task requirements, and when the force requirements

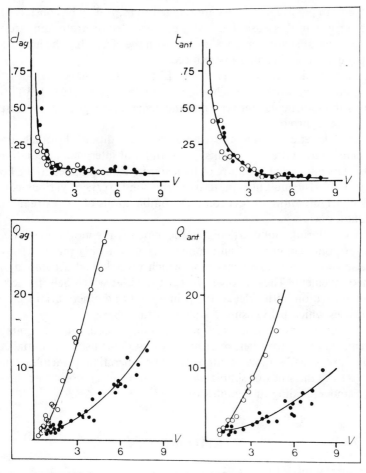

Figure 7 — Timing and excitation level of the antagonist sequences in relation to the peak velocity. d_{ag}: duration of the burst of agonist activity; t_{ant}: time of the onset of antagonist activity (see Figure 5); Q_{ag} and Q_{ant}: excitation level of the burst of agonists and antagonists, respectively (expressed in arbitrary units). Two different inertial loads: I_2 (•) and I_8 (o). Calibration: V: rad/s; d_{ag}, t_{ant}: second; Q_{ag}, Q_{ant}: arbitrary units. (From Lestienne, 1974.)

are low, passive forces are preferentially requested. The nervous organization integrates passive and active muscular forces in such a way as to minimize muscle contractions, and modulates reflex activities in accordance with this general purpose.

Conclusions

1. The elbow movement shows well-defined muscular patterns which

involve both passive and active muscular forces. It may be considered as basic synergy in the sense that EMG patterns demonstrate standardized combination of spatio-temporal orders with the effect that the number of independent motor controls is reduced.

2. Since similar accelerometric and EMG patterns have been described, based on the work of Wachholder and Altenburger (1926 a, b) using various mono-articular movements, the generality of the present results must be recognized.

3. Pure kinetic movements are an extreme condition, the other condition being pure static movements. An infinity of intermediate conditions exists, depending on the nature of the external forces applied to the body limbs. Different muscular patterns correspond to these various conditions which are more or less complex combinations of the extreme fundamental ones.

4. Even during mono-articulated movements, the motor control concerns not only agonist and antagonist muscles crossing the joint. There are also movements of cooperation which precede and accompany the prime movement. They involve synergist muscles which help fix certain joints located on both sides of the prime joint and other various groups of muscles which help ensure a stable postural base.

5. Complex movements are not a simple addition of units of movements acting together, and the task requirements as the initial conditions are usually more composite. The operating principles which emerge from the study of simple movements, however, constitute the key to the understanding of complex ones.

References

BOUISSET, S. 1973. EMG and muscular force in normal motor activities. In: J. Desmedt, (ed.) New Developments in Electromyography and Clinical Neurology, pp. 547-583, Karger, Basel.

BOUISSET, S., and Goubel, F. 1973. Integrated electromyographical activity and muscle work. J. Appl. Physiol. 35(5) :695-702.

BOUISSET, S., and Lestienne, F. 1974. The organization of a simple voluntary movement, as analysed from its kinematic properties. Brain Research 71:451-457.

BOUISSET, S., Lestienne, F., and Maton, B. 1977. The stability of synergy in agonists during the execution of a simple voluntary movement. Electroenceph. Clin. Neurophysiol. 42:543-551.

GELFAND, I.M., Gurfinkel, I.M., Tsetlin, M.L., and Shik, M.L. 1971. Some problems in the analysis of movements. In: Models of the Structural Functional Organization of Certain Biological Systems, pp. 329-345. M.I.T. Press, Cambridge, Mass.

LE BOZEC, S., Maton, B., and Cnockaert, J.C. 1980. The synergy of elbow extensors muscles during dynamic work in man. I. elbow extension. Eur. J. Appl. Physiol. 44:255-269.

LESTIENNE, F. 1974. Programme moteur et mécanismes de l'arrêt d'un mouvement mono-articulaire. These Doctorat d'Etat, Lille.

LESTIENNE, F. 1979. Effects of inertial load and velocity on the braking process of voluntary limb movements. Exp. Brain Res. 34:407-418.

LESTIENNE, F., and Bouisset, S. 1974. Role played by the antagonist in the control of voluntary movement. 4th Int. Symp. on external control of human extremities. Dubrovnik, 1972. In: Gavrilovic and Wilson (eds.), Advances in External Control of Human Extremities, pp. 12-21. Etan Press, Belgrade.

LESTIENNE, F., and Pertuzon, E. 1974. Détermination in situ de la viscoélasticité du muscle humain inactivé. Eur. J. Appl. Physiol. 32:159-170.

MATON, B., and Bouisset, S. 1975. Motor unit activity and preprogramming of movement in man. Electroenceph. Clin. Neurol. 38:658-660.

MATON, B., Le Bozec, S., and Cnockaert, J.C. 1980. The synergy of elbow extensors muscles during dynamic work in man. II. Braking of elbow flexion. Eur. J. Appl. Physiol. 44:271-278.

SHERRINGTON, Sir Charles S. 1906. The Integrative Action of the Nervous System. 1947 ed. Yale University Press, New Haven.

TALBOTT, R.E. 1979. Ferrier, the synergy concept, and the study of posture and movement. In: R.E. Talbot and D.R. Humphrey (eds.), Posture and Movement. pp. 1-11. Raven Press, New York.

WALCHOLDER, K., and Altenburger, H. 1926a. Beiträge zur Physiologie der willkürlichen Bewegung. VIII Mitteilung : über die Beziehungen verschiedener synergisch arbeitender Muskelteile und Muskeln bei willkürlichen Bewegungen (Contributions to the physiology of Voluntary movement. VIII Communication: Concerning the relation of different synergistic working muscle parts and muscles for voluntary movement). Pflügers Archiv. ges. Physiol. 212:666-675.

WALCHOLDER, K., and Altenburger, H. 1926b. Beiträge zur Physiologie der willkürlichen Bewegung. X. Einzebewegungen. (Contributions to the physiology of voluntary movement. X Individual movements). Pflügers Arch. ges. Physiol. 214:642-661.

A.
Running
and Jumping

Energetics of Running in Humans

Shinji Sakurai and Mitsumasa Miyashita
Laboratory for Exercise Physiology and Biomechanics,
Faculty of Education, University of Tokyo
Tokyo, Japan

Since Fenn (1930) presented his first article on running, many arguments have been presented concerning the energetics of human walking and running. Recently, however, Winter (1979, 1980) stated that a variety of analysis techniques and definitions have evolved producing anomalous results and differing conclusions. Then he proposed a difinition that not only accounted for any external work but also for the internal work done by the limbs themselves, and presented the results for over-ground level gait.

The purposes of the present study were to determine the two total body energies at each instant of time during running and to study the relationship between running speed and mechanical work done per unit of time.

Basic Theory

Modeling

In this study, the human body was assumed to be represented by two planar biomechanical models.

1. Multi-segment model: 11 rigid body segments, including upper arms, forearms and hands combined, thighs, shanks, feet, and trunk (with head and neck combined).

2. C.G. model: the mass center of the human body.

Anthropometric data such as center of mass location, segmental weight/body weight ratio and the rotational moment of inertia about the center of mass of the segment were obtained from the study by Dempster (1955).

Mechanical Energy

Mechanical energy at an instant of time was defined as a sum of the potential and kinetic energies.

1. Total mechanical energy of a multi segment model (ME_{tot}): The instantaneous energy ME_i of the i th segment was determined by the following formula:

$$ME_i = m_i g h_i + \tfrac{1}{2} m_i v_i^2 + \tfrac{1}{2} I_i \omega_i^2 \qquad (1)$$

where m is the segment mass in kg, g is gravitational acceleration (9.8 m/sec²), h is the height of the mass center in m, v is the velocity of the mass center in m/sec, I is the rotational moment of inertia about the mass center in kg-m², and ω is the angular velocity of the segment in rad/sec. Therefore, ME_{tot} was given by the formula:

$$ME_{tot} = \sum_{i=1}^{11} (ME_i) \qquad (2)$$

2. Mechanical energy of the C.G. model ($ME_{C.G.}$): $ME_{C.G.}$ was given by the formula:

$$ME_{C.G.} = MgH(=PE) + \tfrac{1}{2} MV^2 (=KE_{C.G.}) \qquad (3)$$

where M is the body mass (kg), H is the height (m) and V is the velocity (m/sec) of mass center of the body.

3. Internal kinetic energy : Internal kinetic energy (KE_{int}) was defined as the difference between the mechanical energy of the multi-segment model and that of the C.G. model:

$$KE_{int} = ME_{tot} - ME_{C.G.} \qquad (4)$$

Methods

Subjects and Experimental Procedures

The subjects were five male university runners, aged 20.5 to 25.0 yrs. Their body weights were 47.1 to 83.5 kg, heights were 161.0 to 183.0 cm, and $\dot{V}O_2$max were 3.05 to 4.09 1/min (see Table 1). The subjects wore running shoes and swimming trunks while running at different speeds on a running course. The subjects were instructed to keep the running speed during each trial as constant as possible. Two force-platforms (Kistler Inc., Switzerland) were inserted in the ground with their surfaces at the level of the running course. Electrical signals obtained by the platforms were amplified and then recorded by an FM magnetic data recorder.

A 16 mm high-speed movie camera (Photo-Sonics Inc., California)

Table 1

Physical Characteristics of Subjects

Subject	Age (yrs)	Height (cm)	Weight (kg)	$\dot{V}O_2$max (1/min)
HT	21.0	181.0	70.5	4.09
SS	25.0	183.0	83.5	4.07
AK	20.5	164.0	55.2	3.81
TK	21.7	166.0	54.4	3.73
IS	24.7	161.0	47.1	3.05

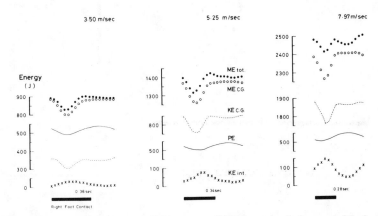

Figure 1—Total mechanical energy of 11 segments (ME_{tot}), mechanical energy of the body's center of mass ($ME_{C.G.}$) and internal kinetic energy (KE_{int}) with kinetic energy ($KE_{C.G.}$) and potential energy (PE) of the body's center of mass in relation to time during one step at three velocities of running.

was positioned so that the focal axis of the lens was perpendicular to the plane of motion. The distance from the camera to the subject was 30 m. The motion pictures were taken at 100.0 frames/sec (exposure time 0.00083 sec) for all 150 running trials. Among them, 37 trials (6 to 9 trials/one subject) were analyzed and used in the results and discussion.

Data Elaboration

In the present study, the analysis was confined to the sagittal plane and completed for only one step of running. The parameters necessary for energy analysis, such as the location and velocity of the body center of gravity, and the linear and angular velocity of each segment, were determined from the films. In order to improve the accuracy, together with the kinematic data obtained by filming, kinetic data from the force-platform measurements were used for the energy analysis. That is to say, the instantaneous horizontal velocities in the C.G. model during foot

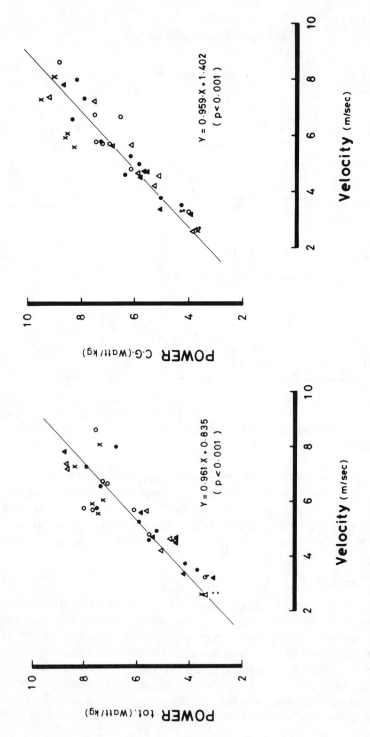

Figure 2—The relationships between running speed and positive work per unit of time in the multi-segmental model (Power$_{tot}$) and in the C.G. model (Power$_{C.G.}$).

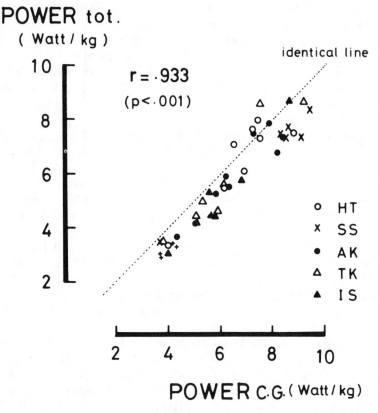

Figure 3—The relationship between the positive power of multi-segment model (Power$_{tot}$) and that of C.G. model (Power$_{C.G.}$).

contact were obtained by A/D conversion (every 0.002 sec) and numerical integration of the force data in direction of the motion.

Results and Discussion

Mean speed of the runners ranged from 2.59 to 8.59 m/sec. Figure 1 shows the two total body energy curves for one of the subjects with KE$_{C.G.}$, PE and ME$_{int}$ at running speeds of: (a) 3.50, (b) 5.25 and (c) 7.97 m/sec. The change in PE ($\triangle$PE) was almost constant (40 J) and independent of the running speed, which indicated that the body center of gravity moved 7 to 8 cm in a vertical plane during each step. On the other hand, the change in KE$_{C.G.}$ ($\triangle$KE$_{C.G.}$) increased with running speed (50, 90, and 190 J).

Both PE and KE$_{C.G.}$ components of C.G. model decreased simultaneously during the first half of the foot contact, and increased during the latter half. These trends were quite similar to the results reported by Cavagna et al. (1964).

In all cases, KE_{int} was almost $180°$ out of phase with $ME_{C.G.}$, i.e., KE_{int} increased during the contact phase and decreased during the release phase. This might have been due to the increase in the kinetic energy resulting from the motion relative to the body center of gravity of each segment during the foot contact phase. These patterns of energy changes were very similar for all subjects.

Since positive work during a step was $\triangle ME_{tot}$ or $\triangle ME_{C.G.}$, the value of positive work was divided by the duration of one step in order to determine the mean positive power ($Power_{tot}$, $Power_{C.G.}$). Figure 2 shows the relationships between the positive power ($Power_{tot}$, $Power_{C.G.}$) and running speed. High linear relationships were found (coefficients of correlation: 0.888 for $Power_{tot}$ and 0.915 for $Power_{C.G.}$, $P < 0.001$).

Figure 3 shows the relationships between $Power_{tot}$ and $Power_{C.G.}$. There was also an almost linear relationship (coefficients of correlation: 0.933, $P < 0.001$). At relatively low speeds, $Power_{C.G.}$ tended to be larger than $Power_{tot}$, but there was no clear difference between them at high speeds. These results did not coincide with the results on walking by Winter (1979).

Winter (1979) reported that instantaneous ME_{tot} was always larger than $ME_{C.G.}$, and that the integral of the energy changes during one stride for the body center gravity was less than that of sum of segment energies. In running, ME_{tot} was larger than $ME_{C.G.}$ at any instant during a step, but $\triangle ME_{tot}$ was less than $\triangle ME_{C.G.}$ at relatively low speeds. This difference between walking and running might be due to the fact that the arms and legs are moving in opposite directions more synchronously especially during contact phase in running than in walking.

References

CAVAGNA, G., Saibene, F.P., and Margaria, R. 1964. Mechanical work in running. J. Appl. Physiol. 19:294-256.

DEMPSTER, W.T. 1955. Space requirements of the seated operator. Wright Patterson Air Force Base, OH:US Air Force.

FENN, W.O. 1930 Frictional and kinetic factors in the work of sprint running. Am. J. Physiol. 92:583-611.

WINTER, D.A. 1979. A new difinition of mechanical work done in human movement. J. Appl. Physiol. 46:79-83.

WINTER, D.A. 1980. Calculation and interpretation of mechanical energy of movement. In: R.S. Hutton (ed.) Exercise and Sport Sciences Reviews. Franklin Institute Press. 6:183-201.

Identification of Critical Variables Describing Ground Reaction Forces During Running

B.T. Bates, L.R. Osternig, J.A. Sawhill, and J. Hamill
Department of Physical Education
University of Oregon, Eugene, Oregon

Ground reaction forces often serve as one of the primary components of evaluation in the analysis of the support phase of running. Miller (1978) presented data for the three components of ground reaction forces (GRF) for a single jogger and Payne (1978) reported corresponding data for a sprinter. Fukunaga et al. (1978) measured two components of the GRF for sprinters running at various speeds while Cavanagh and Lafortune (1980) presented group data on all three components for 17 distance runners. Finally Bates et al. (1980) and Norman (1980) used GRF in an attempt to evaluate shoes. In all of these studies, data have been presented in either the form of a series of continuous curves or by identifying selected descriptive quantities. Although all information is contained in a graphic representation it is difficult to make meaningful comparisons between groups or conditions with this form of data. On the other hand, if one chooses discrete data quantities, which of the characteristics are necessary to completely describe the curves? The purpose of this study was to develop a model made up of a minimum but complete set of unique parameters that would adequately describe the three components of the GRF.

Procedures

The data used in the analysis were obtained from three different groups of runners. All subjects were regular joggers/runners free from injury at the time of testing and running a minimum of 30 km/week. The model was developed using the force time curves obtained from six runners performing in six different shoe conditions (36 runner-shoe combinations). The results were further validated by repeating the procedures using data from a group of five beginning joggers and another group of five experienced distance runners.

The experimental set-up consisted of a Kistler force platform interfaced via a TransEra A/D converter to a Tektronix 4051 graphics system with a sampling rate of approximately 1000 Hz. Running speed was monitored over a 5 m interval using a photoelectric timing system and controlled between .55 and .64 msec^{-1} (6½ to 7½ min/mi).

Force time curves were obtained for 10 successful trials for each subject-condition. When multiple conditions were used, the conditions were systematically varied for each subject and between subjects. Subjects were allowed sufficient practice to assure consistent running speed and a normal foot strike on the force platform. Also, adequate time to rest was allowed between trials and conditions to assure that fatigue would not be a factor.

Based upon previous work on lower extremity function completed in this laboratory (James et al., 1978; Bates et al., 1978; 1980), a set of 43 parameters describing selected temporal and kinetic characteristics of the three GRF components were identified (see Figure 1 and Table 1). All data were normalized by dividing by body mass and the 10 trial averages computed for all variables. Next a Pearson product-moment correlation coefficient matrix was generated for each component (x,y,z) for each of the 10 trial subject-conditions. Average correlation coefficients were then computed from the 36 subject-condition matrices. Using a criterion value of $r = \pm .707$, one of each of the pairs of parameters having on the average 50% or more common variance was eliminated. The remaining variables were then combined into a single set and the process repeated to eliminate any additional variables which might be related between components.

Results and Discussion

The average correlation coefficient matrix generated from the 36 runner-shoe combinations consisted of 318 correlation coefficients of which 80 (25.2%) were significantly different from zero ($r > |.632|$, $p < .05$, df = 8). A summary by GRF components of the number of coefficients exceeding the criterion value of $r > |.707|$ is given in Table 2. Based upon these results, 20 of the original variables were eliminated (Table 3). Examination of the original variable list suggests that a number of variables were in fact dependent and common variance was to be expected. In choosing which variable in a pair was to be retained, preference was given to the value which appeared to be the most meaningful.

With 23 variables remaining, a single average correlation coefficient

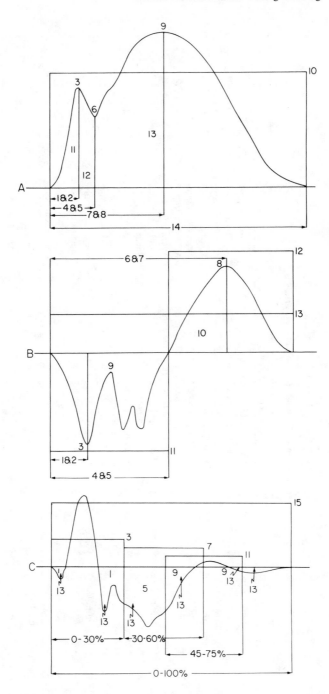

Figure 1—Graphic representation of ground reaction force variables described in Table 1. a = vertical, b = anteroposterior, c = mediolateral.

Table 1

Identification of Ground Reaction Force Variables

Number	Vertical(z)	Anteroposterior(y)	Mediolateral(x)	
1	Abs. time to 1st max. force	Abs. time to max. braking force	Negative impulse	0 to 30%
2	*Rel. time to 1st max. force	*Rel. time to max. braking force	*Algebraic impulse	Support
3	*1st max. force	Max. braking force	Average force	
4	Abs. time to min. force	Abs. time to zero force	Total excursions	
5	*Rel. time to min. force	*Rel. time to zero force	Negative impulse	
6	*Min. force	Abs. time to max. propelling force	Algebraic impulse	30 to 60%
7	Abs. time to 2nd max. force	*Rel. time to max. propelling force	Average force	Support
8	*Rel. time to 2nd max. force	Max. propelling force	*Total excursions	
9	*2nd max. force	*Braking impulse	Negative impulse	
10	*Average vertical force	*Propelling impulse	Algebraic impulse	45 to 75%
11	*Impulse to 1st max. force	Average braking force	Average force	Support
12	*Impulse to 1st min. force	Average propelling force	Total excursions	
13	*Total impulse	Average brake/propel force	Negative impulse	
14			*Algebraic impulse	
15			Average force	
16			*Total excursions	

*Variables retained for model

Table 2

A Summary of the Number of Correlation Coefficients Exceeding
the Criterion Value of r > | .707 |

| r | Component | | | Total |
	z	x	y	
± .707 - .799	1	7	3	11
± .800 - .899	3	9	2	14
± .900 - .999	0	6	39	45
Total	4	22	44	70

Table 3

Variables Eliminated During the Two Phases of Data Processing

| GRF Component | Variables | |
	Phase 1	Phase 2
z	1,4,7	none
y	1,4,6,11,12,13	3,8
x	1,3,4,5,7,9,10,11,12,13,15	6

matrix consisting of 253 values was generated. Fifteen coefficients (5.9%) were significantly different from zero showing considerably greater independence of the variables. This phase of the calculations produced seven coefficients that exceeded the criterion value resulting in the elimination of three additional variables (see Table 3). The final correlation coefficient matrix contained three values that exceeded the criterion value. These were between variables Z3 and Z6, Y5 and Y9, and Y5 and Y10. It was felt that the information provided by these variables was meaningful and should be retained. Consequently the model developed consisted of 10 vertical, 5 anteroposterior and 4 mediolateral variables plus total support time. The variables retained are identified in Table 1.

In order to further validate the model the procedure was repeated on two other sets of data. Data from five skilled female distance runners and five beginning joggers were subjected to analysis. Correlation coefficients greater than + .707 were obtained for four pairs of variables for the beginning group and five pairs for the skilled runners. In these experiments all subjects wore the same type of shoe and it was believed that this common shoe resulted in five of the nine additional significant correlation coefficients (Z2 and Z11, Z5 and Z12, and Z9 and Z10). Possibly two other significant coefficients could have also resulted from this phenomena (X2 and X14). The remaining two appeared to be due to chance.

Realizing that these groups were both small and rather homogeneous, they were combined and the analysis repeated. The results were similar with the three significant coefficients common to both groups in the previous evaluations reappearing (Z5 and Z12, Z9 and Z10, X2 and X14).

The model developed appears to be adequate for the conditions evaluated. The data used to generate and validate the model incorporated both male and female runners of varying skill level and a number of different types of running shoes. All runners were heel-toe runners. The range of running speeds was, however, limited ($.55 - .69$ msec^{-1}) and future investigations are planned to include a greater range of running speeds and possibly different types of footfall patterns.

References

BATES, B.T., James, S.L., and Osternig, L.R. 1978. Foot function during the support phase of running. Running 3(4):24-31.

BATES, B.T., Sawhill, J.A., and Hamill, J. 1980. Dynamic running shoe evaluation. Proceedings of the Special Conference of the Canadian Society for Biomechanics, Human Locomotion I. Canadian Society for Biomechanics, London, Ontario, Canada.

CAVANAGH, P.R., and Lafortune, M.A. 1980. Ground reaction forces in distance running. Journal of Biomechanics 13(5):397-406.

FUKUNAGA, T., Matsuo, A., Yuasa, K., Fujimatsu, H., and Asahina, K. 1978. Mechanical power output in running. In: E. Asmussen and K. Jørgensen (eds.), Biomechanics VI-B, pp. 17-23. University Park Press, Baltimore.

JAMES, S.L., Bates, B.T., and Osternig, L.R. 1978. Injuries to runners. American Journal of Sports Medicine 6(2):40-50.

MILLER, Doris I. 1978. Biomechanics of running—what should the future hold? Canadian Journal of Applied Sport Science 3:229-236.

NORMAN, R.W. 1980. Information content in biomechanical analyses of the effects of shoes on joggers. Proceedings of the Special Conference of the Canadian Society for Biomechanics, Human Locomotion I. Canadian Society for Biomechanics, London, Ontario, Canada.

PAYNE, A.H. 1978. A comparison of the ground forces in race walking with those in normal walking and running. In: E. Asmussen and K. Jørgensen (eds), Biomechanics VI-A, pp. 293-302. University Park Press, Baltimore.

Upper Extremity Contributions to
Angular Momentum in Running

Richard N. Hinrichs and Peter R. Cavanagh
Biomechanics Laboratory
The Pennsylvania State University, University Park, PA

Keith R. Williams
Department of Physical Education
The University of California at Davis, Davis, CA

Despite the large volume of research on the biomechanics of running, the upper extremities have been virtually ignored by researchers. The arm action is commonly thought to play an important role in counteracting excessive rotation of the body caused by the alternate striding of the legs. Although there is some support for this notion in walking (Elftman, 1939), this (balancing) role that the arms play in running has never been shown empirically. The purpose of this study was to investigate the nature of the arm action in running, specifically, to determine what contribution the arms make to the total angular momentum of the body and to more clearly define the balancing role the arms play in the rotation of the body.

Procedures

Male recreational runners (n = 21) ranging in age from 20 to 50 years were filmed while running overground at 3.6 m/sec. A four-camera three-dimensional (3D) cinematography set-up was used (Williams, 1980) with each camera operating at a nominal rate of 100 frames/sec.

A 12-segment mathematical model of the body was used and consisted of the head, trunk, two upper arms, two forearm and hand segments, two thighs, two shanks, and two feet. The body landmarks digitized from films represented the endpoints of each of the segments. Segment

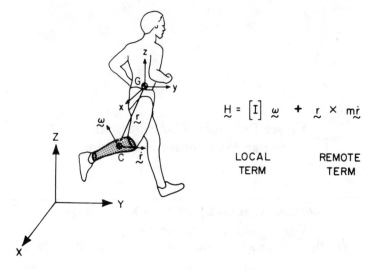

$$\underset{\sim}{H} = [I]\ \underset{\sim}{\omega}\ +\ \underset{\sim}{r} \times m\underset{\sim}{\dot{r}}$$

LOCAL   REMOTE
TERM   TERM

Figure 1—Algorithm for calculating the angular momentum of a body segment relative to the body center of mass.

masses and centers of mass (CMs) were calculated from the mean data of Clauser et al. (1969) and segment moments of inertia were obtained from the data of Whitsett (1963) and were adjusted for height and weight differences according to the method of Dapena (1978).

The segment endpoint data were smoothed using a digital filter with a cutoff frequency of 5 Hz. Linear and angular velocities of each body segment were computed via finite difference differentiation. A computer program was written which calculated from the film data the three components of angular momentum of the body and its segments. The algorithm is described in the following paragraphs.

Consider the body segment shown in Figure 1. The segment possesses two forms of angular momentum, local and remote. The local term is the angular momentum inherent in the segment's rotation about its own CM. Rotations about segment longitudinal axes were ignored, except for the trunk where the rotations were included in the analysis. Dapena (1978) has shown that this introduces little error into the calculations. The remote term is the angular momentum inherent in the movement of the segment CM (point C) relative to the body CM (point G). Referring to Figure 1, the total angular momentum of a given segment is

$$\underset{\sim}{H} = [I]\ \underset{\sim}{\omega} + \underset{\sim}{r} \times m\ \underset{\sim}{\dot{r}} \tag{1}$$

where [I] is the segment inertia tensor, $\underset{\sim}{\omega}$ is the segment angular velocity vector, $\underset{\sim}{r}$ is the vector locating C relative to G, $\underset{\sim}{\dot{r}}$ is the time derivative of $\underset{\sim}{r}$, and m is the mass of the segment.

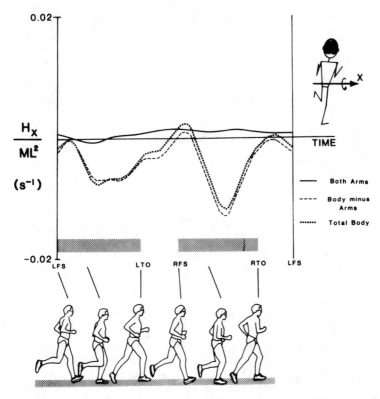

Figure 2—Transverse (x) component of the normalized angular momentum of a typical subject. LFS = left foot strike, LTO = left toe-off, RFS = right foot strike, and RTO = right toe-off.

Once individual segment angular momenta are calculated, the angular momentum of a system of segments, such as an entire extremity or the whole body, can be calculated by adding up the contributions of individual segments. To facilitate comparisons between subjects, the angular momentum was normalized by dividing the absolute values by the mass of the subject (M) and the square of his standing height (L²).

Results

Figures 2 to 4 show the three components of normalized angular momentum for a typical subject. Plotted are the contributions of the arms, the (body-minus-arms), and the total body. The body-minus-arms includes contributions from the head, trunk, and legs. While the body does possess a relatively large amount of angular momentum about the x (transverse) and y (antero-posterior, A-P) axes, it can be seen from Figures 2 and 3 that the arms have very little momentum about these

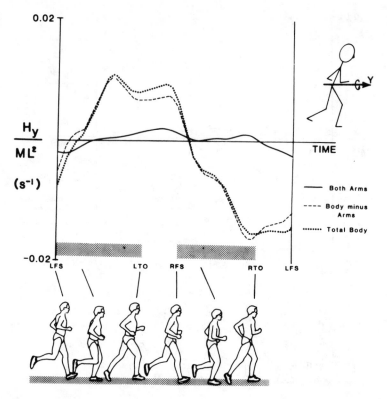

Figure 3—Antero-posterior (y) component of the normalized angular momentum of a typical subject.

axes. It is clear that the arms do not do much to (balance) the rotation of the body about these axes during normal distance running.

The arms do, however, have a substantial effect on the total body angular momentum about the z (vertical) axis, as can be seen in Figure 4. The angular momentum attributed to the body-minus-arms is relatively large about this axis and is derived primarily from the alternate striding of the legs. The arms, however, have comparable angular momentum but in the opposite direction. As a result, the total body angular momentum about the z-axis is small. This result is similar to that found by Elftman (1939). Although the general patterns were similar among subjects there was considerable variation between subjects in the magnitude of the peak normalized angular momentum about the z-axis. Mean peak values in s^{-1} were 0.0088 ± 0.0016 for the arms, and 0.0110 ± 0.0018 for the body-minus-arms, and 0.0034 ± 0.0013 for the total body. Because the peaks were not always time synchronous, the mean peak value for the total body was somewhat larger than the difference between that of the arms and the body-minus-arms.

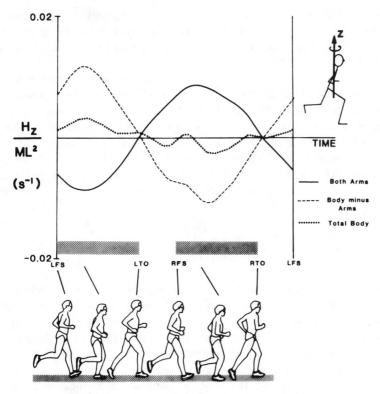

Figure 4—Vertical (z) component of the normalized angular momentum of a typical subject.

On the average, the arms did not totally balance the angular momentum of the rest of the body, but possessed about 80% of the angular momentum of the body-minus-arms in the opposite direction. In some subjects, however, the arms actually possessed larger peak values of angular momentum than did the rest of the body.

Discussion

The arms clearly have little effect on the angular momentum of the body about the x (transverse) and y (A-P) axes. Individually, each arm possesses only a small amount of angular momentum about these axes and each tends to cancel out the other's contribution. About the z (vertical) axis, however, the action of the arms has an additive effect on the angular momentum. A similar pattern is true for the legs but in the opposite direction. Though the arms have only approximately ⅓ the mass of the legs, the center of mass of each arm is situated considerably further from the z axis than is the center of mass of each of the legs. This allows the arms to (compete favorably) with the legs in their remote terms of angular momentum about this axis. The local terms of angular momentum about the z axis were small compared to the remote terms, typically less than 5%.

Because the body is free of external torques during the airborne phase in running (ignoring air resistance) the total body angular momentum between each toe-off (TO) and foot strike (FS) should be constant. The computed total body angular momentum was, in general, not constant during the airborne phase due to errors inherent in the computational process. Hay et al. (1977) have reported similar problems in calculating a constant airborne angular momentum from film data. One probable source of error is the use of potentially inappropriate body segment parameters.

Conclusions

These results show that the arms play an important role in reducing the angular momentum about the vertical axis generated by the legs in normal distance running at 3.6 m/sec. Arm contributions to angular momentum about the other two coordinate axes were, however, minimal. It is unknown at this time whether the relative contributions of the arms vary with different speeds of running.

The variations found between subjects in the magnitude of the arm contributions to angular momentum suggest that measures of angular momentum may be useful in evaluating differences in running style. There are likely to be various combinations of angular momentum from different body segments which would result in a most efficient running style for a given individual.

References

CLAUSER, C.E., McConville, J.T., and Young, J.W. 1969. Weight, volume and center of mass of segments of the human body. Wright-Patterson Air Force Base, Ohio, AMRL-TR-69-70. (NTIS # AD-710-622.)

DAPENA, J. 1978. A method to determine the angular momentum of a human body about three orthogonal axes passing through its center of gravity. J. Biomechanics 11:251-256.

ELFTMAN, H. 1939. The function of the arms in walking. Human Biology 11:529-535.

HAY, J.G., Wilson, B.D., Dapena, J., and Woodworth, G.G. 1977. A computational technique to determine the angular momentum of a human body. J. Biomechanics 10:269-277.

WHITSETT, C.E. 1963. Some dynamic response characteristics of weightless man. Wright-Patterson Air Force Base, Ohio, AMRL-TDR-63-18. (NTIS# AD-412-451.)

WILLIAMS, K.R. 1980. A biomechanical and physiological evaluation of running efficiency. Unpublished Ph.D. Dissertation, The Pennsylvania State University, University Park, Pennsylvania.

Mechanical Energy or Power of Periodic Movements

K. Shibukawa, S. Saito,
T. Yokoi, M. Ae, H. Cho and S. Tada
Institute of Health and Sport Science,
The University of Tsukuba, Ibaraki, Japan

The mechanical efficiency of human movements such as walking and running at uniform velocities is defined as a ratio of output to input, in which the input is usually calculated from a measure of oxygen consumption and the output evaluated as a rate of change in mechanical energy of the body from cinematographical analysis.

Both respiration and locomotion are periodic, but their periodicities are not the same. Oxygen consumption is given only as a mean value within one or two min, which is much longer than one cycle of respiration. On the other hand, mechanical energy can be calculated at each moment sampled with cinematography, and its rate of change may periodically vary within a cycle of locomotion. And the mean value of mechanical work rate or power over one cycle of locomotion will be zero in locomotion on the level at uniform velocity. Therefore, it is necessary to introduce some averaging technique of mechanical power, especially in the evaluation of mechanical efficiency as defined previously.

The present study attempted to introduce the idea of "effective value" as used with alternating electric current which is also periodic.

Effective Value of Periodic Function

Let a function F(t) have the following form:

$$F(t) = Fm + Fp(t,T),$$ (1)

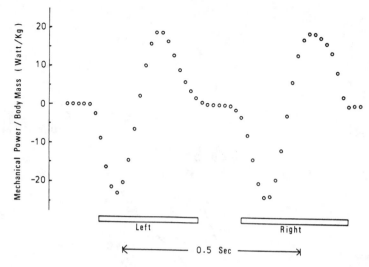

Figure 2—Mechanical power. Subject A, Test A-2.

the mean value, Em, yields the modified total input as follows:

$$Em + Ee = (1 + \sqrt{t_2/t_1}) \, Em \tag{3}$$

This value of Ee, however, should be employed only for the evaluation of the mechanical efficiency as defined previously.

Example

The measures of oxygen intake obtained with the Douglas-bag technique and mechanical power utilizing cinematography were performed for two male subjects running on a treadmill. As the methods used were standard ones, the details were omitted. For the sake of simplicity, the body was assumed to be a particle which was located at the center of gravity of the body.

The wave form of respiration could not be measured, but it was assumed to be rectangular in form and the time ration, $t_1:t_2 = 1 : 2$ because of slow running. An example of mechanical power is shown in Figure 2. (See test A-2 on Subject A). Some numerical results are listed on Table 1, in which R was the ratio of the non-support time to the time for one complete cycle and seemed to have an effect on mechanical efficiency. The efficiency of Subject A seemed higher than that for Subject B (see Figure 3). The efficiency of Subject B was higher than Subject A, at the same R but the ratio R of Subject A was much larger than that of

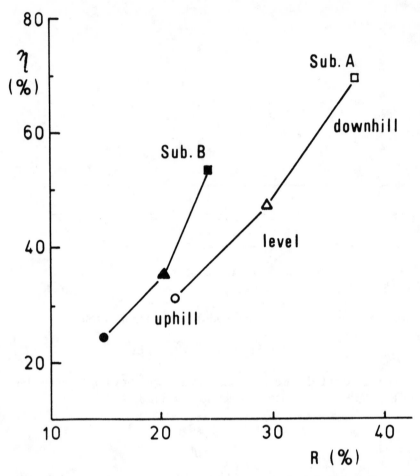

Figure 3—Mechanical efficiency vs non-support time. η = mechanical efficiency, R = ratio of non-support time to time for one cycle.

Subject B. Therefore, large mechanical efficiency may not always reflect running skill.

The mean mechanical power (Pm) and the mean value of input power (Em) were arranged in order of magnitude as follows:

uphill > level > downhill.

On the other hand, the effective value of periodic term of mechanical power (Pe) and the ratio of non-support time to the time for one cycle (R) were arranged as follows:

uphill < level < downhill.

Within the range of this experiment, Pe was much larger than Pm. Finally the mechanical efficiency (η) was arranged in order as follows:

uphill < level < downhill.

Conclusions

1. The concept of effective value of periodic function was introduced to mechanical power in running as an averaging technique.
2. In order to evaluate the mechanical efficiency it was necessary to apply the same averaging technique to both input and output.
3. In order to apply the concept of effective value to oxygen intake, it was necessary to measure the wave form of respiration.
4. The mechanical efficiency may show the effectiveness of energy transformation, but it may not be possible to indicate the skill of running.

Reference

WINTER, D.A. 1979. Calculation and interpretation of mechanical energy of movement. In: R.S. Hutton (ed), Exercise and Sport Sciences Reviews pp. 183-201, Vol. 6 Academic Press.

Rhythmicity of the Standing Triple Jump: Biomechanical Changes During Development

Kiyomi Ueya
Yamanashi University, Kofu, Japan

Rhythm plays an important role in the neuromuscular function of movement. Its formation is especially important during development in children. Successful performance of the standing triple jump, which involves three reciprocal single-leg supports in succession, results from using an appropriate rhythm to efficiently coordinate the forces of hop, step, and jump to produce as long a trajectory as possible. The rhythm for attaining distance in the entire jump, and in each of its phases, is in turn determined by how force is applied during the three periods of contact. This involves the magnitudes, directions, and durations of these forces, their timing in relation to the airborne periods, and the kinesiological characteristics of the form. The success of the performer depends on how well the nervous system has developed, how much the muscular system has matured, and on the form itself. The standing triple jump might thus serve as a general assessment for the development of physical fitness and motor ability in children. The standing triple jump has generally been thought to require the initial push of the standing long jump and the rhythm of the running triple jump, thus assuming an intermediate position between the two in terms of performance criteria, but little research has centered on the standing triple jump.

The purpose of this study was to examine the following: 1) factors related to attaining total distance and the distance of each phase in the activity, 2) forces exerted during each period of contact, 3) temporal relationships of the three phases, 4) form during the activity, and 5) the relationship of the above to development of the standing triple jump in children.

Methods

Three force plates were arranged to measure vertical and horizontal forces for each contact phase in the standing triple jump. A 16mm movie camera was used to record a side view of the activity.

A total of 360 children participated in this study — 20 boys and 20 girls of each year for ages 7 to 15 years. Each child performed the triple jump under five different conditions. Jump A: Land on the left foot to begin the step contact phase. Jump B: Land on the right foot to begin the step contact phase. Jump C: Jump with both arms held to the side. Jump D: Jump with the right arm held to the side. Jump E: Jump with the left arm held to the side. The child freely chose on which foot to land for beginning the step phases in jumps C, D, and E. Jump A or B, depending on which coincided with the child's preferred mode, was taken to be that child's free-style jump. In addition, the leg making final contact in the take-off for a running long jump was recorded, as well as the distance jumped in a standing long jump.

Results and Discussion

Distance Jumped

As shown in Figure 1, the boys steadily increased in distance as a function of age (r = 0.978), whereas, the girls showed relatively sharp increases at ages 11 and 13, but still with a consistently overall increase as a function of age (r = 0.948). The increases for the boys, however, could still be divided into two periods of faster development — (7 to 9 and 13 to 15 years). The first period may reflect primarily neural development, whereas, the second may be due more to muscular growth. Mean increase between 7 and 15 years was 37.48 cm for boys and 27.34 cm for girls. Distance traversed in the standing triple jump correlated highly with that in the standing long jump for boys 9 years and older and girls 8 years and older ($0.59 < r < 0.86$, $p < 0.01$), excepting boys of 10 and 11 years and girls of 12 years of age. The correlation was practically zero for 7-year-old children.

Distances of the Separate Phases

For any age the jump phase assumed the major portion of the distance traversed, the step phase the second largest portion, and the hop phase the least (see Figure 2). As the age increased, so did the proportion of the total distance traversed in the jump phase, whereas, the relative allotments to hop and step phases both decreased. The rates of these changes were dramatic in the earlier years studied, but they tended to

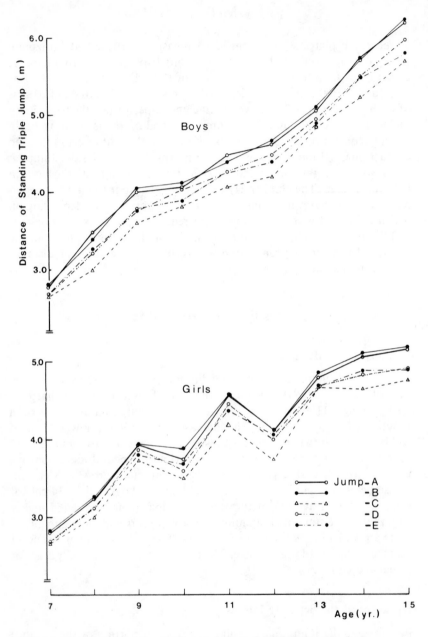

Figure 1—Development of the standing triple jump from 7 to 15 years of age.

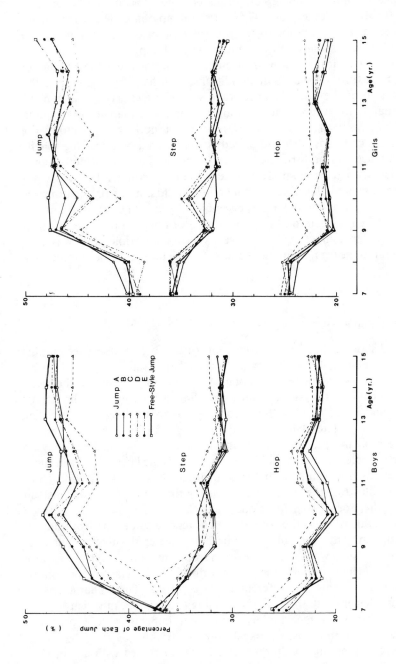

Figure 2 – Developmental changes in jumping patterns.

taper off in the older children. By 15 years the distribution of total distance for most jumping conditions became 22% for the hop phase, 31% for the step phase, and 47% for the jump phase. This might be interpreted as a stabilization of the proper sense of rhythm required for a favorable performance in the standing triple jump.

In jump C, with both arms restrained, the hop phase assumed a larger share than for the other jump conditions, and the jump phase a smaller proportion of the total distance traversed. The free-style jump, on the other hand, involved a relatively longer jump phase than any other jump condition, as well as a comparatively short hop phase. Among the 7-year-old children, 75% had already determined a preferred leg for the step contact, but performance via the preferred pattern did not become significantly better than performance with the contralateral sequence until nine years and older. Restraining one arm did not impair performance more than restraining the other, except in 15-year-old boys, who jumped further when the arm on the step-leg side was held to the side. In most cases the step-contact leg in the free-style standing triple jump was contralateral to the leg making final contact in the takeoff for a running long jump.

Contact Forces

Figure 3 shows a representative example of the force patterns seen in this study. Although all five jumps had basically the same force patterns, a number of points differed among them, notably many of the peak values, details of the patterns during landing or propulsion in the step or jump, and the general patterns for the horizontal lateral forces. The relatively large lateral forces in jump C reflect efforts by the legs to maintain balance, a function in which the arms thus appear to play an important role. Some other differences, such as the lateral forces in jump A as opposed to jump B, or the peak forces in jump D as opposed to jump E, illustrate the significance of using the limbs of one side instead of the other for a given function.

In general, whereas the fore-aft forces tended to decline from hop to step to jump, the vertical forces became greater during the same sequence. Not only was there a transition from a more horizontal to a more vertical tendency, but also the patterns shifted from relatively prolonged high-impulse forms to briefer high-power forms. These characteristics contrast sharply with those of the running triple jump, in which the hop and the jump produce mainly horizontal impacts with a primarily downward step interposed (Fukashiro et al., 1980).

For each jumping condition, peak forces increased with age in boys ($r = 0.902$, $p < 0.001$) and in girls ($r = 0.847$, $p < 0.001$). As an example, the posteriorly directed (F_x) and downward (F_y) peak forces per

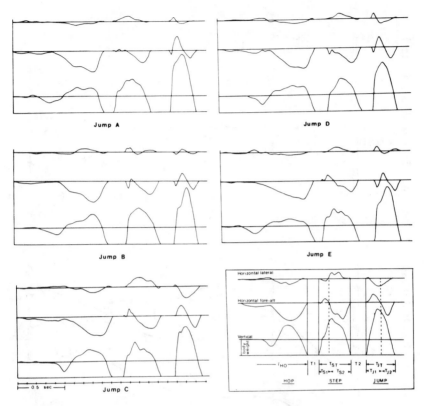

Figure 3—Force patterns in the five jumping conditions performed by a 13-year-old boy who weighed 42.6 kg. Jump A was the free-style jump.

unit body weight of the free-style jump in 7 to 15-year-old boys ranged as follows:

hop	F_x/w: 0.877 - 1.578	F_y/w: 1.570 - 1.940
step	F_x/w: 0.759 - 1.191	F_y/w: 2.028 - 2.354
jump	F_x/w: 0.446 - 1.006	F_y/w: 2.489 - 3.026

Note that the F_y values include body weight. Girls showed a similar trend to boys in changes of these ratios up to about 13 years. Their ratios increased at a smaller rate thereafter, reaching by 15 years only about 75 to 80% of the boys' values for each component of force.

Temporal Relationships

Figure 4 provides an overview of the time spent in different parts of the standing triple jump for various ages in boys. Contact time for the hop

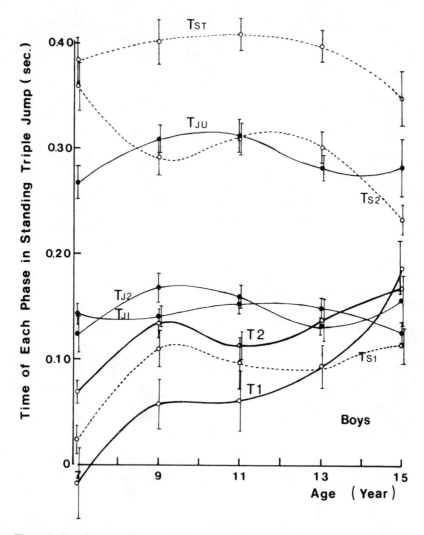

Figure 4 — Development of temporal components in performance of the standing triple jump. Mean values and standard deviations for boys. See Figure 3 for a diagram explaining the symbols.

phase was omitted from this figure because of the difficulty in accurately determining its value. Nevertheless, contact time for the hop always lasted longer than contact time for the step, which in turn always exceeded contact time for the jump. This was in contrast to the relative distances covered in each phase.

Whereas the landing period was much shorter than the propulsive period during contact time for the step phase, these two periods were about equal during contact time for the jump phase. The performers spent a little longer time landing during the contact for the jump than for the step. Temporal relationships of events on the ground did not necessarily change rectilinearly from 7 to 15 years of age. Notably constant were the landing period during contact time for both step and jump from 9 to 15 years, and total contact time for step from 7 to 13 years.

The younger performers spent more time in the air between step and jump contacts than between hop and step contacts, but this gap tended to close as age increased, and by 15 years the relative lengths of these airborne times reversed. The speed of performance, which increased with age, probably accounted for a great part of this change.

Temporal relationships underwent greater changes from 7 to 9 years and from 13 to 15 years than in the intervening ages. Changes for girls paralleled those for boys up to about 13 years of age; thereafter changes were greater in the boys.

Form

Analyzing the films revealed that the following factors increased during development: 1) translation of the body's center of mass during the contact phases, 2) extension of the legs and forward inclination of the trunk in the final moment of each propulsion phase, 3) forward inclination of the body posture in the airborne phases, 4) greater and more efficient use of the arms to help propel the body upward, and 5) greater amplitude in reciprocal limb motion at the hip joints.

The standing triple jump develops from a leg-dominated movement to full-body activity, from a horizontal displacement to a springy trajectory through the air, and from slow movement to sharp speedy movement. These changes begin to occur around 9 years in both boys and girls. Up to then, the standing triple jump looks more like stepping over objects that are in their way. Its development clearly lags behind the developmental processes for single jumps such as the standing long jump or the vertical jump (Tsujino et al., 1974).

The standing triple jump develops into a form quite different from that required for the running triple jump (Muraki, 1970). It demands a unique rhythmicity and form of its own.

Conclusion

Development of the standing triple jump in both boys and girls of ages 7 to 15 years may be divided into three periods, based on kinematic, kinetic, and performance criteria. Rhythmicity begins to emerge, in both boys and girls, as a productive factor at about 9 years of age. Effective use of the arms and legs plays an important role for attaining this rhythm.

References

FUKASHIRO, S., Muraki, Y., Miyashita, M., et al. 1980. Biomechanics of the triple jump: (2) Kinetics. Shin Taiiku 50 (10-11):68-74. (in Japanese)

MURAKI, Y. 1970. Effective limb action in the triple jump. Tokyo Joshi Taiiku Daigaku Kiyo 5:96-106. (in Japanese)

TSUJINO, A., Okamoto, T., Goto, Y., et al. 1974. Changes of motion and power during development: Jumping motions (vertical jump and standing long jump) in Shintai Undo no Kagaku (1): Human Power no Kenkyu, pp. 203-243. Kyorin Shoin, Tokyo. (in Japanese)

Longitudinal Study of Running of 58 Children Over a Four-year Period

Yoshihiro Amano
Aichi University of Education, Aichi Prefecture, Japan

Shiro Mizutani
Mie University, Mie Prefecture, Japan

Tamotsu Hoshikawa
Aichi Prefectural University, Nagoya, Japan

There is ample research on morphological, intellectual, and psychological development in children, but there is little information on movement pattern, especially using longitudinal procedures.

This study was conducted to further analyze selected parameters relating to the development of movement patterns in a manner that would enable the teacher to better understand the role played by these parameters in the maturation of movement patterns.

The purpose of this research was to provide longitudinal information on the development of running patterns, such as step frequency, step length, and leg action with reference to running performance in the same children from kindergarten through second grade.

Methods

In 1977, 58 kindergarten children of both sexes, aged four years, in Nagoya participated as experimental subjects. The subjects were asked to perform a 30 m run with maximum effort. Study variables included movement patterns using a high speed camera, velocity-time curve, step length, step frequency, physique, and power of the leg muscles.

Since 1977, data have been collected longitudinally for each of the four years.

Table 1

Mean Values Measured Longitudinally in Each of Four Years

Items	4 yrs mean s.d.	5 yrs mean s.d.	6 yrs mean s.d.	7 yrs mean s.d.
Body Height	105.7	111.1	118.4	123.5
(cm)	4.8	5.4	5.2	5.7
Body Weight	16.89	18.36	21.13	23.33
(kg)	1.93	2.11	2.98	3.42
Rohrer's Index	143.1	134.1	127.0	123.7
(Index)	12.2	11.7	11.0	13.9
Power of Leg Muscle	1.81	5.13	10.07	18.16
(kg m/sec)	0.55	0.75	1.81	3.07
Average Step Length	89.2		107.8	113.6
(cm)	8.1		7.1	9.4
Average Step Frequency	4.01		3.94	3.89
(steps/sec)	0.27		0.23	0.28
30 m Running Time	8.47	8.08	7.10	6.78
(sec)	0.73	0.58	0.51	0.47

Results

Average values of all measurement variables obtained in this study for four years are given in Table 1.

Running time showed a decrease of 20% as compared to four years ago. Increases of variables related to running performance were 17% for body height, 27% for step length, and ten-fold for leg power, but step frequency decreased by 3% during this period.

Individual characteristics in the developing pattern of running time are shown in Figure 1. As demonstrated in Figure 1, the faster children whose running performances were above the mean plus one standard deviation remained in that range through all four years.

In order to make this trend, more apparent, the H-scores of each variable in children belonging to the two extremes in running performance were computed and are given in Table 2. The children who dropped throughout the four years in the range of mean time plus one standard deviation had lower H-scores in all measurement variables. As for the children whose H-scores decreased with age, there was a tendency toward obesity.

Magnitudes of the contributions of various parameters with reference

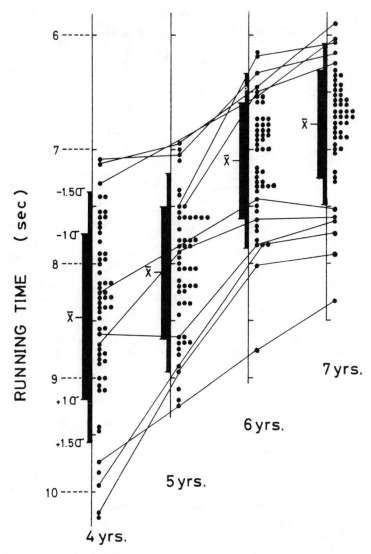

Figure 1—Developmental patterns of running performance for children from 4 to 7 years of age.

to running performance were determined from the slope of a regression equation in each year of the four year period (see Table 3).

Running forms and profiles of subjects who were in one of three different groups with regard to running ability are presented in Figure 2.

Table 2

Various Variables H-scores Measured on the Same Children

Subjects		Faster Children					Slower Children					
		1	2	3	4	5	1	2	3	4	5	6
Items		m.	m.	m.	m.	f.	m.	m.	f.	f.	f.	m.
30m Running Time	4 yrs.	76	72	−	76	−	26	17	−	22	45	53
	5 yrs.	77	−	57	74	64	18	29	37	31	55	−
	6 yrs.	67	65	75	71	66	5	25	30	30	36	44
	7 yrs.	76	72	71	68	65	4	16	22	25	26	28
Body Height	4 yrs.	44	43	−	50	−	46	38	−	35	40	43
	5 yrs.	40	−	62	74	50	−	35	45	32	37	−
	6 yrs.	35	41	65	75	49	44	36	41	31	37	61
	7 yrs.	37	42	66	74	49	43	38	42	35	37	60
Body Weight	4 yrs.	29	44	−	40	−	36	54	−	45	47	47
	5 yrs.	28	−	58	65	51	−	48	51	38	51	−
	6 yrs.	29	48	54	61	45	33	47	43	40	49	82
	7 yrs.	30	53	56	63	51	35	49	72	36	50	99
Rohrer's Index	4 yrs.	30	53	−	35	−	38	77	−	68	63	59
	5 yrs.	36	−	41	32	51	−	73	60	66	75	−
	6 yrs.	37	61	36	32	45	32	68	52	63	70	83
	7 yrs.	37	64	39	38	52	37	64	91	49	66	98
Power of Leg Muscle	4 yrs.	83	41	−	59	−	45	41	34	−	60	71
per Weight	5 yrs.	98	−	60	59	61	−	31	32	43	58	−
	6 yrs.	73	61	44	71	37	64	36	29	46	27	23
	7 yrs.	81	51	70	74	48	28	30	13	35	37	25
Average Step Length	4 yrs.	62	48	−	74	−	24	13	−	32	55	55
	5 yrs.	−	−	−	−	−	−	−	−	−	−	−
	6 yrs.	54	38	74	60	56	19	15	36	31	50	56
	7 yrs.	60	42	74	58	60	34	11	49	42	54	45
Average Step Frequency	4 yrs.	70	83	−	54	−	57	66	−	41	34	46
	5 yrs.	−	−	−	−	−	−	−	−	−	−	−
	6 yrs.	65	82	56	64	62	38	65	39	48	32	35
	7 yrs.	66	82	45	59	54	29	69	26	37	25	37

Faster Children: beyond average values minus standard deviation.
Slower Children: under average values plus 1.5 standard deviations.

Discussion

Running is a phylogenetic movement of which proper function depends upon closely integrated action of reflexes. Also, Asmussen and Nielsen (1955) and Åstrand and Rodahl (1970) reported that running speed is proportional to body height.

Data for the faster children obtained in this study demonstrated that factors of step frequency, step length, and leg muscle power play an important part with respect to running performance. Generally speaking, a

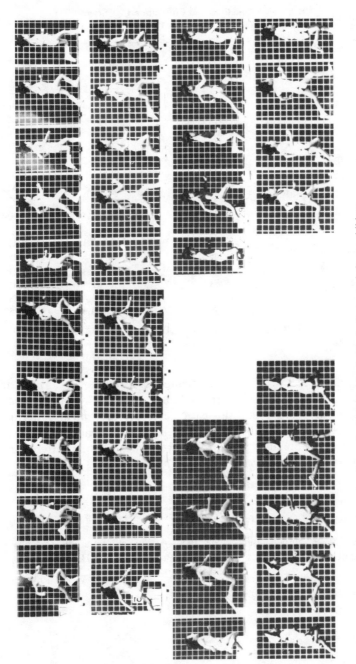

Figure 2 – Successive pictures of running form obtained longitudinally in faster and slower children.

Table 3

**Regression Equations and Correlation Coefficients
between Running Speed and Various Variables**

Variable		Regression Equation	Correlation Coefficient
Body Height	4 yrs	$Y = 0.023\ X + 1.14$	0.363[**]
(cm)	5 yrs	$Y = 0.013\ X + 2.28$	0.256
	6 yrs	$Y = 0.015\ X + 2.46$	0.266
	7 yrs	$Y = 0.016\ X + 2.45$	0.313[*]
Average Step Length	4 yrs	$Y = 0.026\ X + 1.25$	0.692[***]
(cm)	5 yrs		
	6 yrs	$Y = 0.026\ X + 1.44$	0.640[***]
	7 yrs	$Y = 0.015\ X + 2.73$	0.475[***]
Average Step Frequency	4 yrs	$Y = 0.344\ X + 2.19$	0.309[*]
(steps/sec)	5 yrs		
	6 yrs	$Y = 0.571\ X + 2.19$	0.453[**]
	7 yrs	$Y = 0.400\ X + 2.87$	0.386[**]
Power of Leg Muscle	4 yrs	$Y = 0.130\ X + 3.34$	0.238
(kg m/sec)	5 yrs	$Y = 0.179\ X + 2.80$	0.517[***]
	6 yrs	$Y = 0.062\ X + 3.62$	0.387[**]
	7 yrs	$Y = 0.052\ X + 3.49$	0.550[***]

[***]: $p < 0.001$ [**]: $p < 0.01$ [*]: $p < 0.05$

physical performance may be represented as a function of: 1) physical resources, such as geometrical (body build), energetic potential (aerobic and anaerobic power), skill (logical or effective movement pattern), 2) efficiency of the neuromuscular system, and 3) motivation. These variables change in quantity and quality over time, yielding different patterns. Therefore, this investigation will be continued.

References

ASMUSSEN, E., and Nielsen, K.H. 1955. A dimensional analysis of physical performance and growth in boys. J. Appl. Physiol. 7:593-603.

ÅSTRAND, P.O., and Rodahl, K. 1970. Textbook of Work Physiology. McGraw-Hill, Inc., New York.

Race Walking versus Ambulation and Running

M. Marchetti, A. Cappozzo, F. Figura and F. Felici
Laboratory of Biomechanics
Universita degli Studi, Roma-Italy

The efficiency of human walking and running has been scrutinized by a number of authors (for references see Pierrynowski et al., 1980). Two mechanisms of recovery of mechanical energy have been demonstrated in this regard (Cavagna, 1978): the first is due to the potential and kinetic energy exchanges within and between body segments. This mechanism predominates in walking at natural speeds, particularly with the torso acting as a conservative system (Ralston and Lukin, 1969). The second mechanism results from the recovery of some of the work done on the elastic component of the stretched muscles, and it predominates in running, thereby accounting for the high efficiency of runner's muscular work (Cavagna and Kaneko, 1977).

As far as the present authors are aware, the efficiency of race walking has not been investigated. This paper deals with the assessment of the mechanical energy variation in the body segments of a race walking athlete. Based on the data, a comparison of race walking with normal ambulation and running was carried out. The energy exchanges within and between body segments were used to determine what energy recovery mechanism was dominant in race walking. The total body mechanical energy variations and the metabolic energy expenditure were considered in order to determine how the efficiency of race walking compares with that of the other locomotor acts.

Materials and Methods

Four race walking athletes of national level were the subjects for this study. Relevant anthropometric data are reported in Table 1. Previously obtained data on untrained subjects were also utilized in this study (Cap-

Table 1
Anthropometirc Data of the Subjects

Subject	Age (yr)	Body Mass (kg)	Stature (m)
A.Z.	28	65.9	1.76
S.D.	22	66.3	1.86
S.R.	19	54.6	1.66
S.C.	22	52.3	1.65

pozzo et al., 1976). The following body segments were considered: trunk plus head (hereafter referred to as trunk), right and left upper limbs, right and left thighs, and shank plus foot. The 3D kinematics of these segments were obtained through a stereophotogrammetric technique. The details of this technique and its accuracy may be found in Cappozzo (1981). The instantaneous potential (V), rotational and translational kinetic energies (T_r, T_t) were calculated, for each of the aforementioned body segments, during seven ambulation strides, nine race walking strides, and six running strides.

Body segment inertial parameters were calculated using the regression equations provided by Clauser, McConville, and Young (1969).

The total body energy, in a given instant of time, was:

$$E = \sum_{s=1}^{S} (V_s + T_{rs} + T_{ts})$$

The energy variations $\triangle V$, $\triangle T$, $\triangle E$ referred to in the following were obtained by summing the positive increments, within one stride, using the relevant time function energies.

The total positive muscular work (W_p) was calculated as the total body energy variation($\triangle E$).

Metabolic energy expenditure was obtained by conventional indirect calorimetry while the subjects were walking or running on a treadmill.

Results

Energy Expenditure

In Figure 1 the data on metabolic energy expenditure for this study are reported together with the relevant regression curves reported by Zarrough, Todd, and Ralston (1974), with regard to ambulation and race walking, and by Menier and Pugh (1968) with regard to running. The regression curves obtained on our data are remarkably consistent with those of previous researchers.

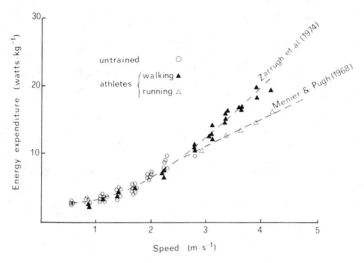

Figure 1—Metabolic energy expenditure rate, per unit body mass, in ambulation, race walking and running at different speeds.

Mechanical Energy Variations

An example of segmental total mechanical energy vs time plots is shown in Figure 2 for race walking and running. All energy components for each body segment are depicted. For ambulation, reference may be made to published data (Ralston and Lukin, 1969; Cappozzo et al., 1976; Winter, 1979). The present results concerning running are in agreement with those of Cavagna and Kaneko (1977). No reference was found reporting these data for race walking.

Rotational ($\triangle T_r$) vs Translational ($\triangle T_t$) Kinetic Energy Variations. The $\triangle T_r$ of the trunk segment was found negligible in every instance. For the lower limbs, the $\triangle T_r$ represented some 3 to 5% of the $\triangle T_t$ during running or race walking. As a percentage, $\triangle T_r$ was more important in ambulation ($\simeq$ 10% at 1.2 m sec^{-1}) than in race walking and running.

The $\triangle T_r$ of the upper limbs represented some 15% of the $\triangle T_t$ in running and 10% in race walking. In ambulation it was negligible.

Potential (V) versus Kinetic (T) Energy. During ambulation, the trunk V and T plots disclosed a typical conservative mechanism: potential and kinetic energies were of approximately the same amplitude and counter-phased. They shifted in phase during running (Cavagna et al., 1964; Ralston and Lukin, 1969). Also, during race walking V and T were in phase. The athletic style of race walkers involved a pronounced tilting and side-to-side displacement of the pelvis aimed to limit vertical excur-

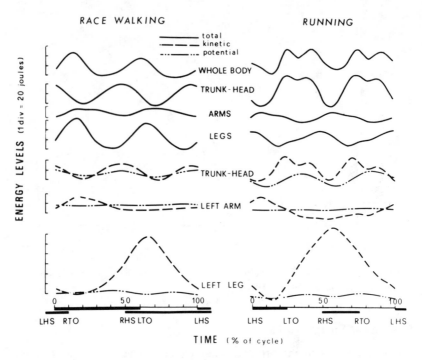

Figure 2—Mechanical energy of body segments and the entire body over one stride period in race walking (2.7 m • sec⁻1) and running (4.2 m • sec⁻1).

sion of the body center of gravity (Elson, 1967; Marchetti et al., 1981). Thus, the trunk $\triangle V$ was maintained at low levels while $\triangle T$ was similar to that observed in running.

Trunk ($\triangle$Et) versus Lower Limbs ($\triangle$El) Energy Variations. During ambulation, El provided the major contribution to the total body energy variations, confirming previous observations (Ralston and Lukin, 1969; Cappozzo et al., 1976). During race walking, as well as in running, the contribution of Et became more relevant because of the lack of a conservative mechanism in the upper body.

Total Body Energy Variations ($\triangle$E). The total energy variations augmented with the increase in walking or running speed. In Figure 3 the mean positive muscular power ($\dot{W}_p$), obtained as the positive muscular work per unit of time during one stride period, is plotted as a function of speed. For ambulation and race walking, the following single regression curve was found:

$$\dot{W}_p = 0.3 \ v^2$$

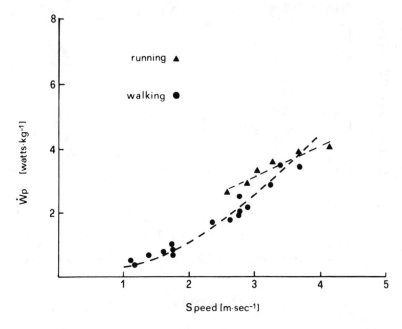

Figure 3—Relationship between positive muscular power (W_p) per unit body mass and walking and running speed.

where $\dot{W}_p$ was measured in watts and v was the speed in m • sec⁻¹. A linear relationship was found to provide the best fit for running $\dot{W}_p$ data:

$$\dot{W}_p = 0.31 + 0.95 \, v$$

Muscular Work Efficiency

Efficiency was calculated as the ratio of $\dot{W}_p$ and metabolic energy expenditure rate minus the standing energy expenditure rate (Cavagna and Kaneko, 1977). This latter rate was assumed at 1.43 watts for all subjects. Hence, for ambulation and race walking

$$\text{efficiency} = 0.3 \, v^2/(2.23 - 1.43 + 1.26 \, v^2)$$

For running, efficiency was calculated in the speed range attained in race walking (2.2 ÷ 4.17 m • sec⁻¹). From efficiency values thus the two curves of Figure 4 were obtained. With regard to ambulation and race walking performed by athletes, the efficiency assumed approximately constant value for speeds above 1.4 m • sec⁻¹. From the findings of Cavagna and Kaneko (1977) this does not appear to apply to untrained walkers, in which case the efficiency shows a maximum at 1.4 m • sec⁻¹ (so-called optimal speed).

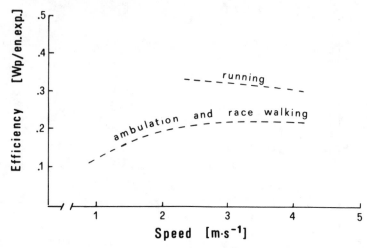

Figure 4—Efficiency in ambulation, race walking, and running as a function of the speed.

Discussion

The efficiency of a locomotor act is a conventional quantity. Its computation implies a number of assumptions that are relative to both the mechanical work done by the muscles and to the metabolic energy expenditure. Regarding this latter energy, should the basal expenditure value, instead of the standing one, be subtracted from the walking expenditure, lower values of efficiency would be obtained (Cavagna and Kaneko, 1977).

Mechanical work done by muscles has been calculated referring either to the positive total body energy variation (Ralston and Lukin, 1969) or to twice this variation (Winter, 1979). In the latter case both positive and negative muscular work were considered. Cavagna and Kaneko (1977) added the absolute values of what they called internal and external muscular work. The former was related to the body center of gravity energy variations, the latter to the variations in energy of the limbs calculated with respect to the body center of gravity.

With regard to efficiency of ambulation, the results of this study are consistent with those found by Ralston and Lukin (1969) and Winter (1979). The efficiency of running was assessed by Cavagna and Kaneko (1977), but their results were not directly comparable with those obtained in this study because of a different definition of muscular mechanical work. By whatever means employed to evaluate race walking efficiency, it is consistently evident that race walking represents a less efficient form of locomotion than running. Thus, the mechanism of energy recovery by means of elastic components of stretched muscles does not operate to the same extent in race walking as it does in running.

References

CAPPOZZO, A. 1981. Analysis of the linear displacements of the head and trunk during walking at various speeds. J. Biomechanics 14:411-425.

CAPPOZZO, A., Figura, F., Marchetti, M., and Pedotti, A. 1976. The interplay of muscular and external forces in human ambulation. J. Biomechanics 9:35-43.

CAVAGNA, G.A. 1978. Aspects of efficiency and inefficiency of terrestrial locomotion. In: E. Asmussen and K. Jørgensen (eds.), Biomechanics VI-A, pp. 3-22. University Park Press, Baltimore.

CAVAGNA, G.A., and Kaneko, M. 1977. Mechanical work and efficiency in level walking and running. J. Physiol. (London) 268:467-481.

CAVAGNA, G.A., Saibene, F.P., and Margaria, R. 1964. Mechanical work in running. J. Appl. Physiol. 19:249-256.

CLAUSER, C., McConville, J., and Young, J. 1969. Weight volume and center of mass of segments of the human body. AMRL TR- 69, 70, Wright-Patterson Air Force Base, Ohio.

ELSON, R. 1967. Race walking. The London Hospital Gazette 70:Supplement IX-XVI.

MARCHETTI, M., Cappozzo, A., Figura, F., and Felici, F. 1982. From walking to race walking. X, Hung. Rev. Sports. Med. 3:177-183.

MENIER, D.R., and Pugh, L.G.C.E. 1968. The relation of oxygen intake and velocity of walking and running, in competition walkers. J. Physiol. (London) 197:717-721.

PIERRYNOWSKI, M.R., Winter, D.A., and Norman, R.W. 1980. Transfer of mechanical energy within the total body and mechanical efficiency during treadmill walking. Ergonomics 23:147-156.

RALSTON, H.J., and Lukin, L. 1969. Energy levels of human body segments during level walking. Ergonomics 12:39-46.

WINTER, D.A. 1979. A new definition of mechanical work done in human movement. J. Appl. Physiol.: Respirat. Environ. Exercise Physiol. 46:79-83.

ZARROUGH, M.Y., Todd, F.N., and Ralston, H.J. 1974. Optimization of energy expenditure during level walking. Europ. J. Appl. Physiol. 33:293-306.

The Effect of Age and Sex on External Mechanical Energy in Running

A. Matsuo and T. Fukunaga
College of General Education
University of Tokyo, Tokyo, Japan

The external mechanical work of the center of gravity of the body in running (W_{ext}) is comprised of three components: 1) the work due to forward velocity changes (W_f), 2) the work done against gravity (W_v), and 3) the work due to lateral displacements of the center of gravity (W_l) (Fenn, 1930; Cavagna et al., 1976; Fukunaga et al., 1980). In previous studies it has been reported that the external mechanical work exerted during 1 sec of running increased linearly with the velocity up to about 12 joule/kg•sec at maximum velocity. The relationship between external work and running velocity may be considered to be effected by age and sex.

The purpose of this study was to determine the effects of age and sex on the external mechanical energy output in running.

Method

The experimental subjects in this study were 10 males and 10 females at each age from 8 to 20 years and athletes (10 male sprinters). All subjects were requested to run normally on an experimental running track, keeping the velocity as constant as possible, and also to step on the force platform naturally. The subjects were able to satisfy these requirements after several practices and then they made several runs at different velocities from slow to a maximum sprint.

The external mechanical energy of the center of gravity of body was measured by means of Kistler force platforms which were inserted in the experimental running track as shown in Figure 1. The average velocity of

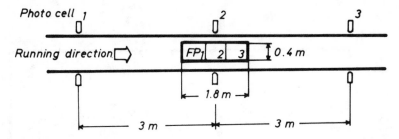

Figure 1—Schematic illustration of experimental apparatus.

the subject passing over three force platforms (V_f) was determined by an electrical signal from photocells placed 3 m apart in front of and behind the force platforms. The signals from the force platforms and photocells were recorded on magnetic tape and subsequently read every msec by an analog-to-digital converter. Using these digital values, the mechanical energy of the body exerted in forward, vertical, and lateral directions was calculated by a computer.

The forward energy (E_f) of the center of gravity was represented by the following equation:

$$E_f = \frac{m}{2} \ (V_f + \frac{1}{m} \ \int F_f dt)^2 \tag{1}$$

where m is body weight, V_f the forward velocity measured by photocells, and F_f the forward force. The vertical energy was defined by

$$E_v = \frac{m}{2} \ V_v^2 + m \cdot g \cdot h \tag{2}$$

where $V_v = \frac{1}{m} \ \int (F_v - m \cdot g) \ dt$, $h = \int V_v \ dt$, $g = 9.8 \ m/sec^2$, and F_v was the vertical force. The lateral energy (E_1) was calculated from the equation:

$$E_1 = \frac{m}{2} \ (\frac{1}{m} \ \int F_1 dt)^2 \tag{3}$$

where F_1 was the lateral force. The external mechanical energy of the subject was represented by the following equation:

$$E_{ext} = E_f + E_v + E_1 \tag{4}$$

Mechanical work in one step (W) was calculated as the increment of the energy in one step (W_f, W_v, W_1, W_{ext}). The external work rate per second ($\dot{W}$) was calculated from:

$$\dot{W} = W \cdot S_f$$

where S_f was the step frequency per sec.

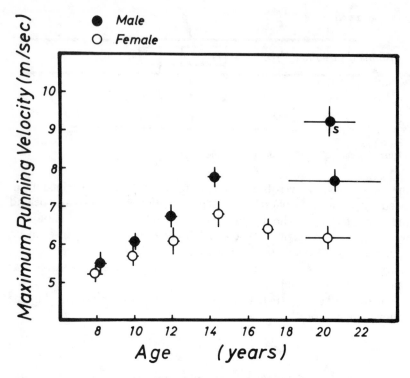

Figure 2—The relation between age and maximum running velocity. Values are means ± S.D. Symbols: closed = males, open = females, s = sprinters.

Results and Discussion

In Figure 2, maximum running velocities of each age and sex are represented by the mean and standard deviation. The maximum velocity of running increased progressively from 5.5 m/sec at 8 years to 7.8 m/sec at 14 years in male subjects and also from 5.2 to 6.8 m/sec at 14 years in females. Ikai (1968) reported that maximum running velocity increased progressively with age up to 7.8 m/sec for 16-year-old males, and up to 6.2 m/sec for 11-year-old females. The tendency of the maximum velocity to increase with age in the present study was in good agreement with the results of Ikai (1968).

Of all subjects through every age and sex, it was observed that the external mechanical work, $\dot{W}_f$ and $\dot{W}_{ext}$ exerted in a given time of running increased rectilinearly with the velocity while $\dot{W}_v$ was independent of the velocity. Maximum values of $\dot{W}_f$ and $\dot{W}_{ext}$ were obtained at the maximum running velocity for each age and sex. The external mechanical work per kg body weight, $\dot{W}_f, \dot{W}_v$, and $\dot{W}_{ext}$ at maximum velocity for each age and sex are indicated by means and standard deviations in Figure 3.

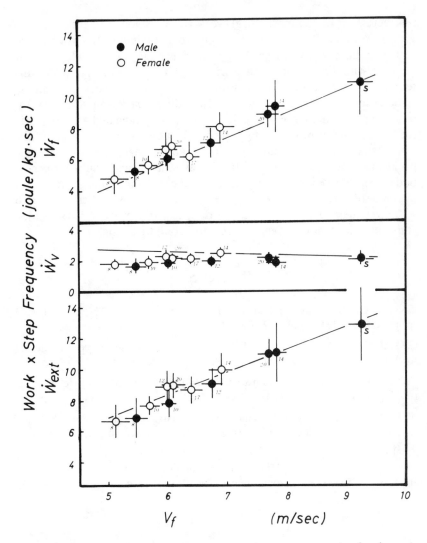

Figure 3 — Work per kg body weight/sec at maximum running velocity ($\dot{W}_f$, $\dot{W}_v$, $\dot{W}_{ext}$) as a function of the running velocity (V_f). Values are means ± S.D. Numbers near the plots present the age of the group, s = sprinters. The solid lines indicate the regression of the mechanical work on the running velocity when sprinters run at various velocities up to a maximum.

$\dot{W}_f$ at maximum velocity increased with age from about 5 to 9 joule/kg • sec in males and from 5 to 8 joule/kg • sec in females. On the other hand, $\dot{W}_v$ indicated approximately constant values of 2 joule/kg • sec independent of age, sex, and athletic background. Maximum values of external work ($\dot{W}_{ext}$) increased with the maximum running velocity which increased with age. The solid lines in Figure 3 indicate the regression of the mechanical work on the velocity in the case of a sprinter running at various velocities up to a maximum. It was observed that maximum $\dot{W}_f$, $\dot{W}_v$, and $\dot{W}_{ext}$ at each age and sex were plotted near the regression line obtained for sprinters. There were no significant differences in the regression equations of mechanical work to running velocity for subjects of any age, sex, and athletic background. For example, at a given running velocity of 5 m/sec, $\dot{W}_{ext}$ indicated approximatly the same value of 6.6 joule/kg • sec in all subjects in spite of great differences in step length and frequency among age, sex, and athletic background groups. Consequently, the external mechanical work per kg body weight in a given time of running indicated almost constant values, in spite of the differences in age, sex, and body dimensions, and whenever the run covered a certain distance, the external work yielded constant values independent of age and sex and training.

References

CAVAGNA, G.A., Komarek, L., and Mazzoleni, S. 1971. The mechanics of sprint running. J. Physiol. 217:709-721.

CAVAGNA, G.A., Thys, H. and Zamboni, A. 1976. The sources of external work in level walking and running. J. Physiol. 262:639-657.

FENN, W.O. 1930. Work against gravity and work due to velocity changes in running. Am. J. Physiol. 93:433-462.

FUKUNAGA, T., Matsuo, A., Yuasa, K., Fujimatsu, H., and Asahina, K. 1980. Effect of running velocity on external mechanical power output. Ergonomics. 23(2):123-136.

IKAI, M. 1968. Biomechanics of sprint running with respect to the speed curve. In: J. Wartenweiler (ed.), Biomechanics I. pp. 282-290. Karger, Badel/New York.

B.
Throwing and Kicking

Biomechanical Analysis of Throwing in Twins

Shintaro Toyoshima and Tamotsu Hoshikawa
Aichi Prefectural University, Nagoya, Japan

Toshiaki Goya
Aichi University of Education, Aichi Prefecture, Japan

Hideji Matsui
Nagoya University, Nagoya, Japan

Throwing is a very typical movement for human beings, because it comprises a series of rotations of the body in the sagittal plane. Therefore, a throwing movement can be performed by a biped, but not by a quadruped and is a rather ontogenetic movement which requires a lot of learning or training.

Manifestations of various human physical capacities are related to intrinsic (heredity) and extrinsic (learning and training) factors. Although there is ample evidence as to the effects of both factors on morphological, intellectual, and psychological aspects of character, there is little information on motor ability or movement pattern. It is necessary to determine the magnitude of the influence of both factors relative to motor ability or movement patterns, in the construction of a physical education curriculum and the selection of sport talent.

The purpose of this study was to estimate to what extent intrinsic components may account for existing individual differences of the throwing pattern and its performance, employing monozygous and dizygous twins as subjects.

Methods

There were 18 pairs of monozygous and dizygous twins of both sexes and

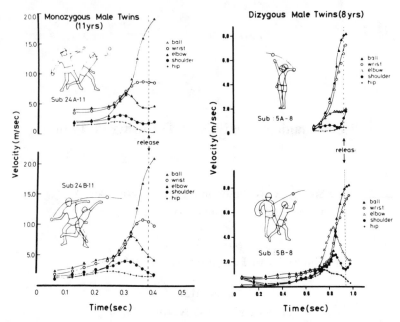

Figure 1—Changes in velocities of hip, shoulder, elbow and wrist joints determined from films taken from the side.

one set of female identical triplets employed as subjects in this study. The ages ranged from 7 to 12 years.

The data for this experiment were obtained by photography. Photographs were taken from two directions, the top and side, using two high speed cameras, simultaneously, and the films were analyzed by means of a Graf-pen system connected directly to a computer.

The subjects, who were appropriately marked with white targets at the wrist, elbow, shoulder, and hip, threw a hard rubber ball weighing 125 g at maximum effort.

Comparison of the differences between monozygous and dizygous twins was made by intrapair difference computed through the following equation:

$$\frac{2|(A - B)|}{A + B} \times 100 = \text{Intrapair difference (\%)}$$

where A and B are twin brothers/sisters, respectively.

Results

Figures 1 and 2 show the data obtained from the film which was taken from the top and the side, simultaneously. It was found that there were

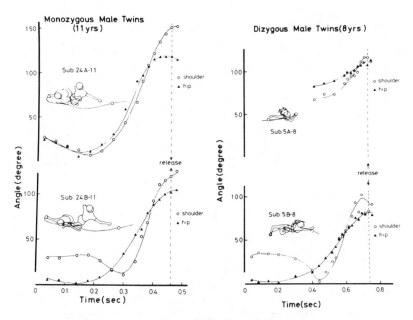

Figure 2 — Changes in angle formed between the lines joining the acromions processes and both iliums as determined from films taken from overhead.

some distinctive differences in temporal patterns, displacements, and velocities in the hip, shoulder, elbow, and wrist joints.

Such differences become more apparent when the pair of values were plotted against each other on a system of X-Y coordinates, as can be seen in Figure 3a-3f for ball velocity, stride length, and velocities of the hip, shoulder, elbow, and wrist joints in the horizontal direction, respectively. However, the values of the monozygous and dizygous pairs yielded no systematic distribution and the two groups had an almost equal scatter. Furthermore, there is no significant difference in percent of intrapair differences between monozygous and dizygous twins.

In order to investigate the body rotation with respect to sagittal line, the angle formed between lines joining both acromion processes and both iliums and the direction of the throw was analyzed. Comparison of the magnitude of the body rotation was made at points of 80, 60, 40 and 20% of the throwing motion with respect to a duration of 100%. A significant difference was found at 80 and 60% but the differences were not statistically significant at 40 and 20% between the two types of twins.

Discussion

In previous research on twins' physical characteristics, it has been stated

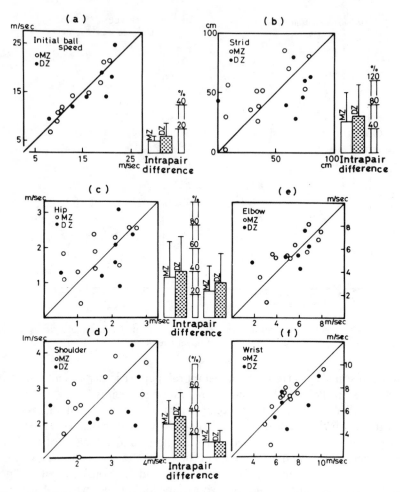

Figure 3—Differences in (a) initial ball velocity, (b) distance of the forward step, (c) horizontal velocities of the hip, (d) shoulder, (e) elbow, and (f) wrist joints between twin partners during throwing.

that monozygous twins resemble each other more than dizygous, also that the resemblances are greater for twins than for non-twins.

Although it may be assumed that there is a similarity even in movement patterns there is no information to support this assumption. In this study the similarity of throwing patterns for twins was determined from intrapair differences.

Data obtained in side views show that there were no differences in the velocity of the ball, and the moment when the velocities of the hip, shoulder, elbow, and wrist joints reached their maximums.

Even from the top view, there was no significant difference apparent in the magnitude of the body rotation between monozygous and dizygous twins.

Wild (1938) has divided the action of throwing into four stages, and has suggested that certain characteristics appear in the step in the throwing direction and the body rotation. Furthermore, in the analysis of throwing, Leme and Shambes (1978) have suggested that twisting body movements do not develop naturally.

Our study offers evidence that the percentage of intrapair differences in trunk movement in relation to stride length and body rotation is higher even in monozygous twins as compared to that of wrist, elbow, and other joint movements. That is to say, the results indicate that the influence of heredity on the throwing ability was found to be small and training or learning is most important to effective utilization of movement in trunk.

References

LEME, A.S., and Shambes, M.G. 1978. Immature throwing pattern in normal adult women. J. Human Movement Studies 4:85-93.

WILD, M.R. 1938. The behavior pattern of throwing and some observations concerning its course of development in children. Res Quart. 9:20-24.

Interaction Between Segments During a Kicking Motion

Carol A. Putnam
School of Physical Education
Dalhousie University, Nova Scotia, Canada

During the initial phase of a kicking motion, thigh rotation appears to dominate the activity while there is very little rotation of the shank relative to the thigh. Throughout the remainder of the movement, thigh rotation decreases considerably, while that of the shank increases. Little is known about why this particular pattern of segment motions is adopted or how it contributes to the quality of performance of a kick.

Several biomechanical principles have been formulated in an attempt to explain this and similar patterns of segment motions. These include the principles of summation of speed, summation of force, and transfer of angular momentum proposed, for example, by Bunn (1972). Plagenhoef (1971) explained these segment motion patterns in terms of the influence that the angular acceleration of one segment has on that of an adjacent segment. These explanations, however, have been stated in somewhat vague terms and have rarely been substantiated by vigorous investigations of those movements to which they were meant to apply.

If movement patterns similar to that found in kicking are to be understood, one must first gain an appreciation of the manner in which the segments interact with each other during the course of the activity. Initial attempts at this have been made by Phillips et al. (1978), Wahrenberg et al. (1978) and Winter and Robertson (1978). The purpose of the present study was to examine the interaction between the thigh and shank during a punt kick in an effort to explain the observed pattern in their motions. In particular, attention was focused on the decrease in thigh angular velocity which occurs during the latter part of the movement.

Procedures

Theoretical Considerations

The kicking leg was modelled as a planar, two-segment system, linked by a frictionless pin joint at the knee. The two segments included the thigh and shank, where the foot was treated as a rigid extension of the shank. Newtonian equations of motion were written for the system and modified so that the angular acceleration of each segment was expressed as a function of the proximal joint torque acting on that segment and kinematic variables of the system. These equations have the following form.

Equation for shank:

$$(I_{CS} + r_S^2 m_S)\ddot{\theta}_S = T_K - (r_S \cos\phi \, m_S \, l_T \ddot{\theta}_T) - (r_S \sin\phi \, m_S \, l_T \dot{\theta}_T^2) \qquad (1)$$

$$+ (r_S \sin\theta_S m_S \ddot{x} - r_S \cos\theta_S m_S \ddot{y}) - (r_S \cos\theta_S m_S g)$$

Equation for thigh:

$$(I_{CT} + r_T^2 m_T)\ddot{\theta}_T = T_H - (r_S \cos\phi \, m_S \, l_T + l_T^2 m_S)\ddot{\theta}_T - (r_S \sin\phi \, m_S \, l_T \dot{\theta}_T^2) \qquad (2)$$

$$- (l_T \cos\phi \, m_S \, r_S + r_S^2 m_S + I_{CS})\ddot{\theta}_S + (l_T \sin\phi \, m_S r_S \dot{\theta}_S^2)$$

$$+ (r_T \sin\theta_T m_T + l_T \sin\theta_T m_S + r_S \sin\theta_S m_S)\ddot{x}$$

$$- (r_T \cos\theta_T m_T + l_T \cos\theta_T m_S + r_S \cos\theta_S m_S)\ddot{y}$$

$$- (r_T \cos\theta_T m_T g + l_T \cos\theta_T m_S g + r_S \cos\theta_S m_S g)$$

Where:
θ, $\dot{\theta}$, and $\ddot{\theta}$ are the orientation angle, angular velocity and angular acceleration, respectively, of the shank (subscript S) or thigh (subscript T); l is the segment length; r is the distance between the proximal joint and mass center of the segment; m is the segment mass; I_{CS} and I_{CT} are the transverse centroidal moments of inertia of the shank and thigh, respectively; T_K and T_H are the joint torques at the knee and hip, respectively; ϕ is the knee angle or $(\theta_S - \theta_T)$; $\ddot{x}$ and $\ddot{y}$ are the horizontal and vertical acceleration, respectively, of the hip; and g is the acceleration due to gravity.

All terms on the right hand sides of Equations (1) and (2) have the units, Newton-meters. Each may be regarded as a moment about the center of gravity of the shank (Equation 1) or thigh (Equation 2). With the exception of the joint torques, these moments are motion-dependent

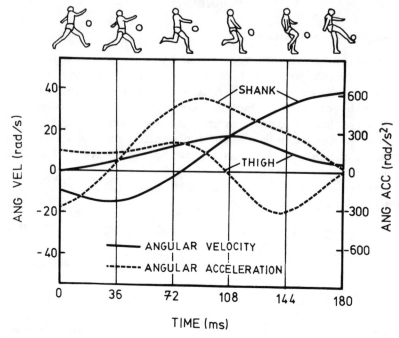

Figure 1—Angular kinematic data.

and therefore fictitious. For the purpose of this study, however, they were treated as applied moments which influence the angular accelera- tion of the shank (Equation 1) or thigh (Equation 2). Those moments which are functions of angular motion variables are hereafter referred to as angular motion-dependent moments. The interaction between the angular motions of the thigh and shank was investigated by considering the influences that these angular motion-dependent moments have on the angular accelerations of the thigh and shank.

Experimental Data

Eighteen subjects were used in this study. Segment inertial parameters for these subjects were estimated using data presented by Dempster (1955) and Dempster and Gaughran (1967). Two punt kicks performed by each subject were filmed with a Locam camera operating at 300 frames/sec. Displacement-time data for the hip, knee, and ankle of the kicking leg were determined from the film and smoothed with cubic spline functions. Segment kinematic and joint torque data were subse- quently calculated.

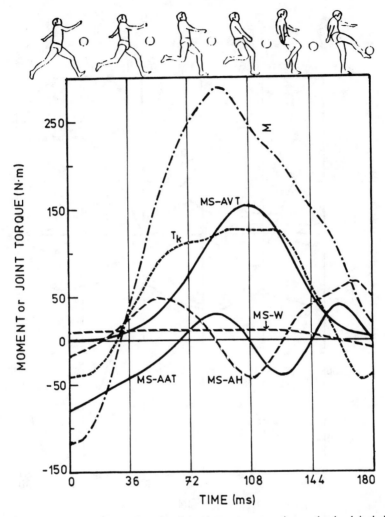

Figure 2—Knee torque (T_K) and motion-dependent moments acting on the shank including those which are functions of the angular velocity and acceleration of the thigh (MS-AVT and MS-AAT, respectively), the acceleration of the hip (MS-AH) and the weight of the shank (MS-W). Σ represents the sum of the knee torque and the motion-dependent moments, that is, the left-hand side of Equation (1).

Results

The results presented are those of the trial with the highest foot speed at ball contact. These results are representative of all trials analyzed. Thigh and shank angular velocity- and angular acceleration-time curves are shown in Figure 1. Positive thigh rotation is in the direction of hip flexion while positive shank rotation is in the direction of knee extension.

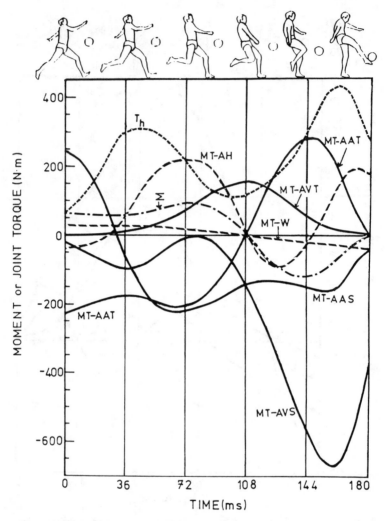

Figure 3—Hip torque (T_H) and the motion-dependent moments acting on the thigh including those which are functions of the angular velocity and acceleration of the thigh (MT-AVT and MT-AAT, respectively), the angular velocity and acceleration of the shank (MT-AVS and MT-AAS, respectively), and acceleration of the hip (MT-AH) and the weight of the thigh and shank (MT-W). Σ represents the sum of the hip torque and the motion-dependent moments, that is, the left-hand side of Equation (2).

The proximal joint torques and all motion-dependent moments influencing the angular accelerations of the shank and thigh are shown in Figures 2 and 3, respectively. The solid curves in these figures represent the angular motion-dependent moments.

Discussion

The angular motion of the thigh influenced that of the shank in such a manner as to cause the shank to be angularly accelerated in the positive direction throughout most of the kick (MS-AVT and MS-AAT in Figure 2). MS-AVT was particularly influential in angularly accelerating the shank, having a magnitude approximately equal to that of the knee torque. The magnitude of this moment changed primarily as a function of fluctuations in the angular velocity of the thigh. Therefore, when the thigh started to slow down during the latter part of the kick, the magnitude of MS-AVT decreased. As a result, the effect that the angular motion of the thigh had in positively accelerating the shank was never as great, during the later part of the kick, as it had been when the thigh was rotating at or near its maximum angular velocity. Furthermore, when the angular velocity of the thigh started to decrease, MS-AAT changed from assisting in the positive angular acceleration of the shank to tending to angularly accelerate the shank in the negative direction. This moment continued to act in the negative direction until the knee had extended beyond 90°, an event which occurred approximately midway between the time that the thigh started to slow down and the foot contacted the ball. It appears, therefore, that it is disadvantageous to decrease the angular velocity of the thigh, at least in regard to the influence that the thigh's motion has on the shank.

The reason for the observed decrease in the thigh's angular velocity during the latter part of a kicking motion is understood by examining the influence of the shank's motion on that of the thigh (MT-AVS and MT-AAS in Figure 3). Both MT-AVS and MT-AAS were negative throughout most of the kick and tended to decrease the angular velocity of the thigh. MT-AVS was especially influential in decreasing the thigh angular velocity during the latter part of the kick when the shank was rotating very rapidly. As a result, the thigh was accelerated in the negative direction despite a high magnitude of positive hip torque.

Conclusion

It was concluded that the decrease in the thigh's angular velocity observed during the latter half of a kicking motion does not serve to increase the angular velocity of the shank, but rather, this decrease occurs as a result of the influence of the shank's angular motion on the thigh.

Acknowledgments

Sincere appreciation is extended to Dr. J.G. Hay and Professor J.G. Andrews for their guidance and assistance throughout this study. This research and its continuation are being supported by NSERC Operating Grant #A7788.

References

BUNN, J.W. 1972. Scientific Principles of Coaching. Prentice-Hall, Englewood Cliffs, N.J.

DEMPSTER, W.T. 1955. Space requirements of the seated operator. WADC Technical Report 55-159, Wright-Patterson Air Force Base, Ohio.

DEMPSTER, W.T., and Gaughran, G.R.L. 1967. Properties of body segments based on size and weight. Amer. J. Anat. 120:33-54.

PHILLIPS, S.J., Roberts, E.M., and Huang, T.C. 1978. Intersegmental relationships in human lower extremity swing motions. A paper presented at the second annual meeting of the American Society of Biomechanics, University of Michigan, Ann Arbor.

PLAGENHOEF, S.G. 1971. Patterns of Human Motion: A Cinematographic Analysis. Prentice-Hall, Englewood Cliffs, N.J.

WAHRENBERG, H., Lindbeck, L., and Echolm, J. 1978. Knee muscular moment, tendon tension, force and EMG during a vigorous movement in man. Scand. J. Rehab. Med. 10:99-106.

WINTER, D.A., and Robertson, D.G.E. 1978. Joint Torques and energy patterns in normal gait. Biological Cybernetics 29:137-142.

Analysis of Powerful Ball Kicking

Toshio Asami
College of General Education,
University of Tokyo, Tokyo, Japan

Volker Nolte
Institut für Biomechanik, Deutsche Sporthochschule Köln

Some researchers have analyzed the mechanical phenomenon of ball kicking. Togari et al. (1972) reported that though there was a high correlation between foot velocity immediately before impact and kicked ball velocity, skilled players kick the ball with higher velocity than unskilled players even though foot velocity was the same, and suggested that the rigidity of the foot during impact was an important factor for a powerful kick as well as foot velocity. Shibukawa (1973) considered the body of the kicker at impact as a system of rigid bodies connected at three joints, the hip, knee, and ankle, and theoretically calculated the effect of foot velocity and joint fixation on ball velocity. The results of Shibukawa (1973) were consistent with the experimental data obtained by Togari et al. (1972). He concluded that ball velocity was greatly influenced by joint fixation of the kicking leg as well as foot velocity at impact.

Plagenhoef (1971) measured foot velocity before and after impact, time of impact, and striking mass of the foot and concluded that the ability to attain body rigidity during impact is the key to better performance in a hard kick of a soccer ball.

But no research was found to measure quantitatively the rigidity of the foot during impact. The purpose of this study was to determine the mechanical phenomena occurring during impact of the foot and ball in powerful kicking and to analyze quantitatively the effects of foot velocity and rigidity of foot to the ball velocity.

Method

Four German professional footballers, including one international player, and two amateur footballers were used as subjects.

They were requested to kick a placed ball four times as powerfully as possible, aiming at the center of a handball goal placed 10 m from the ball. The ball was an officially approved ball by FIFA, 430 g in weight, 68 cm in circumference and with 0.65 kg/cm^2 air pressure. Two 16 mm high speed cameras operating at 500 and 100 frames/sec were used to film the kick, one from the side and one from behind the kicker, respectively. A Kistler force platform set under the surface of ground was used to measure the force of the supporting foot during the kick and force curves were recorded on magnetic tape with a signal synchronized with the film.

Measurement of the displacement and the velocity of the hip, knee, ankle, foot, and ball, and the angular displacement and velocity of the knee, ankle, and foot joints were taken from the film using an HP 9845/HP9887A Digitizing System. Mean force during the impact and the striking mass were calculated from the measured data. Impact time of the ball and foot was calculated from the numbers of frames in which ball and foot were in contact, the distance between the ball and foot in the frame just before and after impact and the velocity of the ball and foot at that moment.

The vertical force of the supporting foot at impact was measured from the force curve.

Results and Discussions

Since leg movement during a powerful kick is not a planar motion, the lateral displacement was obtained from the film which was taken from behind the kicker. The results showed that the lateral displacement was less than 2% when the analysis of movement was limited to the period immediately before impact to immediately after impact. Therefore, this factor was disregarded.

The mean values, standard deviations, minimum and maximum values of ball velocity (V_B), foot velocity immediately before (V_{Fb}) and after (V_{Fa}) impact, ratio between V_B and V_{Fb} (V_B/V_{Fb}), striking mass (SM), impact time (IT), mean force applied between foot and ball during impact ($\overline{F}$), maximum angular displacement of the ankle ($\triangle \angle A$) and foot ($\triangle \angle F$) during impact, vertical force of supporting foot to ground at the moment of impact (F_Z) and F_Z per body weight (F_z/W), and the coefficient correlations between V_B and other parameters are shown in Table 1.

Maximum V_B of 34.0 m $\bullet$ sec^{-1} in this study was higher than the 32.8

Table 1

Mean, Standard Deviation, Minimum and Maximum Values of Each Variable, and Coefficient Correlations Between Ball Velocity and Other Variables

Variable	n		mean	S.D.	Min.	Max.	r
Ball velocity (V_B)	19	(m sec⁻¹)	29.9	2.9	21.5	34.0	
Foot velocity before impact (V_Fb)	19	(m sec⁻¹)	28.3	1.6	25.8	31.1	0.738***
Foot velocity after impact	19	(m sec⁻¹)	15.5	1.8	12.7	18.6	0.645**
V_B/V_{Fb}	19		1.06	0.07	0.81	1.15	0.822***
Striking mass	19	(kg)	1.02	0.14	0.80	1.26	0.720***
Impact time	19	(msec)	12.0	1.6	10.1	16.5	−0.886***
Mean force	19	(N)	1100	221	560	1419	0.952***
Maximum angular displacement of ankle	19	(degrees)	19.9	6.7	9	32	−0.409
Maximum angular displacement of foot	19	(degrees)	34.6	31.1	4	108	−0.805***
Vertical force of supporting foot at impact (F_Z)	16	(N)	1820	354	1340	2350	0.275
F_Z/Body weight	16		2.39	0.44	1.54	3.29	0.336

**p < 0.01, ***p < 0.001.

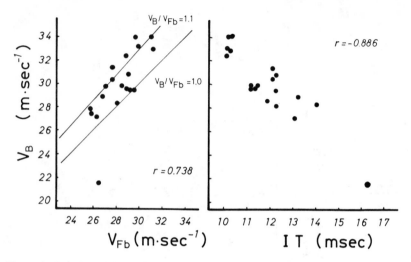

Figure 1—Relation of foot velocity immediately before impact and impact time to ball velocity.

m • sec^{-1} of a top Japanese player reported by Togari (1970), and 29.1 m • sec^{-1} by Plagenhoef (1971). The mean value of 29.9 m • sec^{-1} was also higher than 27.4 m • sec^{-1} by Zernicke and Roberts (1978).

Mean V_B/V_{Fb} of 1.06 was consistent with Togari's (1970) data from a skilled player, but lower than 1.25 by Plagenhoef (1971) and slightly lower than 1.16 by Shibukawa (1973) which was calculated theoretically with the ankle joint completely fixed and knee joint completely free.

The mean striking mass was 1.02 kg with a range from 0.80 to 1.26 kg which was quite inconsistent with the 3.90 kg (2.59 to 4.81) reported by Plagenhoef (1971). This difference was mainly dependent on the difference in the foot deceleration during ball impact. In this study, the foot velocity decreased from 28.3 to 15.5 m • sec^{-1} on the average and differed very much from Plagenhoef's data which decelerated from 24.1 to 21.0 m • sec^{-1} during impact.

Mean impact time of 12.0 msec was not so much different from 15 msec reported by Robert and Metcalfe (1968), but a little bit longer than the 8 msec Plagenhoef (1971) found. These inconsistencies with Plagenhoef's data may have been partially due to the changes in 1975 in the rules of football which altered the pressure of the ball from 1.0 to 0.6-0.7 kg/cm^2.

As shown in Table 1, all of the mechanical parameters were highly correlated to the ball velocity. Figure 1 shows the relationship of V_{Fb} and IT to V_B.

Maximum angular displacement of the ankle ($\triangle\angle A$) and the foot ($\triangle\angle F$) during impact were measured as described in Figure 2 which shows the relationship between these parameters and ball velocity. The

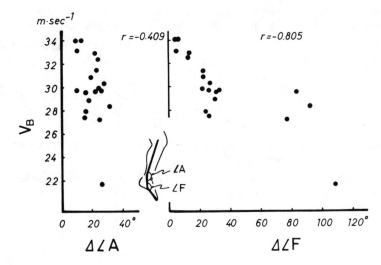

Figure 2 — Relation of maximum angular displacement of ankle and foot during ball impact to ball velocity.

coefficient correlations of $\triangle\angle A$ and $\triangle\angle F$ to V_B were -0.409 and -0.805, and to IT were 0.489 and 0.942, respectively. These results suggest that rigidity of the foot is more important than ankle fixation for a powerful kick, although the latter affects ball velocity to a certain extent.

From the film it was also observed that the passive plantar flexion of the forefoot during impact measured by $\triangle\angle F$ was quantitatively large when the ball was struck by the metatarsus and/or phalanges which form the anterior part of foot.

The mean vertical ground reaction force of the supporting foot at the movement of impact was 1820 N, 2.4 times body weight. No correlation was found between ground reaction force and ball velocity and no special pattern of force curves was found in relation to ball velocity.

Acknowledgments

Appreciation is extended to Dr. W. Bauman, Mr. P. Galbierz and all the other staff members of the Institut für Biomechanik in the Deutsche Sporthochschule, Köln, for their help and assistance throughout the study.

References

PLAGENHOEF, S. 1971. Patterns of Human Motion: A Cinematographic Analysis. Prentice-Hall, Englewood Cliffs, New Jersey.

ROBERT, E.M., and Metcalfe, A. 1968. Mechanical analysis of kicking. In: J. Wartenweiler and E. Jokl. (eds.), Biomechanics, pp. 315-319, Karger, Basel.

SHIBUKAWA, K. 1973. Effect of joint fixing when kicking a soccer ball. Bulletin of the Institute of Sport Science, The Faculty of Physical Education, Tokyo University of Education. 11:81-83.

TOGARI, H. 1970. Studies on the velocity of kicked ball and its relation to kicking form. The Proceedings of the Department of Physical Education, College of General Education, University of Tokyo. 5:5-12.

TOGARI, H., Asami, T., and Kikuchi, T. 1972. A kinesiological study on soccer. Jap. J. Physic. Educ. 16(5):259-271.

ZERNICKE, R.F., and Roberts, E.M. 1978. Lower extremity forces and torques during systematic variation of non-weight bearing motion. Med. Sci. Sports 10(1):21-26.

II.
SPORT

Keynote Lectures

Exercise Bioenergetics:
The Analysis of Some Sport Activities

F. Saibene, P. Cerretelli and P.E. di Prampero
Centro Studi di Fisiologia del Lavoro Muscolare
del C.N.R., Milano, Italy
Département de Physiologie de l'Université
Genève, Switzerland

Exercise physiologists seem much more concerned about the amount of energy a human subject expends to perform a given movement, rather than about the way he does it. Perhaps this depends on the fact that only energy consumption is easily measured, especially during repetitive movements, like those occurring in most sport activities, while the measurement of mechanical work often represents a formidable task. Overall energy expenditure, however, is by no means a reliable criterion for evaluating the skill of a movement or for inferring potential improvements in performance upon training. In fact, it does not provide analytical measurements of the various work components and their relative cost.

In many activities, work is mainly done against gravity, as in jumping and in weightlifting, or in accelerating the body or part of it, for example, in sprint running and throwing. By contrast, in other activities, most of the effort is sustained in overcoming the resistance of the fluid (air or water) surrounding the body, as in cycling or in swimming. Furthermore, not all work performed by muscles originates from simultaneous degradation of chemical energy, but energy can also be derived from mechanical energy previously stored by elastic components when the contracted fibers of muscles are forcibly stretched by an external force. In the latter case, the calculated mechanical efficiency of an exercise (η) may even exceed 0.25, the conventional value for the transformation of chemical energy into mechanical work, as determined (in vivo) in intact muscles.

Biomechanics not only deals with the description of the kinematics of human movements, but also with the analysis of the dynamics. Thus, the amount and the direction of all forces acting on the body should be assessed. The knowledge of the work performed and of the corresponding amount of energy expended allows one to calculate η for any particular movements or sport activities, and possibly to improve the performances.

About a century ago, Marey and Demeny (1885) made an estimate of the work components activated in running, the most common and natural of all sporting activities. Fifty years later, Fenn (1930) repeated those measurements in sprint running analyzing the various components of the total energy expended and their mechanical counterparts. The results are presented in the following diagram:

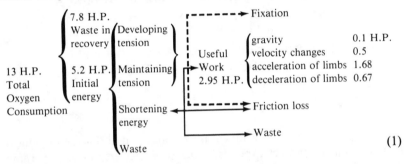

$$(1)$$

The first five items of the right hand column represent all the work actually performed, not only in running, but in all movements that lead to a displacement of the center of mass of the body. Actually, the first one, fixation, does not directly provide any displacement of the body center of mass, but it helps to position the different parts of the body properly to better perform useful work.

Another, perhaps more common classification is based on the division of total work performed by the body (W_{tot}) into its external (W_{ext}) i.e. the work resulting from the interaction between the body and its surroundings (ground, air, water), and internal (W_{int}) components. In activities requiring progression of the body in a forward direction, W_{ext} is performed: 1) to lift its center of mass (W_v); 2) to increase its forward speed (W_f); and 3) to overcome frictional drag (W_D) (air, water). W_{int} is performed: 1) for the movements of the limbs which do not lead to a displacement of the body center of mass; 2) to overcome internal friction or viscosity; and 3) for stretching the series elastic components during isometric contractions. All these components except the last two, can be measured, and their relative role in locomotion evaluated. It should be considered that not only the type of movement, but also its speed can determine the amount of energy expended for some of the work components. Thus gravity, drag, changes of forward speed, and acceleration/deceleration of the limbs relative to the trunk, may become in turn the main factors opposing progression. Moreover, there are instances in which one kind of energy can be converted into another, e.g. potential into kinetic, and vice verse, and the actual work performed can be less than the sum of its components.

The advantages of a combined approach, i.e. measurement of the energy expended and of the mechanical work performed, to the analysis of human movements will become apparent from the following sections, where different sport activities will be analyzed. Most of the work done in the past deals with aerobic repetitive events, that is, activities in which a steady state is attained. Under such conditions, the energy expenditure is readily measured and work can be evaluated from the mechanical analysis of a single movement cycle representative of the whole movement sequence. By contrast, the energy expenditure of short anaerobic events cannot be easily measured and the calculation of mechanical work often requires the analysis of a complete sequence of movements, as each of them may be different from the one preceding.

Walking and Running

Walking and running are the most natural forms of human locomotion. Despite the fact that the transition from walking to running takes place spontaneously, both the energetics and mechanics of walking and running, at least at physiological speeds, are very different from one another.

The energy cost of human locomotion is quite well known from the work of Chauveau (1901) and the extensive work of Margaria (1938) and of Margaria et al. (1963). These researchers showed: 1) that in walking there is an optimal speed at which the cost per unit distance (E) is minimal for any given incline of the ground and 2) that in running at aerobic speeds the energy cost (E) is almost independent of the speed when allowance is made for the minor fraction due to air drag (D). The latter at maximal aerobic speed ($6 - 7$ m $\cdot$ sec^{-1}) is 6% of the overall energy expenditure (Pugh, 1970).

In walking and running, gravitational and inertial forces account almost entirely for mechanical external work (W_{ext}). As clearly pointed out by the extensive work of Cavagna and his associates (1963, 1964, 1966, 1976), W_{ext} is imposed by the changes of kinetic (W_f) and potential (W_v) energy taking place during each step as a consequence of the bipedal mechanics that characterizes human gait. The mechanics of walking is pendulum-like so that kinetic and potential energy are almost 180° out of phase. Walking involves an alternate exchange of the two forms of energy: the contracting muscles only restore a small part of the energy that is not recovered at each step. Incidentally, the maximum percent recovery $(W_f + W_v - W_{ext})/(W_f + W_v)/100$ (Cavagna et al., 1976), occurs approximately at a speed at which W_{ext} is minimal.

In running, the mechanics is spring-like, so that kinetic and potential energy are almost completely in phase, which leads to a greater energy dissipation during each step. Part of this energy, however, can be stored by stretching the elastic elements of the previously contracted muscles, and recovered during the following cycle.

Internal power ($\dot{W}_{int}$) increases approximately with the square of the speed both in walking and in running (Cavagna and Kaneko, 1977). At a given speed, W_{int} is higher in walking than in running. In walking, W_{int} is always greater than W_{ext} while in running, W_{int} exceeds W_{ext} only at speeds higher than 5-6 m • sec-1. Total work per unit distance (W_{tot}) increases with speed in both forms of locomotion. As for the efficiency, in walking η is highest at about 1.2 m • sec-1, a finding that can be explained on the basis of the properties of the contractile components of the muscle, as described by its force-velocity relationship. The optimal speed of walking, therefore, is set by both a mechanical (exchange and recovery of mechanical energy) and a metabolic factor (efficiency of contraction). On the contrary, in running, η increases steadily with speed from 0.45 at 3 m • sec-1, to 0.75 at 6 m • sec-1. This depends on the ability of the muscles to store elastic energy (Cavagna and Citterio, 1974), particularly when this is preactivated before the contact of the foot with the ground during each running cycle (Komi, 1981).

The changes of W_{tot}, and of its fractions (W_v, W_f, W_{ext}, W_{int}) the energy cost per unit distance (E) and the overall efficiency (η) are plotted in Figure 1 as a function of the speed during walking (A) and running (B).

Competitive Walking

A particular type of locomotion is competitive walking. Measurements of energy expenditure made by Menier and Pugh (1968) and recently confirmed by our group (unpublished data) indicate that at speeds of 2 m • sec^{-1} or greater, competition walking is relatively less expensive than normal walking. The energy expended per unit time ($\dot{E}$), however, at the maximum speed attained in competition (4 m • sec^{-1}) is still 20% higher than in running at the same speed (see Figure 2). Cavagna and Franzetti (1981) observed that this type of locomotion is characterized by different phases: up to 3 m • sec^{-1}, kinetic and potential energy changes are most or less 180° out of phase as for normal walking. From 3 m • sec^{-1} onwards, kinetic and potential energy changes are almost in phase as in running and W_{ext} necessarily increases while percent recovery decreases. Since W_{int} at a given speed can be assumed to be the same for both normal and competition walking (Cavagna and Franzetti, 1981), it follows

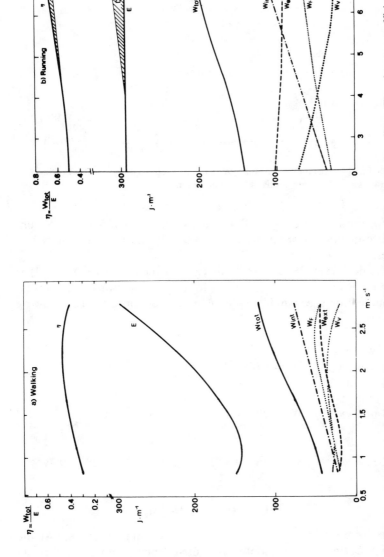

Figure 1 – Mechanical work (W_{tot}), together with its components (W_{int}, W_{ext}, W_v, W_f), energy expenditure (E) and apparent efficiency (η) as a function of the speed (v) in walking (A) and in running (B). The hatched area represents the maximum extra expended to overcome air resistance (D) and the related effect on the efficiency. (Partially redrawn from Cavagna et al., 1976-1977).

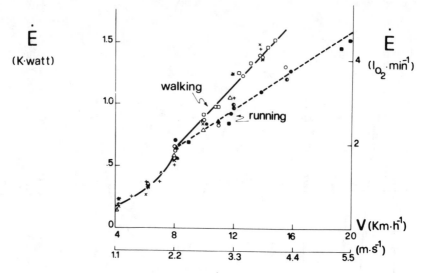

Figure 2—Energy expenditure per unit time ($\dot{E}$) as a function of the speed (v) for normal walkers (4-8 km • h^{-1}), competitive walkers and runners.

that W_{tot} must also increase. Observed η values reach 0.5 to 0.6 for running, presumably because of the recovery of elastic energy stored in muscles.

Mechanical work (W, J • m^{-1}) and energy expenditure (E, J • m^{-1}) as a function of the speed (v, m • sec^{-1}) for a 70 kg man walking at normal and competition speed and/or running may be obtained by the following equations:

normal walking
(1.4 m • sec^{-1}) : W = 63.7   E = 140

comp. walking
(> 3.3 m • sec^{-1}) : W = 29.7 + 35v   E = 123 + 63v + $0.7v^2$ (2)

running
(> 2.2 m • sec^{-1}) : W = $46v^{-1}$ + 83.3 + 19.9v   E = 280 + $0.7v^2$

Cross-Country Skiing

Another form of locomotion which is widely used in northern and Alpine countries during winter months is cross-country skiing. Elite cross-country skiers are characterized by the highest maximal aerobic power ever recorded. This finding is likely due to the involvement of large muscle groups of both the upper and lower limbs.

In a recent study from our laboratory (Saibene et al., 1981) on a group of cross-country skiers, the friction of the ski (F) as well as the energy expenditure (E) of the subject were measured when performing with two different styles, the so called "diagonal pole" in which the subject pushes alternatively on the poles at each stride and the "double pole" in which arms push simultaneously. The results indicate that (see Figure 3): 1) in cross-country skiing, at speeds below 5.5m $\cdot$ sec^{-1}, E is less than in running; 2) the double pole is more economical than diagonal pole technique; 3) the energy expended to overcome friction is a major component of the overall E; and 4) the difference in maximal aerobic speed attained between elite (athletes characterized by the highest aerobic power) and average skiers is comparatively smaller ($\triangle_{CC}$) than the homologous difference between runners ($\triangle_R$). This is due to the fact that in cross-country skiing, the increase in the cost of transport (E) with an increase in the speed is greater than in running.

The cost of level cross-country skiing (double pole) for a 70 kg man is given by the following equation:

$$E = 38 + 3.3F + 2.8v^2 \qquad (3)$$

where: E is J $\cdot$ m^{-1}; F is N; and v is m $\cdot$ s^{-1}

The efficiency of this exercise has not been calculated so far, since W was not measured.

All locomotion activities reported in the previous paragraphs deal with speeds lower than 6 m $\cdot$ sec^{-1} where the energy expenditure for the work performed against air resistance is only a minor fraction of the total (see for example D in Figure 3) and most of the work performed is against gravitational and inertial forces. However, there are other forms of locomotion in which maximal aerobic speed can be higher. Under these conditions, air resistance, which increases with the square of the speed, could represent a significant fraction of the whole energy cost.

Speed Skating

When racing on skates, aerobic speed may attain almost 11 m $\cdot$ sec^{-1} (e.g., in the 10,000 m race). From measurements of energy expenditure (E) at different skating speeds and from measurements of air drag (D) in a wind tunnel at different air velocities (di Prampero et al., 1976) it was calculated that at the speed of 11 m $\cdot$ sec^{-1} air resistance accounts for half of the energy output. The equation for calculating the cost of skating per unit distance for a 70 kg man is:

$$E = 73 + 0.8v^2$$

where: E is J $\cdot$ m^{-1} and v is m $\cdot$ sec^{-1} \qquad (4)

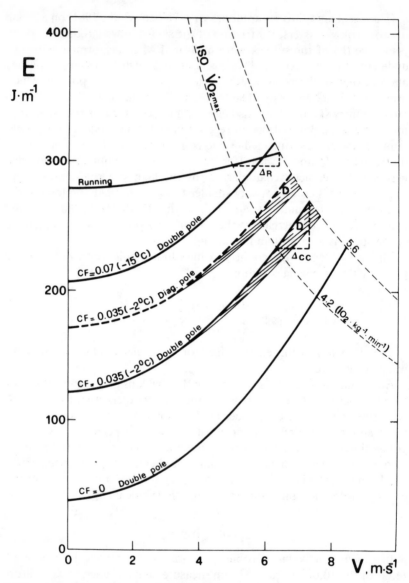

Figure 3 — Energy expenditure (E) as a function of the speed (v) in cross-country skiing at different C_F's and using different techniques, and in running. The hatched area represents the maximum energy expended to overcome the air resistance (D). The interrupted curves represent different iso-$\dot{V}_{O_2}$ max values. $\triangle$ R and $\triangle$ CC are, respectively, the difference in maximal speed expected between top class ($\dot{V}_{O_2}$ max = $5.6\ell \cdot min^{-1}$) and average athletes ($\dot{V}_{O_2}$ = $4.2\ell \cdot min^{-1}$) in running and in cross-country skiing (see text).

The first term at the right side of the equation is constant, independent of speed, and represents all of the energy expended except the fraction due to D. This term is ¼ of the corresponding term for running, which indicates that in these two forms of locomotion the external and internal work components are substantially different. As a matter of fact, due to the particular type of progression, the speed is more likely to be uniform in skating and the variations of kinetic energy $(W_F = \frac{m}{2}(v_1^2 - v_2^2)$, where m is the mass of the subject and v_1 and v_2 the maximal and minimal speed with each stroke) correspondingly smaller. The squared term of the equation, which is due to D, depends on the projected area (Ap) of the subject's body. The skater, in order to achieve his highest possible speed, should reduce to a minimum his Ap. This happens to be the case since in skaters measured Ap is only $0.34m^2$, a figure to be compared with homologous figures of 0.63 m^2 in walkers and $0.47m^2$ in runners. Should a skater assume the posture of a middle-distance runner (max Ap), the cost of skating at 11 m • sec^{-1} would increase by more than 20% (see Figure 4). Since the body surface area (S.A.) increases approximately in proportion to two-thirds power of the weight, it turns out that heavier subjects are comparatively better suited for skating than lighter ones for equal muscular power per unit b.w. Thus, while high body weight taxes the performance of walkers, runners, and particularly cross-country skiers, it appears to be advantageous for skaters.

Cycling

The bipedal mechanism of human locomotion, such as in walking and running and, with some differences, in level skiing and speed skating, leads to a vertical displacement of the center of mass of the body and to simultaneous changes of its forward speed with each step. If the alternate movements of the legs can be transformed by an appropriate system of levers into a rotary movement, as on a bicycle, the vertical displacement of the body center of mass as well as the changes of its forward speed, become negligible, provided the force exerted by the muscles on the pedals is kept constant. In the latter situation, W_{ext} is performed mostly against frictional forces. In cycling, the resistance to motion (R) is represented by the sum of the rolling resistance (Rr) and air resistance or drag (D). Rr depends on the characteristics of the road and tire surfaces, on the inflated pressure of the tires and on the weight of both subject and bicycle and is independent of speed. At a given speed (v), D depends on Ap, as already mentioned, on the drag coefficient (C_D), and on air density (ρ). D increases as the square of the air velocity, the latter being equal to the forward speed (v) in calm air, according to the equation:

$$D = 0.5 \, C_D \, Ap \, \varrho \, v^2 \qquad (5)$$

By towing a number of subjects on racing posture on a bicycle over a flat track, R was measured at different constant speeds and the external mechanical work per unit distance was calculated as (di Prampero et al., 1979):

$$W_{ext} = 3.1 + 0.2v^2$$

where W_{ext} is in J $\cdot$ m^{-1} and v is m $\cdot$ sec^{-1} (6)

These results are similar to those obtained by Pugh (1974).

The internal power ($\dot{W}_{int}$ developed in cycling was calculated by di Prampero et al. (1979) using the method described by Fenn (1930) for sprint running, who found it depended on the pedalling frequency (f) according to the equation:

$$\dot{W}_{int} = 7 \, F^3$$ (7)

where f is expressed in sec^{-1}. If the ratio between the number of revolutions of the pedal shaft and that of the wheels is kept constant at all speeds, frequency and velocity are linearly related, thus allowing one to express $\dot{W}_{int}$ as a function of v. Under these conditions, W_{tot} can be calculated from the previous two equations. For instance, should the development of the pedalling cycle be 8 m, f would be v/8 and total work:

$$W_{tot} = 3.1 + 0.21 \, v^2$$ (8)

It can be seen that at the maximal speed (about 14 m $\cdot$ sec^{-1}) W_{int} accounts only for about 6% of the total work. Pedalling frequency affects only slightly the mechanical efficiency of cycling (Seabury et al., 1977): assuming, therefore, an efficiency value of 0.25 also E can be calculated for all speeds according to:

$$E = 12.5 + 0.8 \, v^2$$ (9)

where the units are the same as for the previous equations. It may be observed that while the value attributable to air resistance is very close to the corresponding value for skating, the first value is about 5 times smaller (see Figure 4).

As already mentioned, air resistance to motion is linearly related to ρ and hence, for a given temperature, to barometric pressure. The present 1 hr world record for cycling was in fact achieved in Mexico City, at a barometric pressure of about 580 mmHg, so that D was reduced to about 75% of its sea level control value. In Figure 5, the net energy consumption ($\dot{E}$) during cycling is plotted as a function of the speed (v). The fami-

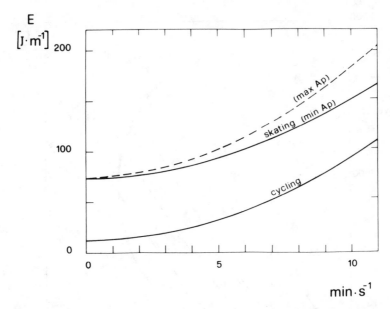

Figure 4—Energy expenditure (E) as a function of the speed (v) in cycling and in speed skating. The interrupted curve represent the energy expenditure calculated for a skater having the same posture of a middle-distance runner.

ly of curves refers to different altitudes and the thick line indicates the average relationship between maximal aerobic power at any given altitude and maximal cycling speed for an athlete whose $\dot{V}_{O_2}$ max at sea level is 5.34 1 • min^{-1}, i.e., that of the present record holder. It may be noted that the ideal altitude for the best performance appears to be somewhere between 3500 and 4000 m above sea level.

Aquatic Locomotion

During aquatic locomotion, the speed of progression is very low, yet the frictional drag represents the major component of the whole cost of transport. This depends on the fact that the density and the viscosity of the medium through which the body moves are greatly increased (800 and 50 times respectively). In this section swimming and rowing (man-powered aquatic locomotion) will be analyzed.

Swimming

In swimming, external mechanical work is performed essentially to overcome water resistance. In fact: 1) the speed oscillations with each stroke are negligible at least for the crawl (Faulkner, 1968), and 2) the work per-

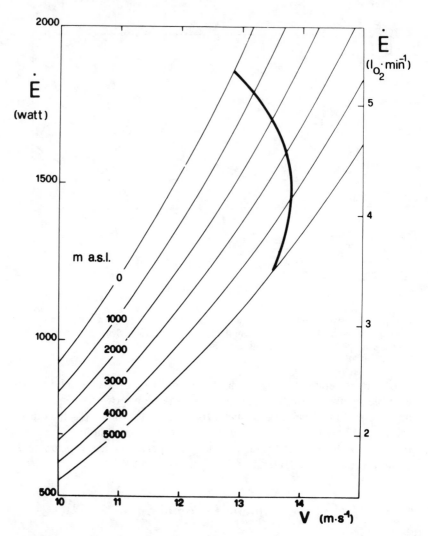

Figure 5 — Energy expenditure per unit time ($\dot{E}$) as a function of the speed (v) in level cycling at various altitude above sea level (a.s.l.). The thick curve indicates the average reduction of $\dot{V}_{O_2}$ max due to altitude for an athlete whose $\dot{V}_{O_2}$ max is 5.34 1 • min^{-1} at sea level.

formed against gravity is very minor. Mechanical power output is given by the product of body drag (D) times the speed of progression (v):

$$\dot{W} = D \cdot v \qquad (10)$$

During actual swimming, D can be assessed by adding, or subtracting, known extra drag loads to, or from, swimmers moving at constant speeds (see Figure 6) (di Prampero et al., 1974). Under these conditions,

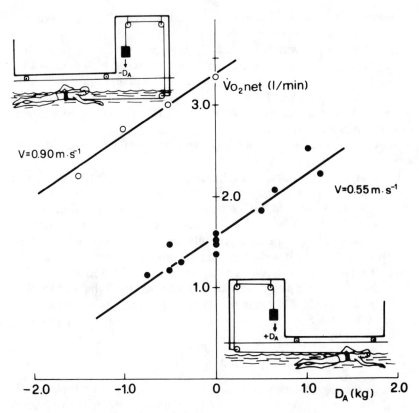

Figure 6—Oxygen consumption above resting as a function of the added drag in one subject swimming at two different speeds (0.90 m • sec^{-1}, open circle; 0.55 m • sec^{-1}, filled circles). The insert shows the experimental setup for adding (or subtracting) a known drag (see text) (Redrawn from di Prampero et al., 1974).

the swimmer's energy expenditure ($\dot{E}$) is linearly related to the added drag. Extrapolation of this relationship to the subject's resting $\dot{V}_{O_2}$ yields the force that, if applied to the swimmer in the forward direction, would allow him to progress at that given speed without any increase in energy expenditure. This force is assumed to be equal and opposite to the swimmer's body drag (D) at the considered speed. Furthermore, if only negative added drags are used (i.e., if the swimmer's body drag is reduced), this method allows one to calculate the energy expenditure for speeds greater than that corresponding to the subject's $\dot{V}_{O_2}$ max. In this case, back-extrapolation of the added drag/$\dot{V}_{O_2}$ relationship to added drag = 0 yields the energy expenditure that would allow the swimmer to swim free at the considered speed. From the above data (i.e., D, v, $\dot{E}$) the overall efficiency of progression can be easily calculated as:

$$\frac{\dot{W}}{\dot{E}} = \frac{D \cdot v}{\dot{E}} = \eta \tag{11}$$

Di Prampero et al. (1974) and Pendergast et al (1977) have shown that in swimming the front crawl:

1) D (N) increases with v (m $\cdot$ sec^{-1}): D $= 56 \cdot v^{1.4}$, at all speeds. It is about twice that observed for subjects passively towed in water, presumably because of the movements of the head and limbs.

2) E amounts to 0.98 kJ $\cdot$ m^{-1} at speeds up to 1.1 m $\cdot$ sec^{-1} and increases at higher speeds to 1.7 kJ $\cdot$ m^{-1} at 1.8 m $\cdot$ sec^{-1}(the maximal speed in the studies cited).

3) η increases linearly with the speed from 0.04 at 0.6 m $\cdot$ sec^{-1} to 0.08 at 1.8 m $\cdot$ sec^{-1}.

In addition, it was observed (Pendergast et al., 1977) that in women having similar technical ability, when expressed per unit of surface area of the body (S.A.) E, is about 0.8 that of man, independent of the speed. The latter finding, showing that females are naturally better swimmers than males, can be attributed essentially to the lower torque (Tq) of the female body. In fact, it was observed that, when lying horizontally in the water (supported by a rigid beam) the product of the weight at the feet times the distance nipples to feet for women is about ½ that of men (7 N $\cdot$ m in women vs. 14.4 N $\cdot$ m in men) Pendergast and Craig, 1974). In addition, E per unit S.A. (J $\cdot$ m^{-1} $\cdot$ m^{-2}) is linearly related to Tq (N $\cdot$ m) as described by:

$$E = 400 + 23 \cdot Tq \qquad (12)$$

for both men and women. This shows that, at least under these conditions, the difference in energy cost of swimming between the two sexes can be explained entirely in terms of S.A. and torque.

Kayaking

Using the same procedure described above for measuring body drag on swimmers, the drag of a loaded boat (D$_{LB}$), the energy cost of progression (E), and its mechanical efficiency (R) were determined for kayaking. The results are summarized in the following table (Pendergast et al., personal communication, 1981):

Table 1

Energy Expenditure for Kayaking

Speed (m $\cdot$ sec^{-1})	D$_{LB}$ (N)	E (kJ $\cdot$ m^{-1})	η
1.2	12	0.20	0.06
1.7	20	0.25	0.08
2.0	30	0.28	0.11
2.3	48	0.41	0.12

Rowing

A schematic representation of the forces acting externally on the oar in rowing is shown in Figure 7 (di Prampero et al., 1971; Celentano et al., 1974). Such forces are: 1) the water reaction against the blade of the oar $\vec{p}$. This is the resultant of two components, $\vec{p_a}$ acting in the same direction as the shell and useful for its progression and $\vec{p_t}$, perpendicular to $\vec{p_a}$, causing a strain on the shell and therefore wasted; 2) the force exerted by the rower (T) which as a first approximation acts along the same direction as the boat; 3) the resistance to the progression of the shell (D) which is applied to the oarlock pin of each oar; and 4) the oarlock reaction (P_t) to the force $\vec{p_t}$.

The work performed by the propulsive force F ($= T + p_a$) during a number of complete rowing cycles to travel a given distance (s) was measured by means of strain gauges mounted on the oarlock pin as:

$$W = \int_0^s F \cdot ds \simeq \bar{v} \cdot \int_0^t F \cdot dt \qquad (13)$$

where t = the time required to cover the distance (s) and $\bar{v}$ = the mean speed.

Since, neglecting deformation of the oar and the shell, the work performed by F (the propulsive force) and by D (the resistance to progression) over the distance (s) are equal, it necessarily follows that;

$$\bar{D} = \frac{W}{s} = \frac{\bar{v}}{s} \cdot \int_0^t F \cdot dt \qquad (14)$$

where $\bar{D}$ = the mean resistance to progression.

$\bar{D}$ (N) was therefore calculated and found to increase approximately with the square of v (m $\cdot$ s^{-1}) as described by:

$$\bar{D} = 4.7 \, \bar{v}^{1.95} \qquad (15)$$

Hence, the mechanical power ($\dot{W}_{ext}$, W) necessary to maintain the progression of the boat is given by:

$$\dot{W}_{ext} = \bar{D} \cdot \bar{v} = 4.7 \cdot v^{2.95} \qquad (16)$$

The relationship between W_{ext} and v is shown in Figure 8. The individual $\dot{V}_{O_2}$ value were obtained indirectly from the heart rate/$\dot{V}_{O_2}$ relationship, previously determined during simulated rowing in a basin on the same group of oarsmen. In the speed range from 3 to 4.5 m $\cdot$ sec^{-1}, the apparent efficiency of rowing varies between 0.10 and 0.16. The value of η is influenced, among other factors, by rowing frequency and by the fact that part of the mechanical work of the rower (W_{ext}-W_{useful}) is lost in the deformation of the shell and in the water friction of the blades of the oars.

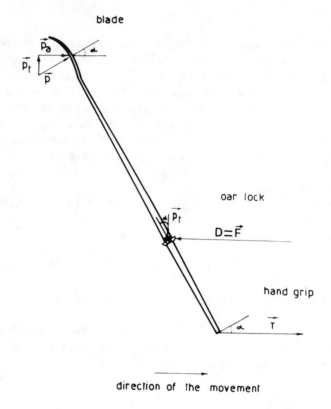

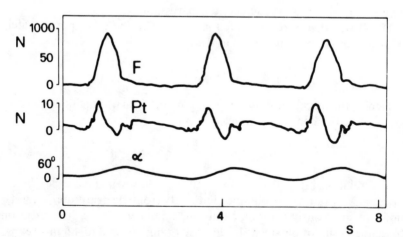

Figure 7—Schematic representation of the forces acting externally on the oar and actual record of axial force (F), transverse force ($\overrightarrow{p_t}$) and oar position angle (α) as a function of time (see text) (Redrawn from Celentano et al., 1974).

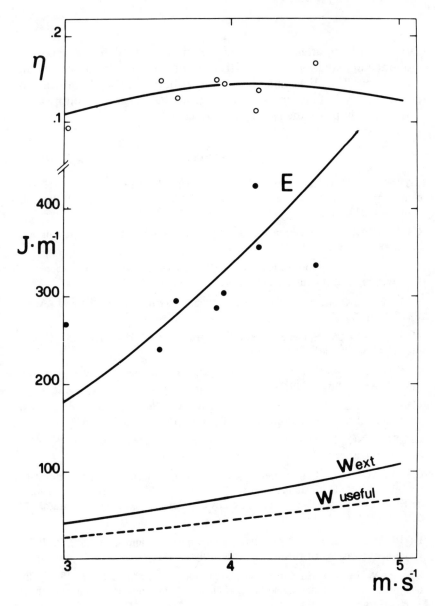

Figure 8—Mechanical work performed (W_{ext}) and work utilized for progression (W_{useful}) energy expenditure (E) and apparent efficiency (η) of rowing as a function of the speed (v) (Data from di Prampero et al., 1971).

Conclusions

In this article an attempt was made to analyze: 1) the forces opposing motion; and 2) the energy cost of transport on different forms of human locomotion. The above analysis can be profitably applied in practice to predict the level of performance, to estimate the energy expenditure and, perhaps, to improve the skill of the athlete.

References

CAVAGNA, G.A., and Citterio, G. 1974. Effect of stretching on the elastic characteristics and the contractile component of frog striated muscle. J. Physiol. (London), 239:1-14.

CAVAGNA, G.A., and Franzetti, P. 1981. Mechanics of competition walking. J. Physiol. (London) 315:243-251.

CAVAGNA, G.A., and Kaneko, M. 1977. Mechanical work and efficiency in level walking and running. J Physiol. (London) 268:467-481.

CAVAGNA, G.A., and Margaria, R. 1966. Mechanics of walking. J. Appl. Physiol. 21:271-278.

CAVAGNA, G.A., Saibene, F., and Margaria, R. 1963. External work in walking. J. Appl. Physiol. 18:1-9.

CAVAGNA, G.A., Saibene, F., and Margaria, R. 1964. Mechanical work in running. J. Appl. Physiol. 19:249-256.

CAVAGNA, G.A., Thys, H., and Zamboni, A. 1976. The sources of external work in level walking and running. J. Physiol. (London) 262:639-657.

CELENTANO, F., Cortili, G., di Prampero, P.E., and Cerretelli, P. 1974. Mechanical aspects of rowing. J. Appl. Physiol. 36:642-647.

CHAVEAU, A. 1901. La dépencse énergétique qu'entraînent respectivement le travail moteur et le travail résistant de l'homme qui s'élève ou descend sur la zone de Hirn. Evaluation d'après l'oxygène absorbé dans les échanges respiratoires. C.R. Acad. Sci. 132:194-301.

DI PRAMPERO, P.E., Cortili, G., Celentano, F., and Cerretelli, P. 1971. Physiological aspect of rowing. J. Appl. Physiol. 31:853-857.

DI PRAMPERO, P.E., Cortili, G., Mognoni, P., and Saibene, F. 1976. The energy cost of speed-skating and the efficiency of work against the air resistance. J. Appl. Physiol. 40:584-591.

DI PRAMPERO, P.E., Cortili, G., Mognoni, P., and Saibene, F. 1979. Equation of motion of a cyclist. J. Appl. Physiol. 47:206-211.

DI PRAMPERO, P.E., Mognoni, P., and Saibene, F. 1979. Internal power in cycling. Experientia 35:925.

DI PRAMPERO, P.E., Pendergast, D.R., Wilson, D.W., and Rennie, D.W. 1974. Energetics of swimming in man. J. Appl. Physiol. 37:1-4.

FAULKNER, J.A., 1968. Physiology of swimming and diving. In: H. Falls (ed.), Exercise Physiology, pp. 415-446. Academic Press, New York.

FENN, W.O., 1930. Frictional and kinetic factors in the work of sprint running. Amer. J. Physiol. 92:582-611.

FENN, W.O. 1930. Work against gravity and work due to velocity changes in running. Amer. J. Physiol. 93:433-462.

KOMI, V.P. 1981. Inegrative approach of biomechanics and physiology in the study of locomotion. Int. Conf. Sports Med., Utrecht.

MAREY, J., and Demeny, G. 1885. Mesure du travail mécanique effectué dans la locomotion de l'homme. C.R. Acad. Sci. 101:905-909.

MAREY, J., and Demeny, G. 1885. Variation du travail mécanique dépensé dans les différentes allures de l'homme. C.R. Acad. Sci. 101:910-915.

MARGARIA, R. 1938. Sulla fisiologia e specialmente sul consumo energetico della marcia e della corsa a varie velocità ed inclinazioni del terreno. Atti Reale Accad. Naz. Lincei 7:299-368.

MARGARIA, R., Cerretelli, P., Aghemo, P., and Sassi, G. 1963. Energy cost of running. J. Appl. Physiol. 18:367-370.

MENIER, D.R., and Pugh, L.G.C.E. 1968. The relation of oxygen intake and velocity of walking and running, in competition walkers. J. Physiol. (London) 197:717-731.

PENDERGAST, D.R., and Craig, Jr., A.B. 1974. Biomechanics of floating in water. The Physiologists 17:305.

PENDERGAST, D.R., di Prampero, P.E., Craig, Jr., A.B., Rennie, D.W., and Wilson, D.W. 1977. Quantitative analysis of the front crawl in men and women. J. Appl. Physiol. 43:475-479.

PUGH, L.G.C.E. 1970. Oxygen intake and treadmill running with observations on the effect of air resistance. J. Physiol. (London) 207:823-835.

PUGH, L.G.C.E. 1974. The relation of oxygen intake and speed in competition cycling and comparative observation on bicycle ergometer. J. Physiol. (London) 241:795-808.

SAIBENE, F., Cortili, G., Colombini, A., and Magistri, P. 1981. Energy cost of cross-country skiing. Int. Conf. Sports Med., Utrecht.

SEABURY, J.J., Adams, W.C., and Ramey, M.R. 1977. Influence of pedalling rate and power output on energy expenditure during bicycle ergometry. Ergonomics 20:491-498.

Application of Biomechanics Research to Sport

W. Baumann
Institut für Biomechanik
Deutsche Sporthochschule Köln, Köln, FRG

Biomechanics has developed rapidly in the last decade throughout the entire world. Substantial progress has been made in the areas of technical instrumentation and theoretical methodology. More and better qualified people are working in more and better equipped laboratories than ever before. If this development is used as an indicator of the actual importance and capacity of the discipline, biomechanics has been very successful. Within this article, the considerations will be restricted to biomechanics of sport and focussed mainly on its application.

The tasks of biomechanics of sport can be specified as follows:

1. Characterization of the mechanical structure of the human body and of sports techniques.

2. Identification of factors influencing performance.

3. Improvement of sports techniques and of the efficacy of training processes.

The first task yields a quantitative description of the human body and its motions in terms of mechanical properties, kinematics and kinetics, which are fundamental prerequisites for the solution of most of the problems. The extent and accuracy of this description depend largely on the basic assumptions and idealizations and on the capacity of the available methods of data acquisition and data processing. The second point provides the conditions for a systematic and efficient improvement of performance. This can be achieved by defining the formal relations between the influencing factors and performance or by casual explanations of the phenomena. Problems connected herewith are mainly concerned with the field of modelling, the theoretical approach. The last item emphasizes the area of application, the prerequisites of which might be provided perhaps by very sophisticated theoretical optimization procedures or in a relatively simple manner: objectification of the input-output relations of the variables in training, feedback to the process and systematic experiments in the training process. In whatever way the approach is carried out the scientists must be concerned about the practical application of their results otherwise they cannot prove their hypotheses.

Since the methods of data acquisition and processing as well as the modelling concepts are of decisive importance for the correctness and applicability of the results, a critical examination of some points seems to be justified.

In the area of data acquisition and processing the technical equipment is generally accepted to satisfy most of the requirements. Steadily improved technical specifications of the measuring instruments and the irresistible promulgation of digital computers indeed have contributed to substantial improvements in the measuring technology. These include minimizing the size and weight of complex measuring systems, improved frequency response especially of force-platforms, semi-automatic and automatic numerical evaluation of optical images, on-line processing of synchronized kinematic and kinetic data, and interactive processing of large amounts of data via digital computer, all of which offer wonderful possibilities for practical and theoretical experiments for modelling and simulation.

In practical applications, however, we are often near the limits of the capacity of our methodology. This can be illustrated by some examples from the areas of measurement methodology and modelling.

Measuring Methods

The analysis of sports techniques requires measuring methods with least possible interference with the athlete, high accuracy of measurement and high measuring frequency. First, we must realize that the quasi-static human gait and fast sport motions are very different with regard to intensity and signal frequency. Therefore, one cannot simply apply the many reliable findings and refinements in measuring methodology from gait studies (e.g., Winter et al., 1974) to the much faster events in sports.

Kinematic Data

Concerning the recording of kinematic data such as displacement as a function of time, cinematography and videorecording are the only noninterfering measuring instruments which can be applied without the knowledge of the subject under test. The time resolution of cine recording is determined by the frame rate which should be selected in relation to the movement to be examined. The frame-by-frame analysis provides the

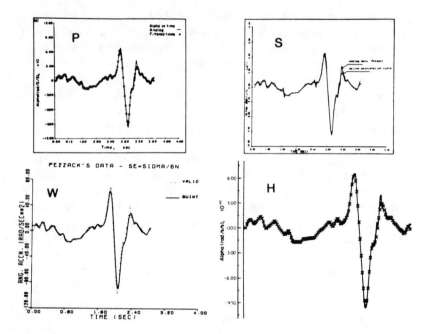

Figure 1—Four different derivative determining techniques applied to data of Pezzack et al. (1977). P = Pezzack, W = Wood and Jennings, 1979; S = Soudan and Dierckx, 1979; H = Hatze, 1981.

basic displacement-time data of the motion from which positions of the joints and body angles can be calculated. From these data the corresponding velocity-time and acceleration-time functions can be inferred. Measurement errors and the properties of the numerical differentiation process require exceptional care in the data analysis. Special attention has been devoted to that problem in recent years. Pezzack et al. (1977) presented the results of an assessment of derivative determining techniques where a horizontal arm movement was recorded simultaneously by an electrogoniometer, accelerometer and cinefilm. The experimental design took into account only the contamination of the data with noise, however. Other authors using the experimental data of Pezzack et al. (1977) proposed different techniques for data processing (Wood and Jennings, 1979; Soudan and Dierckx, 1979; Hatze, 1981). Comparing results of the different authors (see Figure 1), it is clear that all procedures yield good agreement with the analog acceleration signal. Now we apply a simple 9-point parabola fit to the same data and again, the resulting acceleration is acceptable (see Figure 2). Obviously in this case the sampling frequency was so high that the data processing technique could hardly influence the result. Moreover, the measurement errors are purely stochastic with such high frequencies, so that the low pass characteristics of all the methods used smoothed them out in a very similar way.

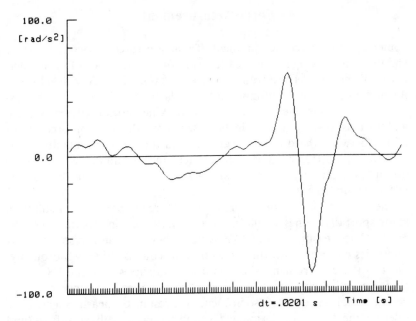

Figure 2—Nine-point parabola fit applied to Pezzack's data.

So why do we have so many problems with the derivation of acceleration of a long jumper's center of gravity (C.G.)? First, because our main errors are not random, secondly because they usually are of lower frequency than the assumed noise, and third because we do not know the frequency spectrum of our signals. The above mentioned procedures do not solve our problems in the analysis of sports movements. An experimental test could help to clarify the situation: 3-dimensional accelerometers could be fixed at the ankle and the knee joint respectively, and filmed during a jump or sprint stance. If the assumption of a rigid tibia is true, the accelerations in the direction of the tibial axis at both points must be equal. If the movements of two points are so closely related to each other, it is quite appropriate to take this into account in the numerical treatment of the coordinates and not to do so as if the accelerations are independent. If we do not want to continue selecting arbitrarily the frame rate and the differentiation procedures, we must obtain reliable information about the frequencies of our measuring signals. Their maximum is determined by the movement of the foot at the beginning of the stance phase. Until we have these data, the previously mentioned different methods actually cannot be assessed and the resulting data may very well be overestimated.

Force Measurements

Sometimes the direct measurement of reaction forces between a surface and the human body is used as a criterion for the accuracy of the accelerations of the c.g. derived from displacement-time data. A critical look at the measuring capacity of modern force platforms will show to what extent this is reasonable. Together with the kinematics of the movement and the mass distribution of the body segments, the internal forces transmitted between the different segments and the net moments about the axes through different joints can be calculated. Again in gait studies the use of force platforms offers many fewer problems than in the analysis of sports movements.

The comfort of direct measurements of force as a decisive quantity in many sport disciplines should not lead to a careless interpretation of the results. The measuring accuracy, particularly the dynamic response characteristics of force platform data should be considered very thoroughly. Most of these instruments show natural frequencies of about 100 to 200 Hz (or even below that), varying with the different components. Inevitably cross-talking effects also influence the results of measurement. Considering the frequently examined support phases with very large and rapidly changing forces, we are in a similar situation as in the kinematic analysis. There is no information about the frequency spectrum of the force signal. When we examine the force graphs of powerful support phases, the force rates often allow the interpretation as resonance effects.

Hochmuth and Knauf (1979) have shown in theoretical calculations of models and the transfer function of force platforms, how different the actual applied force and the measuring signal can be. Our own model simulations and corresponding experiments with two force platforms show that in reality we have appreciable errors, depending on the time history of the applied force and the properties of the surface material of the platform. An example is given in Figure 3 for a step-jump. The difference in the peak force is 900 Newtons, or 35%. In the region of slower force variations, however, the differences between both curves can be disregarded. In agreement with Hochmuth's and Knauf's calculations the total impulse—inferred by integration of the force-time function—can be determined very exactly (Baudzus, 1980).

With similar models it has been shown (Hochmuth and Knauf, 1979) that the effect of the elasticity of the whole body on the measuring result is negligible. That means the measured force represents the force applied to the center of gravity of the body and determines its motion quite precisely. But where is the C.G.? We use individually adjustable models to calculate this fictitious point as a function of the spatial coordinates of the body joints which connect the body segments, the latter are assumed to be rigid (except for the trunk which sometimes can be modified

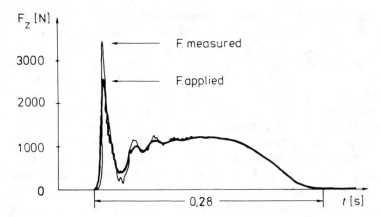

Figure 3—Force measurement errors from an experiment with two piezoelectric platforms.

according to the hip/shoulder distance). This may be correct and can be checked experimentally for selected body positions. In case of a sudden impact force applied to the body via the lower extremity as in the beginning of the stance phase of a jump, the situation changes to a large degree, however. The passive movement apparatus, the skeleton, moves approximately according to the movement defined by the joint axes. Most of the soft tissues, however, especially the inner organs but also the muscles are not coupled tightly to the skeleton and move, but these movements are not considered. It is probably a low estimate if we assume that wandering masses of the order of 10 to 15% of total body mass are neglected in nearly all of our kinematic analyses. This applies most notably in sudden impact situations of fast sports events.

Considering the measurement errors of the force platform and the uncertainty of the c.g. derived from the kinematics, it would be very unusual if there was any close agreement between force and the product of mass and acceleration. In a comparison, the direct force measurements are not so well suited as outer criterion as sometimes assumed. Single force values should not be compared and are very questionable in the impact phase of the stance. Since no mechanical laws are suspended, however, the total impulses should coincide much better.

Modelling

In addition to the measuring methodology, the type of models used in biomechanics is of fundamental importance. Models generally are more or less minor drastic idealizations of the real object which, being extremely complex, in an analysis cannot be represented exactly.

Structural and functional properties of the human body are represented by mostly simple mathematical models based on typical assumptions. The human body consists of rigid segments connected by joints with up to three rotational degrees of freedom. According to individual segmental dimensions and the total body mass, reliable information about segmental masses and their respective distributions can be obtained. The possible shortcomings of the models used in dynamic applications already have been considered briefly.

The process of identification of the essential factors constituting the performance at different levels is one of the prerequisites of effective teaching and coaching. Two different methods to obtain valuable information in that direction have been used: 1) the empirical/statistical approach, and 2) the theoretical/mechanical approach.

For the most part, the empirical/statistical approach is selected. Based on empirical data about kinematic and kinetic variables of many single events, attempts are made to infer general principles or laws. This is carried out by different analysis of variance techniques and multiple regression analysis. The method is an inductive one. The application of statistical procedures yields mainly formal relations and is characteristic for situations where insufficient knowledge about a complex object comes together with too generally formulated questions. The advantage of an arbitrary selection and combination of variables stands against some inherent risks of the approach. These risks include: 1) neglect of mechanical laws whether by not using them or by interpretations inconsistent with them, 2) loss of information by giving up the time history of the variables, 3) application of an unscientific (shotgun) method in selecting the variables, and 4) ignoring the conditions of applicability.

Despite the advantages of this approach in a first screening, it will probably not contribute that much to an explanation of the phenomena.

Since human movement is governed to a considerable extent by mechanical laws, it appears promising to describe and to explain the movement in terms of mechanics. This approach is deductive and utilizes fully the mechanical laws, the classical procedures of solving the equations of motion, and all the available knowledge about the mechanical structure and function of the movement apparatus. When we are analyzing simple problems, this model is extremely useful. Facing the problem of identifying the essential factors contributing to an actual performance, which requires a complete causal explanation of the event, the system to be studied becomes extremely complex and turns out to be mechanically undetermined. The number of unknowns exceeds the number of equations.

In order to overcome this calamity (the overshooting degrees of freedom) one has to establish criteria according to which the biological system selects a special solution among the unlimited possibilities. Assuming that nature in the course of evolution has taken into account

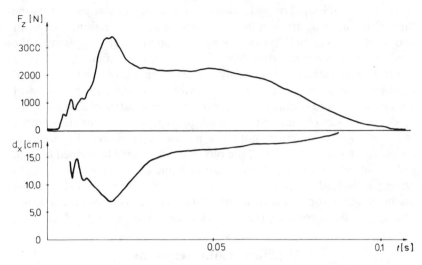

Figure 4 — Stance phase of a sprinter. F_z = vertical force, d_x = horizontal distance of the point of application of F_z to the ankle joint in the direction of the long axis of the foot.

some physical concepts such as efficiency and economy of body construction and motion, a number of different criteria have been introduced e.g., minimizing forces in muscles and ligaments (Cappozzo et al., 1975; Seireg and Arvikar, 1975) and minimizing energy expenditures (Nubar and Contini, 1961; Becket and Chang, 1968; Hatze, 1976).

Of course, there is much evidence that in long distance walking, energy can be minimized and it seems reasonable to reduce the forces of the movement apparatus in the tissues, inasmuch they do not cause atrophy as a consequence of underload. In sports, however, it is very unlikely that summed up energies or forces are minimized. Except for endurance disciplines, in most sports movements the total power has to be maximized and consequently the forces in muscles, joints, and ligaments will increase. Useful criteria may rather be connected with the efficiency of muscular effort and with the uniformity of relative stress distribution with respect to the ultimate stresses in the different biological materials during violent sports activities.

For static cases, Pauwels (1965) demonstrated convincingly that the construction features of the bones and joints as well as the function of the one-and two-joint muscles follow the principle of reduction of tension in the parts of the the movement apparatus with the most stress. A dynamic example from our own studies is given in Figure 4. In the stance phase of the sprint, the higher the force, the more the Achilles tendon is loaded. To the same extent, the force increases as the horizontal distance of the point of application from the ankle joint is decreased, thus loading the joint and preventing the tendon from becoming overloaded. Not considering the additional imperfections and unsolved problems such as the difficulties in the determination of antagonistic muscle action, the role of two-joint muscles and the viscoelastic properties of the tissues, the mathematical treatment of a 15-or 17-link system in 3-dimensional space is not an easy task even when only the net moments of the muscles and ligaments are to be determined. The effort to optimize the motion of such a system is so prohibitive that even further drastic reductions of the model such as confinement to a planar movement, reducing the number of body links and others do not allow a reasonable solution for practical purposes (Bauer, 1980). Thus, the physical reality may be pushed back in order to permit a mathematical solution. Remembering the difficulties in getting reliable data of actual motions, the discrepancies between the mathematically sophisticated model of the physically poor phenomenon and the far more complex reality is considerable.

Practical Considerations

Obviously the biomechanics of sport is not ready to offer closed solutions of complex problems. Neither do we have the adequate instrumentation to investigate violent sports movements with sufficient accuracy, nor do we have sufficiently valid concepts to explain and to interpret the phenomena. Consequently, we must try 1) to overcome these deficiencies and 2) to answer less complicated questions. In the practice of sports there are numerous problems which can be solved only in cooperation with biomechanics.

Without great theoretical effort, the instrumentation of biomechanics can be used very effectively in the training process for: 1) objectification and comparison of the movement technique in training and competition, 2) feedback of selected variables to the coach/athlete, and 3) comparison of technical and conditional elements of training exercises and competitive events.

In this connection the comprehensive work of Farfel (1977) and his school must be mentioned. In numerous applications they proved the advantage of biomechanical (and other) rapid information systems in the process of learning and controlling movements at different levels of performance and in many different situations.

Many more engineering analyses of simple partial systems should be carried out. For empirically obtained sports techniques, mechanical models have to be constructed and with parameter variations, so the influence of different variables can be studied separately. An excellent example in this approach is the recently published simulation study of Morawski et al. (1977) of pole vaulting in which consideration was given to the influence of basic parameters of the athlete and the pole on the reached height, which resulted in very successful individual selection of poles.

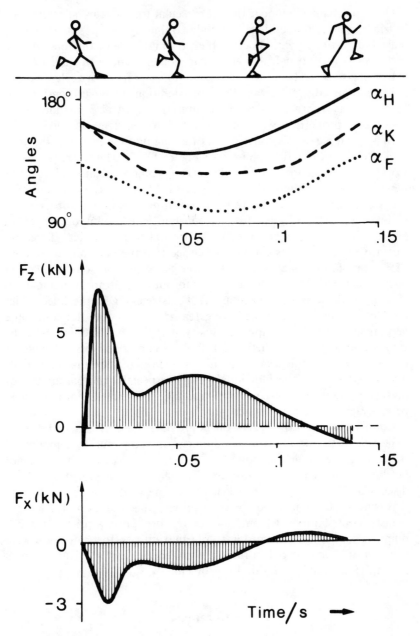

Figure 5—Take-off phase in long jumping, horizontal and vertical forces, time histories of the angles at the hip, knee, and ankle joints, respectively. (E. Kollath, 1980)

Using mechanical models, the net muscular torques can be determined in training and competition. The indeterminancy of the system will be mastered partially by considering the working mode of the musculature involved. If we only calculate muscular moments and then enter selected values in a statistical analysis we are still far away from an explanation of the technique. But if we take into consideration the length-tension and force-velocity characteristics of the muscles which also includes some viscoelastic properties, we make one step into the right direction.

In an example of the stance phase of a long jump, we can see the knee joint flexing in the first phase (see Figure 5). The knee extensors are working in an eccentric mode and thus can produce maximum force. For most of the stance phase, the knee angle is held constant, and the muscles are working isometrically, thus transmitting maximum forces while maintaining the ankle/hip distance constant which is important for an effective energy transfer from the horizontal to the vertical direction. Tentatively, hypotheses can be established that the knee angle of about 135° is an optimum between the possible generation of isometric tension due to the length of the muscle and the torque of the transmitted load increasing with decreasing knee angle. The stretching in the knee under concentric contraction of the knee extensor muscles does not contribute very much to the total impulse. This type of interpretation seems to be very useful in judging whether a technique is more or less effective. It concerns both technical and conditional factors influencing the performance. Of course, the individual ability, especially the parameters of muscular power and muscular strength, under different conditions, must be analyzed and taken into account.

Cooperation with the practitioners provides the only chance to prove hypotheses, to modify the conditions, to carry out systematic experiments. The empirical knowledge of the experienced coach should not be underestimated in that process. Together with a reduction of the discussed methodological deficiencies and the vital necessity of a standardization of our methods, we can provide the conditions for a more realistic contribution of biomechanics to sport. From a practical standpoint, it seems more promising to put maximum effort into the useful solution of a small problem than in a perfect but for practical purposes insignificant solution of a major problem.

References

BAUDZUS, W. (1980). Vergleichende Untersuchungen der mechanischen Eigenschaften künstlicher Bodenbeläge. (Comparative investigations of mechanical equality of artificial floor surfaces) Diplomarbeit, Deutsche Sporthochschule Köln.

BAUER, W.L. (1980) Mathematical modelling and optimization and their influence of sports movements. In: W. Baumann (ed.), Biomechanics and performance in sport. Proc. Int. Symp. Köln. (in print)

BAUMANN, W., and Stucke, H. 1979. Belastung des Bewegungsapparates aut Sportstättenboden. (Load on the movement apparatus on artificial surfaces.) Forschungsber. No. BII-1/79. des Instituts für Biomechanik, DSHS Köln.

BECKET, R., and Chang, K. 1968. An evaluation of the kinematics of gait by minimum energy. J. Biomechanics 1:147-159.

CAPPOZZO, A., Leo, T., and Pedotti, A. 1975. A general computing method for the analysis of human locomotion. J. Biomechanics 8:307-320.

FARFEL, W.S. 1977. Bewegungssteuerung im Sport. (Movement direction in Sport) Sportverlag Berlin.

HATZE, H. 1976. The complete optimization of a human motion. Math. Biosci. 28:99-135.

HATZE, H. 1981. The use of optimally regularized Fourier series for estimating higher-order derivatives of noisy biomechanical data. J. Biomechanics 14:13-18.

HOCHMUTH, G., and Knauf, M. 1979. Zu einigen Fragen der Dynamometrie bei biomechanischen Untersuchungen im Sport. (Several questions of dynamometry for biomechanical investigations in Sport.) In: G. Marhold Biomechanische Untersuchungsmethoden im Sport. Int. Symp., Karl-Marx-Stadt, pp. 81-99.

KOLLATH, E. 1980. Fur Kinetic des Weitsprungs unter besonder Berücksditigung der Gelenkbelastung. (On the kinetics of long jump under special considerations of the load on joints.) Dissertation DSHS Köln.

MORAWSKI, J.M., Buczek, M., Wiklik, K., Lasocki, B., Mizikowski, W., and Seiwinkski, M. 1977. Pole vault simulation studies. Rep. MINS/PNT, No-s: 1/77, 8/77, and 12/77, Institute of Sport, Warsaw.

NUBAR, Y., and Contini, R. 1961. A minimal principle in biomechanics. Bull. Math. Biophys., 23:377-391.

PAUWELS, F. Gesammelte Abhandlungen zur funktionellen Anatomie des Bewegungsapparates. (Total treatises of functional anatomy of the movement apparatus.) Berlin, Heidelberg, New York, 1965.

PEZZACK, J.C., Norman, R.W., and Winter, D.A. 1977. An assessment of derivative determining techniques used for motion analysis. J. Biomechanics 10:377-382.

SEIREG, A., and Arvikar, R.J. 1975. The prediction of muscular load sharing and joint forces in the lower extremities during walking. J. Biomechanics 8:89-102.

SOUDAN, K., and Dierckx, P. 1979. Calculation of derivatives and Fourier coefficients of human motion data, while using spline functions. J. Biomechanics 12:21-26.

WINTER, D.A., Sidwall, H.G. and Hobson, D.A. 1974. Measurement and reduction of noise in kinematics of locomotion. J. Biomechanics 7:157-159.

WINTER, D.A., Quanbury, A.O., Hobson, D.A., Sidwall, H.G., Reiner, G.,

Trenholm, B.G., Steinke, T., and Schlosser, H. 1974. Kinematics of normal locomotion — A statistical study based on T.V. data. J. Biomechanics 7:479-486.

WOOD, G.R., and Jennings, L.S. 1979. On the use of spline functions for data smoothing. J. Biomechanics 12:477-479.

A.
Track
and Field

A Biomechanical Analysis of the Segmental Contribution to the Take-off of the One-leg Running Jump for Height

M. Ae, K. Shibukawa, S. Tada, and Y. Hashihara
Institute of Health and Sport Science,
The University of Tsukuba, Ibaraki, Japan

There have been many studies on the segmental contributions to the take-off of the high jump. In track and field, this event is performed under optimal conditions in order to clear the crossbar at a maximum height (Dyatchkov, 1968; Hay, 1975). On the other hand, few studies relate to and discuss the contributions of the body segments in the one-leg running jump for height as a basic human movement, which is performed in various situations (Kinpara et al., 1966). The purpose of this study was to investigate the changes in the segmental contribution (role and degree) to the take-off in the one-leg running jump for height due to differences in the length of approach run, focusing on the momentum generated by the body segments.

Procedures

Experiment

Five male high jumpers performed a running jump for maximum height using a one-leg take-off from a straight approach run of 1, 3, 5, and 7 steps without clearing a crossbar. They were filmed at 250 frames/sec with a Milliken high speed camera synchronized with ground reaction force data which were obtained using a Kistler force platform.

Data Reduction

The x and y coordinates of the segmental endpoints on the subjects were obtained in every frame covering the take-off phase (touchdown to lift-off of the take-off foot) with a graph-pen system. These data were then filtered at 24.3 Hz cutoff frequency with a Butterworth digital filter (Winter, 1979). Body segment parameters of Chandler et al. (1975) were used to estimate the segmental centers of gravity.

Segmental momenta of the arms, head plus trunk (shortened to trunk), free leg, and take-off leg were computed and were subdivided as shown in the equations (1) to (4), considering their movements relative to the joints and/or the ground.

$$m_a V_a = m_a V_{a/s} + m_a V_{s/h} + m_a V_h \qquad (1)$$

$$m_t V_t = m_t V_{t/h} + m_t V_h \qquad (2)$$

$$m_{fl} V_{fl} = m_{fl} V_{fl/h} + m_{fl} V_h \qquad (3)$$

$$m_{tl} V_{tl} = m_{tl} V_{tl} \qquad (4)$$

(m_i = mass; V_i = velocity; $V_{i/j}$ = velocity of i relative to j; a = arms; s = shoulder; h = hip; t = trunk; fl = free leg; and tl = take-off leg.)

Some of these momenta in the right sides of the equations (1) to (4) seemed to be generated not only by the single segment, but also by the other segments. For instance, the momentum of the arms is the sum of the momenta generated by the arms ($m_a V_{a/s}$), the trunk $m_a V_{s/h}$), and the take-off leg ($m_a V_h$). Based on this idea, the momenta of the segments could be computed by the equations (5) to (8) and defined as generated momentum (GM).

$$GM_a = m_a V_{a/s} \qquad (5)$$

$$GM_t = m_a V_{s/h} + m_t V_{t/h} \qquad (6)$$

$$GM_{fl} = m_{fl} V_{fl/h} \qquad (7)$$

$$GM_{tl} = (m_a + m_t + m_{fl}) V_h + m_{tl} V_{tl} \qquad (8)$$

Segmental generated impulses in each frame were also computed by subtracting generated momenta at touchdown of the take-off foot from those of each frame, and then the mean percent contributions of the segments were obtained by dividing total impulse of each segment over the take-off phase by the whole body impulse (the absolute value). Because of the calculation, some mean percent contributions exceeded minus 100% which meant decelerative effect had occurred.

Results and Discussion

By increasing the number of approach steps, i.e., the length of approach run, approach velocity was also increased. Consequently, the jumps in this study could be referred to as Slow (S), Medium-slow (MS), Medium-fast (MF), and Fast (F) one-leg running jumps with respect to the approach velocities.

Mean Percent Contribution (See Figure 1)

With respect to the horizontal component, the largest and the smallest mean percent contributions under every condition were provided by the trunk (positive) and the take-off leg (negative), respectively. With an increase in approach velocity, the trunk contribution decreased, and that of the take-off leg increased. The mean percent contribution of the arms was larger than the free leg under the conditions S and MS, and the order was inverted under the conditions, MF and F, although the contributions of the arms were much smaller in an absolute sense than those of the trunk and take-off leg. These results reveal that all the body segments except for the trunk decelerate the velocity of the C.G., and that the arms and trunk play a more decelerative role in fast approach runs than in slow ones.

The largest mean percent contribution in the vertical component was shown by the take-off leg, which was followed by the arms, trunk and free leg under the conditions MS, MF, and F, although the trunk and free leg were reversed in the condition S. As approach velocity increased, the contribution of the take-off leg gradually increased, and the contributions of the other segments decreased. This means that the take-off becomes more dependent on the take-off leg from the standpoint of the generated impulsed, as approach velocity increases.

Each segment except the take-off leg still contributed 5 to 10% to the whole body impulse, however, even in the cases of fast approach run. The body segments were also expected to play different roles from the generation of impulse or momentum. Consequently, in order to correctly identify the contribution of the body segments, it was necessary to observe the change in momentum which was the source of the generated impulse.

Change in Generated Momentum

Generated momentum divided by the mass of the subject was normalized to the duration of the latter half of the condition S, and then was averaged for generalization. Therefore, Figures 2 and 3 show the horizontal and vertical velocity curves of the body segments for the mean of five subjects, which are equivalent to the generated momenta. The

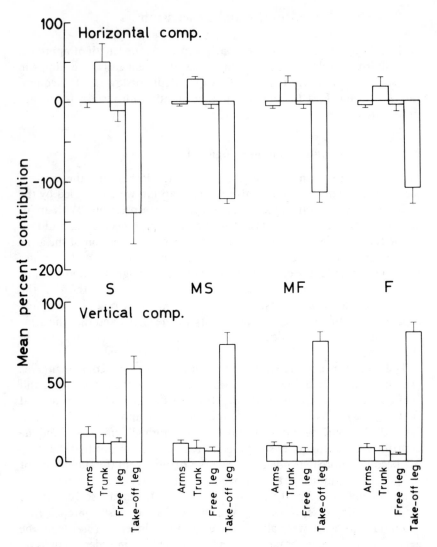

Figure 1—Changes in mean percent contributions of the body segments to the whole body impulse for five subjects during the take-off.

former half was defined as the duration from the take-off foot touchdown to the instant when the distance between the C.G. and the toe of the take-off foot was shortest, and the latter half was from the end of the first half to take-off foot release.

Horizontal Component (See Figure 2). The decrease in the whole body momentum was larger in the first half than the latter half, indicating that the horizontal velocity of the whole body decreased markedly and rapidly during the first half of the jump.

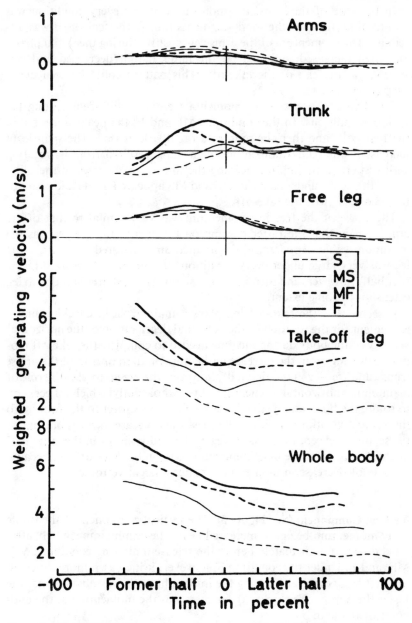

Figure 2—Changes in horizontal generated momenta of the body segments. Mean of five subjects during the take-off.

In the case of the arms, the momentum under every condition was smaller than for the other body segments (notice the scale used was different). The momentum of the arms was positive during the initial phase, and then increased, although in the latter half it decreased and was negative at the end of the take-off. This pattern could be seen under every condition.

The changes in the trunk momenta were very different among the various conditions. In the conditions MF and F, the peaks appeared in the first half, and then the momenta rapidly decreased. The significant increase in the trunk momentum might have resulted from its rotation with respect to the hip in checking the linear movement. On the other hand, the peaks under conditions S and MS appeared at the mid-point or the second half of the take-off.

The curve of the free leg momentum was very similar to that of the arms, except that the faster the approach velocity, the greater was the maximum value. Apparently, the momentum generated by the take-off leg was very large under every condition. It noticeably decreased in the first half, however, and then increased slightly in the latter half due to leg extension to some extent.

It seems that the take-off leg plays a major role in checking linear movement for the rotation of the body and deceleration of the horizontal velocity of the body during the time needed to complete the take-off motion. Inferring from the increase in the horizontal momenta of the arms, trunk, and free leg in the first half, however, they seem to play the role of maintaining horizontal momentum of the whole body, which is necessary to rotate the body and incline it backward with respect to the foot up to the vertical position at release. Besides, the decrease in the momenta of the segments except the take-off leg at the mid-point or in the latter half might also be effective in lightening the load on the take-off leg, because the rate of decrease in momentum creates a negative force.

Vertical Component (See Figure 3). The vertical component of the whole body momentum began to increase linearly beginning immediately after touchdown to immediately before the release under the conditions MS, MF, and F. Under the condition S, however, it increased more slowly in the first half, especially during the initial phase. It was obvious that the faster the approach velocity, the greater was the momentum at the mid-point of the take-off.

Arm momentum was quite small. It was negative initially, but it did not decrease in the first phase. Then it became positive at the middle of the first half and decreased in the latter half. These changes in the momentum show that the arms are swung down relative to the shoulders, and that they play a role in unweighting not in the initial phase but in the latter half. Trunk momentum was also small throughout the take-off. It

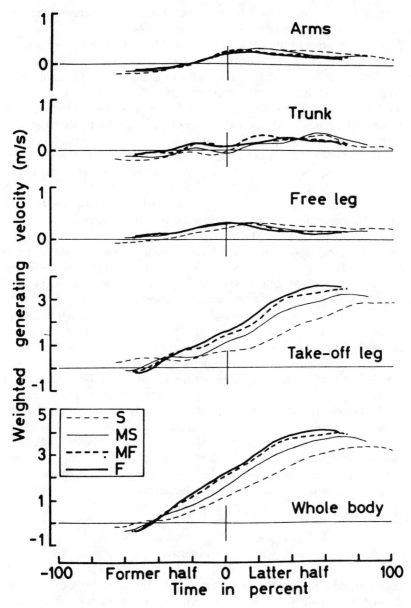

Figure 3 — Changes in vertical generated momenta of the body segments. Mean for five subjects during the take-off.

was negative during the initial phase, but it increased to reach the maximum around the middle of the latter half of the take-off.

Free leg momenta under the conditions MS, MF, and F were positive throughout the take-off while that of the condition S was negative during the first third of the first half. After increasing in the first half, they began to decrease at the mid-point or during the initial phase of the second half. Contrary to the expectation that the free leg was swung down at the initial phase of the take-off, no noticeable swinging down relative to the hip could be seen under the conditions of faster approach velocity.

The take-off leg generated a very large momentum under every condition, producing evidence of the large mean percent contribution shown in Figure 1. As seen in the whole body momentum, the faster the approach velocity, the greater the momentum of the take-off leg is determined by three factors: 1) the angular velocity of the take-off leg, 2) the rotational radius, and 3) the sine of the angle between the take-off leg and the vertical, assuming that the leg extensors eccentrically contract to produce a large force (Åstrand and Rodahl, 1970) as the take-off leg is flexed in the first half of the take-off. Therefore, the large momentum of the take-off leg at the mid-point under the conditions MF and F would be generated by the greater angular velocity (faster approach velocity), longer radius (more extended take-off leg) and larger backward lean than under the S and MS conditions.

Vertical momentum of the whole body seems to be generated principally by the rotation of the take-off leg in flexion, and by the swinging of the free leg and arms in the first half, with the momenta from the take-off leg extension and the lifting up of the shoulders added during the latter half. In addition, it is assumed that the vertical component as well as the horizontal component helped the take-off leg by lightening the load as a result of the decreases in the arms, free leg, and trunk momenta during the latter half. When the extensors of the take-off leg strive to contract, overcoming the load on it, the deceleration of the other segments will assist them in explosive extension of the take-off leg.

Conclusion

Based on these results, the roles, and contributions, of the body segments during the take-off of one-legged running jump may be categorized simply into 1) direct role such as generation of impulse or momentum, and 2) indirect role such as lightening the load on a certain segment, and 3) maintaining momentum, a part played especially by the trunk. Each body segment seems to play different roles at different phases of the take-off. The degree of contribution of the segment would also change so that the faster the approach velocity, the more the jumper is dependent on the lower limbs.

References

ÅSTRAND, P.O., and Rodahl, K. 1970. Textbook of Work Physiology McGraw-Hill.

CHANDLER et al. 1975. Investigation of Inertia Properties of Human Body. Aerospace Medical Research Laboratory, U.S.A.

DYATCHKOV, V.M. 1968. The high jump. Track Technique 34:1059-1075.

HAY, J.G. 1975. Biomechanical Aspect of Jumping. In: J.H. Wilmore and J.F. Keogh (eds.), Exercise and Sport Sciences Reviews, Vol. 3, pp. 135-161.

KINPARA et al. 1966. Studies on the fundamental technique for improving jump force (Part 3) — high jump utilizing approach run. Bulletin of Institute of Sport Science, The Faculty of Physical Education, Tokyo University of Education. 4:32-50.

WINTER, D.A. 1979. Biomechanics of Human Movement. John Wiley and Sons.

Foot to Ground Contact Forces of Elite Runners

Andrew Howard Payne
Physical Education Department
University of Birmingham, Birmingham, England

The apparently simple, but basic, part of running which concerns the athlete's foot contact with the ground receives scant, or at most, contradictory treatment by coaches and biomechanists in the literature. For example Snyder (1956), coach of Jesse Owens, recommended that a sprinter should ". . . land high on the toes, rock down on the foot to get good ankle flexion but not so low that the heel touches the track." Wilson (1963) said that: ". . . long distance runners land heel first during the major part of their running." Doherty (1963) considered all running distances and maintained that sprinters run on their toes, whereas endurance runners' foot contact depends upon speed, distance, and individual style, with the middle distance runner dropping lightly onto the heel after initial ball-of-foot contact, and the long distance runner settling down more on the heel. Doherty implied that it was a fault to run flat-footed or heel first. Nett (1964) wrote: "In the 100 metre and 200 metre runs, the ground is contacted first on the outside edge of the sole, high on the ball. . . . In the 400 metre run, which is run at a somewhat slower pace, the contact point lies a bit further toward the heel; the foot plant is now somewhat flatter . . . . In the further course of motion . . . even in the case of sprinters . . . the heel contacts the ground." Dyson (1970) agreed with Nett: '(The foot) takes the full weight of the body at a point which varies with the runner's speed; in sprinting, well up on the ball of the foot, almost flat-footed at very low speeds. As the body passes over the foot, the heel touches the ground lightly." The East German experts (editor Schmolinsky, 1978) prefer the older version by Snyder: "International class sprinters . . . balance their body weight in such a manner that only the ball of the foot touches the ground in the mid-support phase." Hay (1978) quotes Nett and emphasizes even more strongly: 'In the course of this process (and despite the views occasionally expressed by coaches that this does not or should not occur) the heel of the foot is lowered to the track."

The apparent disagreement among some of the authors is probably due to the short time periods involved in foot contact with the ground and most motion cameras give only a blurred picture when viewed frame by frame. The biomechanics researchers and coaches are also probably more concerned with what is happening higher up in the athlete's body and tend to attach lesser importance to the foot/ground contact. Whatever the reasons, the athlete's only means of evoking force for running is through foot contact with the ground, and conflicting, sometimes dogmatic, advice is not likely to be of much assistance to the athlete concerning this part of a running technique. In an attempt to clear up this confusion the author observed international class athletes in outside laboratory conditions and during actual competition.

Methods

A force platform system developed by Payne (1973) was used to measure the forces at the feet of elite runners. A total of 18 athletes who were specialists from the whole range of running events were tested when running over the double force platform system located in an outdoor sports field with the platforms sunk into a pit so that the tops were level with the surrounding ground. Speed of running was measured by means of photoelectric beam timers situated along the run up. Consistency of speed was checked by means of further timers situated on the far side of the force platforms. Each run was filmed side on by a Hulcher 35 mm sequence camera at 45 frames/sec. with each exposure at $\frac{1}{650}$ sec or less.

In order to check the observations of the above trials, 90 athletes were filmed during international competitions. The Hulcher sequence camera was again used under similar conditions except that some recordings were made at 25 frames/sec and some were panned.

Results

The force-time curves for two athletes are shown in Figures 1 and 2. These two examples were chosen to show the extremes of the range of foot/ground contact that were recorded. Figure 1 is the recording of an athlete who has competed at the Olympic level, using a sprinting style abhorred by many of the experts, i.e. landing heel first. Figure 2 is the recording of a 400 meter specialist (also of Olympic caliber) who runs with no heel contact at all. A characteristic of this type of running is the absence of the first vertical force peak and smoother patterns in the horizontal forces. Figures 3 and 4 show examples of athletes using different ground contact methods in actual international competition. The

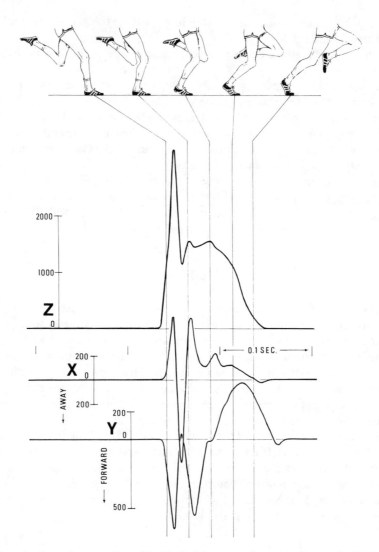

Figure 1—Force-time recording of heel/ball-of-foot type of running. Z = vertical force, X = transverse horizontal force, and Y = in-line-of-run horizontal force. (Newtons) Mass = 63 kg, Velocity = 9.5 m/sec.

athlete in Figure 4 would show a force-time pattern similar to that in Figure 2.

Table 1 shows the results of the classification according to type of foot/ground contact of 90 randomly-chosen athletes running in international competitions. No attempt was made to obtain significant numbers of subjects in the jumping events since the technical requirements of these events call for other attributes as well as running

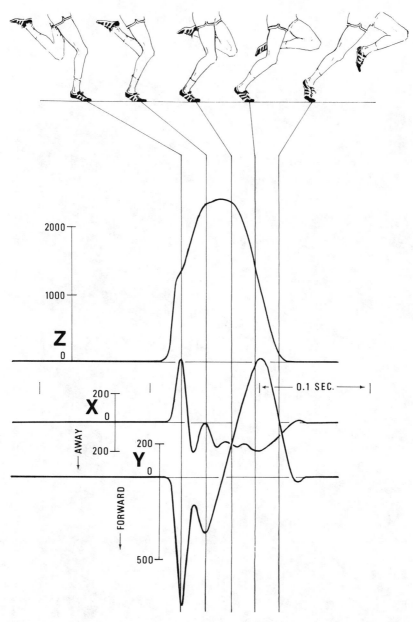

Figure 2—Force-time recording of ball-of-foot-only type of running. Scales as in Figure 1. Mass = 80 kg, Velocity = 9.2 m/sec.

Figure 3 – The athlete leading in this 5000 meter race uses a heel/ball-of-foot ground contact whereas the runner behind uses a ball-of-foot/flat method.

Figure 4 – This athlete runs entirely on the ball of the foot in the final sprint of a 1500 meter race.

Table 1
Types of Foot Contact in Competitive Running

Events	Heel and ball-of-foot	Flat	Ball-of-foot/Flat	Ball-of-foot-only	Total number in sample
Sprints up to 200 meters	28	11	56	6	18
400 meters to 1500 meters	27	32	27	15	41
5000 meters and upwards	64	29	7	0	31
Overall (not sample size weighted)	40	27	26	8	90

(Percentages of athletes using each method in sample)

style, but preliminary results indicate a similar wide variation in types of foot placement. It should be noted that in many cases the distinction between classifications is not clear cut, but a 4-category division was chosen to emphasize the variability of foot/ground contact in runners. Similarly, the dividing line between presence/absence of the first peak in the vertical force pattern is difficult to define precisely, lying somewhere in the ball-of-foot/flat classification, depending also on the shock absorbing nature of other movements by the athlete.

Conclusions

The following conclusions are based on a simplified two-dimensional view of running as seen side-on from a position at right angles to the line of running. The support foot, of course, moves in 3 dimensions and there are many nuances of rolling from the outside of the foot as well as sideways movements of the heel, but these are not considered here.

1. International class athletes range over the whole spectrum of foot/ground contact in running, from the heel-first to the ball-of-foot-only methods.

2. Sprinters and middle distance runners tend to make first contact on the ball of the foot followed by whole foot contact.

3. Long distance runners tend to make first contact on the heel followed by whole foot contact.

4. Observed less frequently are those athletes who run entirely on the front of the foot, but from the results of this study they appear to be mainly 400 to 800 meter specialists.

5. Running with the body weight mainly on the ball of the foot allows for smoother force/time patterns to be evoked from the ground and presumably should be considered mechanically more efficient than those

that cause high impact forces while the center of mass of the athlete's body is in a vertical line behind the support foot.

However, running on the ball of the foot is physiologically more demanding especially when long distance endurance is considered, and is probably a difficult skill in the all-out effort of a sprint run. Mechanical efficiency must sometimes (give way) for the sake of physiological and neuromuscular efficiency.

References

DOHERTY, K. 1963. Modern Track and Field. Bailey Bros. and Swinfen, London.

DYSON, G. 1970. Mechanics of Athletics. University of London Press, London.

HAY, J.G. 1978. The Biomechanics of Sports Techniques. Prentice Hall, Englewood Cliffs, N.J.

NETT, T. 1964. Foot plant in running. Track Technique 15:462-463.

PAYNE, A.H. 1974. A force platform system for biomechanics research in sport. In: R.C. Nelson and C.A. Morehouse (eds.), Biomechanics IV, pp. 502-509. University Park Press, Baltimore.

SCHMOLINSKY, G. (ed.) 1978. Track and Field. Sportverlag, Berlin.

SNYDER, L. 1956. The training of Jesse Owens. International Track and Field Digest. Ann Arbor, Michigan.

WILSON, H. 1963. In: G.F.D. Pearson (ed.), Athletics. Thomas Nelson and Sons, Edinburgh.

Mechanical Efficiency of Sprinters and Distance Runners during Constant Speed Running

M. Kaneko, T. Fuchimoto, A. Ito and J. Toyooka
Osaka College of Physical Education, Osaka, Japan

From the view point of muscular energetics, the efficiency, as well as aerobic power, seems to be of prime importance in distance runners, whereas explosive power may be the crucial factor affecting the performance of sprinters. It has been shown that distance runners' muscles are characterized by a higher percent of slow-twitch (ST) fibers, and sprinters' muscle by a predominance of fast-twitch (FT) fibers (Gollnick et al., 1972; Costill, 1976). It has also been stated that the ST fibers contribute to more economical energy conversion at slow speeds (Awan and Goldspink, 1972; Wendt and Gibbs, 1974), and that the FT fibers produce more powerful and speedy action (Thorstensson et al., 1976; Komi and Bosco, 1977). The biomechanical studies of distance runners have focused on the motion (Miura et al., 1973; Nelson and Gregor, 1976; Cavanagh et al., 1977) and efficiency (Gregor and Kirkendall, 1977). However, whether the distance runners can run more economically than sprinters has not been established. The present study aimed to compare the efficiency of endurance and sprint runners by measuring energy supply and mechanical power output during various 'constant' speeds of level running.

Method

The subjects were three distance runners and three sprinters, aged 20 to 22 years. To measure mechanical work the subjects ran at 6 to 10 different speeds (212 to 366 m/min) as constantly as possible at each speed. The horizontal and vertical forces (F_f and F_v) were recorded by a force platform (1m x 1m) implanted in the running track. Immediately after the trial the forward velocity was produced by integrating F_f electrically to check whether the speed was constant or not. The forces were also stored in an A/D converter to calculate instantaneous velocities (V_f and V_v) and the verical displacement of the body's center of mass (c.o.m.). The external mechanical work (W-ext) to accelerate the body's c.o.m. was calculated by Cavagna's method (Cavagna, 1975), where the total external work was obtained by adding the work due to kinetic energy changes (W-ext • f) and the work due to potential energy changes (W-ext • v) at each step, since the energy transfer between those two should be almost nil in running (Cavagna et al., 1976). The subject's side view from the left was filmed at 200 frames/sec by a high speed movie camera. Analyzing the angles made by limb segments with the horizontal, the kinetic energy (E_k) was calculated, at 0.0lsec intervals, by summing translational and rotational energies at each instant as described by Fenn (1930). The sum of increments in E_k-time curves of all segments was taken as the internal mechanical work (W-int) to accelerate the limbs relative to the trunk. Then, the total work(W-tot) was obtained as W-ext + W-int. For measuring the net energy cost (E-net) the subjects ran on a level treadmill at four selected speeds (200 to 342 m/min), 10 min at each speed. The average VO_2 above resting during the final two minutes was taken as aerobic energy used. The anaerobic energy by glycolysis was also calculated from blood lactate at the fourth minute after exercise and then added to the aerobic energy to obtain the 'maximum' net energy cost (Thys et al., 1975). By converting all values into the same unit of power the mechanical efficiency was calculated as

$$\text{Mechanical efficiency} = \frac{\dot{W}\text{-tot}}{\dot{E}\text{-net}} = \frac{\dot{W}\text{-ext} + \dot{W}\text{-int}}{\dot{E}\text{-net}} \qquad (1)$$

Results

The total mechanical power ($\dot{W}$-tot) was plotted against running speed (V_f) in Figure 1. The regression equations for distance runners (D) and sprinters (S) were

$$\text{D: } \dot{W}\text{-tot} = 0.049V_f - 2.394 \ (r = 0.920) \qquad (2)$$

$$\text{S: } \dot{W}\text{-tot} = 0.050V_f - 3.034 \ (r = 0.899) \qquad (3)$$

where $\dot{W}$-tot is in watt/kg, V_f in m/min, and r is the correlation coefficient. The regression equations are valid for the speed range of about 220 to 340 m/min. The $\dot{W}$-tot in D and S increased with speed in a similar manner (slightly D > S); from about 8.4 to 14.3 watt/kg in D and 7.9 to 13.8 watt/kg in S, and there was no significant difference between the two groups at any speed. The external power ($\dot{W}$-ext) and the internal power ($\dot{W}$-int) of D and S also increased similarly with increasing speed (see Figure 1). The regression equations were as follows (the units of power and speed were same as previously given).

$$D: \dot{W}\text{-ext} = 0.020V_f + 1.072 \ (r = 0.841) \tag{4}$$

$$S: \dot{W}\text{-ext} = 0.020V_f + 1.155 \ (r = 0.819) \tag{5}$$

and

$$D: \dot{W}\text{-int} = 0.029V_f - 3.466 \ (r = 0.893) \tag{6}$$

$$S: \dot{W}\text{-int} = 0.030V_f - 4.189 \ (r = 0.818) \tag{7}$$

The percent of $\dot{W}$-ext to $\dot{W}$-tot decreased with speed from about 65 to 55%, while the $\dot{W}$-int increased with speed by as much as 10%. Neither in $\dot{W}$-ext nor in $\dot{W}$-int was a significant difference observed between the D and S groups, though slightly D < S in $\dot{W}$-ext and D > S in $\dot{W}$-int. The distance runners (D) did appreciably less W-ext/step, however, and about the same W-int/step, resulting in significantly less W-tot/step as compared to the sprinters (S). Looking further into W-ext • f/step and W-ext • v/step, both were considerably less in D than S, resulting in the smaller W-ext/step of D mentioned previously. The gap between $\dot{W}$-tot and W-tot/step is necessarily due to the time occupied in one step; e.g., at 300 m/min the time of one step was about 0.31 sec in D and 0.36 sec in S, corresponding to the stride frequencies of 195 and 167 steps/min, and to the stride lengths of 1.54 and 1.80 meters in D and S respectively. The net energy cost ($\dot{E}$-net) was plotted against speed (V_f) in Figure 2 for distance runners (D) and sprinters (S). The regression equations obtained were:

$$D: \dot{E}\text{-net} = 0.103V_f - 11.175 \ (r = 0.962) \tag{8}$$

$$S: \dot{E}\text{-net} = 0.087V_f - 2.237 \ (r = 0.988) \tag{9}$$

where $\dot{E}$-net is in watt/kg, V_f in m/min. Taking the speed range of 220 to 320 m/min, the energy cost of D changed from 11.5 to 22.0 watt/kg (equivalent to 0.755 to 0.983 kcal/kg/km at the speeds above), whereas that of S changed from 17.0 to 25.8 watt/kg (equivalent to 1.109 to 1.153 kcal/kg/km). The differences between the two groups become smaller with increasing speed, but the distance runners consumed significantly less energy than sprinters at all speeds of running. The mechanical efficiencies estimated from the regression lines of $\dot{W}$-tot and $\dot{E}$-tot are also drawn in Figure 2. The efficiency of distance runners decreased with increasing speed from about 72% to 59%, whereas that of the sprinters re-

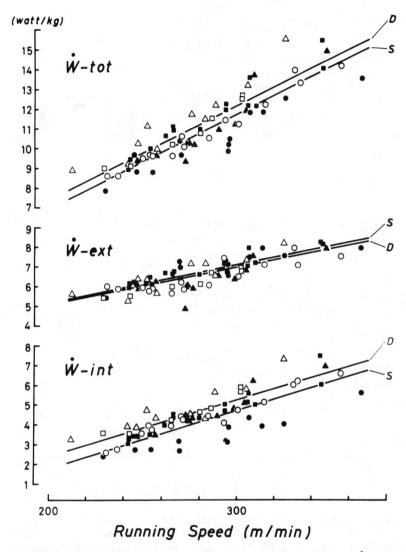

Figure 1—Mechanical power in relation to speed of level running: Total power (Ẇ-tot), external power (Ẇ-ext) and internal power (Ẇ-int) of distance runners (D; open symbols) and sprinters (S; closed symbols). The regression equations are included in the text.

mained almost constant at 47 to 48%—appreciably lower than the distance runners.

Discussion

Gregor and Kirkendall (1977) reported the efficiency of female marathon

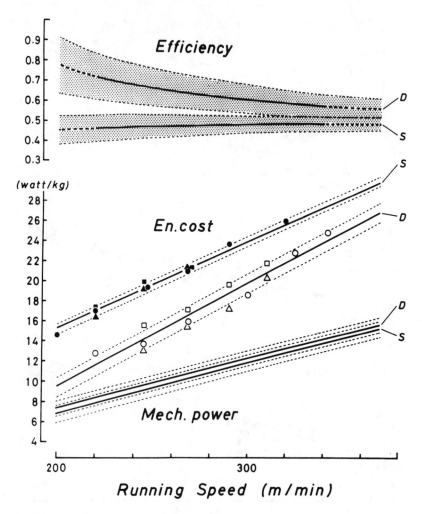

Figure 2—Net energy cost (Ė-net), and mechanical efficiency estimated from the regression equations of Ẇ-tot and Ė-tot for distance runners (D; open symbols) and sprinters (S; closed symbols) at various speeds. The regression lines of Ẇ-tot were also illustrated from Figure 1. The dotted lines indicate the standard error.

runners to be about 25 to 30%. This, however, is not directly comparable to the present data, since their total work was obtained in a different way (Norman et al., 1976, Winter and Quanbury, 1976). Cavagna and Kaneko (1977) have shown that the efficiency increased with increasing speed from 45% at 1.7m/sec to 70% at 9.2m/sec. The efficiency range in the present study is in general agreement over the range of speeds studied (3.3 to 6.2m/sec). However, the efficiency of the distance runners in this study decreased with increasing speed despite the fact that the changes of W-tot, W-ext, and W-int with speed were virtually the same. This discrepancy is due to the net energy cost whereas Cavagna and Kaneko (1977) assumed the E-net was 1 kcal/kg/km for all speeds, in the present study E-net was found to increase considerably with speed from 0.755 to 0.983 kcal/kg/km in the distance runners. The elastic energy recoil theory (after Cavagna et al., 1964) may be a possible mechanism to explain the high efficiency value obtained. Is it possible, however, to explain the different efficiency levels between distance runners (D) and sprinters (S) by the same theory? Looking further into the muscle action during foot contact with the ground, the time and the vertical displacement during the negative phase are D ≃ S, but D < S during the positive phase. In addition the W-tot/step of D is less than S with a shorter stride length at a given speed. Taking these into consideration, it might be that, despite their higher efficiency, the distance runners' muscles stored less elastic energy with a milder spring during the negative phase, and then did less work in the following positive phase. This theory does not favor reuse of elastic energy. For this reason, the higher efficiency observed in distance runners, as compared to sprinters, cannot be explained by elastic energy recoil.

The most probable reason for this may exist in the metabolic response of the muscles. In regard to muscle fiber composition, distance runners are characterized by a higher percent of slow-twitch (ST) fibers (70-75%) and the sprinters by a higher percent of fast-twitch (FT) fibers (Gollnick et al., 1973; Costill et al., 1976). The efficiency of the ST fibers in mammals is higher than the FT fibers if the contraction rate is low (Awan and Goldspink, 1972; Wendt and Gibbs, 1974). It was also shown that the efficiency of men who had a predominance of ST fibers decreased with increasing cycling speed (Suzuki, 1979). From these reports it seems that the distance runners in the present study might have more ST fibers than the sprinters, and thus could work more efficiently at such relatively slow speeds as in endurance running rather than in sprinting.

Conclusion

Under various constant speeds of running the mechanical efficiency of distance runners was found to be appreciably higher (59% to 72%) than the sprinters' (47% to 48%), which resulted from the less net energy cost and about equal mechanical work at a given speed.

Acknowledgment

This study was supported, in part, by a grant for scientific research (C) from the Ministry of Education, Japan (No. 56580091).

References

AWAN, M.Z., and Goldspink, G. 1972. Energetics of the development and maintenance of isometric tension by mammalian fast and slow muscles. J. Mechanochem. Cell Motility 1:97-108.

CAVAGNA, G.A. 1975. Force plates as an ergometer. J. Appl. Physiol. 39:174-179.

CAVAGNA, G.A., and Kaneko, M. 1977. Mechanical work and efficiency in level walking and running. J. Physiol. 268:467-481.

CAVAGNA, G.A., Saibene, F.P., and Margaria, R. 1964. Mechanical work in running. J. Appl. Physiol. 19:249-256.

CAVAGNA, G.A., Thys, H., and Zamboni, A. 1976. The sources of external work in level walking and running. J. Physiol. 262:639-657.

CAVANAGH, P.R., Pollock, M.L., and Linda, J. 1977. A biomechanical comparison of elite and good distance runners. Ann. N.Y. Acad. Sci. 301:328-345.

COSTILL, D.L., Daniels, J., Evans, W., Fink, W., Krahenbohl, G., and Saltin, B. 1976. Skeletal muscle enzymes and kinetic factors in male and female tract athletes. J. Appl. Physiol. 40:149-154.

FENN, W.O. 1930. Frictional and kinetic factors in the work of sprint running. Am. J. Physiol. 92:583-611.

GOLLNICK, P.D., Armstrong, R.B., Sanbert, IV, C.W., Piel, K., and Saltin, B. 1972. Enzyme activity and fiber composition in skeletal muscle of untrained and trained men. J. Appl. Physiol. 33:312-319.

GREGOR, R.J., and Kirkendall, D. 1977. Performance efficiency of world class female marathon runners. In: E. Asmussen and K. Jørgensen (eds.), Biomechanics VI-B, pp. 40-45. University Park Press, Baltimore.

KOMI, P.V., and Bosco, C. 1977. Utilization of elastic energy in jumping and its relation to skeletal muscle fiber composition in man. In: E. Asmussen and K. Jørgensen (eds.), Biomechanics VI-A, pp. 79-85. University Park Press, Baltimore.

MIURA, M., Kobayashi, K., Miyashita, M., Matsui, H., and Sodeyama, H. 1973. Experimental studies on biomechanics on long distance runners. In: H. Matsui (ed.), Our researches, 1970-1973, pp. 46-56. Dept. of Physical Education, University of Nagoya, Nagoya, Japan.

NELSON, R.C., and Gregor, R.J. 1976. Biomechanics of distance running: A longitudinal study. Res. Quart. 47:417-428.

NORMAN, R.W., Sharratt, M.T., Pezzack, J.C., and Noble, E.G. 1976. Reexamination of the mechanical efficiency of horizontal treadmill running. In: P.V. Komi (ed.), Biomechanics V-B, pp. 87-93. University Park Press, Baltimore.

SUZUKI, Y. 1979. Mechanical efficiency of fast- and slow-twitch fibers in man during cycling. J. Appl. Physiol. 47:263-267.

THORSTENSSON, A., Grimby, G., and Karlsson, J. 1976. Force-velocity relations and fiber composition in human knee extensor muscles. J. Appl. Physiol. 40:12-16.

THYS, H., Cavagna, G.A., and Margaria, R. 1975. The role played by elasticity in an exercise involving movements of small amplitude. Pflugers Arch. 354:281-286.

WENDT, I.R., and Gibbs, O.L. 1974. Energy production of mammalian fast- and slow-twitch muscle during development. Am. J. Physiol. 226:642-647.

WINTER, D.A., and Quanbury, A.O. 1976. Instantaneous energy and power flow in normal human gait. In: P.V. Komi (ed.), Biomechanics V-A, pp. 334-340. University Park Press, Baltimore.

A 3-Dimensional Cinematographical Analysis of Foot Deformations During the Take-off Phase of the Fosbury Flop

**Y. Muraki, T. Sakamoto, S. Saito,
M. Ae, and K. Shibukawa**
Institute of Health and Sport Science
The University of Tsukuba, Ibaraki, Japan

One of the most prominent characteristics of the Fosbury flop technique arises from the curved run-up before the take-off. This curved run-up has enabled many jumpers to get into an ideal preparatory body positioning for the take-off easily and naturally and to jump successfully in the initial stage of athletic development without advanced mastery of technique. At the same time there have been heard frequent complaints of foot trouble by Fosbury jumpers because the take-off foot is planted primarily in an abducted position and not in the direction tangential to the run-up (Krahl et al., 1978; Muraki, 1979).

Krahl et al (1978) discussed this problem based on the observation of take-off foot deformations in the Fosbury technique using 2-dimensional cinematography. Since 1978, we have endeavored to design special spiked shoes for Fosbury flop jumpers which have cross-sectional tilted soles for the curved run-up and take-off, and have intended to contribute to a rational take-off movement and to avoid injuries caused by abnormal foot deformations which are most frequently found medially at the ankle joint.

This investigation was intended to determine the degree of foot deformations and the effects of different shoe models by applying a newly developed 3-dimensional cinematographical technique called Direct Linear Transformation Method (Shapiro, 1978; Walton, 1979).

Table 1

Symbols and Characteristics of the Spiked Shoes Used in this Study

Cross sectional views of the shoe soles	Symbols of the models	Characteristics of the shoes	Number of spikes	
			Forefoot	Heel
L R	R	Normal sprint spiked shoes on the market (Asics PAW DS-3400)	6	0
	F	High jump shoes with flat soles reinforced on the heel caps	6	2
4	T*	High jump shoes with tilted soles of 4 mm differences between both edges, and reinforced on the heel cap	6	2
10	Tx*	High jump shoes with tilted soles of 10 mm differences between both edges, and reinforced on the heel caps	6	2

*Whole of the soles were tilted toward the center of the run-up curve, and of which the thickest edge of the soles were limited within 13 mm in accordance with the IAAF rules. The arrangement of spikes and the materials of the soles in each model were the same, excepting the R-model.

Materials and Method

Three Fosbury flop jumpers were employed as subjects, all of whom used the left foot for the take-off. Their best personal records were 2.25m, 2.10m, and 2.05m for Subject A, Subject B, and Subject C, respectively. The experimental jump trials with their own full run-ups were performed on a synthetic all-weather surface with the jumpers wearing four kinds of spiked shoe models (see Table 1). For three months before the experiment, the jumpers practiced using the prepared shoes to eliminate any sense of insecurity. The height of the bar cleared ranged from 90% to 98% of their personal best records.

The take-off foot was filmed during the take-off phase by two high-speed 16mm cine-cameras (1PL and DBM-5) running at 250 frames/sec. Each camera was placed 5 m from the center and 80 cm above the base of the control area, a cubic meter, and at an angle of 45° between them. The X and Y coordinates of the indentifying marks painted on the take-off foot of each subject (see Figures 2 and 3) were obtained in every frame of film. The positions of these marks in the control space were

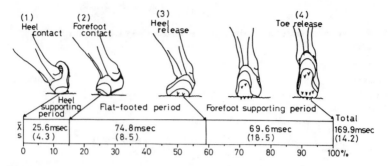

Figure 1—Sequence of the takeoff foot action during the Flop jump take-off phase.

determined by using the Direct Linear Transformation Method through the following three stages: 1) time-matching of the digitizer-coordinates, 2) application of an analytical model, and 3) smoothing of coordinates produced by that model (Shapiro, 1978; Walton, 1979). The parameters shown in Figures 2 and 3 were determined in two views to indicate foot deformations.

Observations and Discussion

General View of the Take-off Foot Movement During the Take-off Phase (See Figure 1)

The sequence of take-off foot action during the take-off phase starts at the instant of heel contact with the ground (1), and ends at the instant of the toe release from the ground (4), passing through the moments of forefoot contact (2) and of heel release (3). The mean take-off time in this study was 169.9 ± 14.2msec ($\bar{X}$ ± SD) and the mean percentage of heel contact period (1 to 2) to the entire take-off time (1 to 4) was 15.1% (25.6 ± 4.3msec). The flatfooted period (2 to 3) was 44.0% (74.8 ± 8.5msec) and the forefooted period (3 to 4) was 40.9% (69.6 ± 18.5msec).

The take-off time and the ratios did not show any definite relationships with respect to the differences in various models of spiked shoes. It was observed, however, that foot deformations mainly appeared during the first half of the take-off phase, and that those foot deformations coincided with the appearance of maximum force measured with the force platform (Kuhlow, 1973).

The major foot deformation frequently observed was forced pronation which compounded forced eversion, abduction, and dorsiflexion in ankle and foot joints. This appears to be consistent with the observations of Krahl et al. (1978).

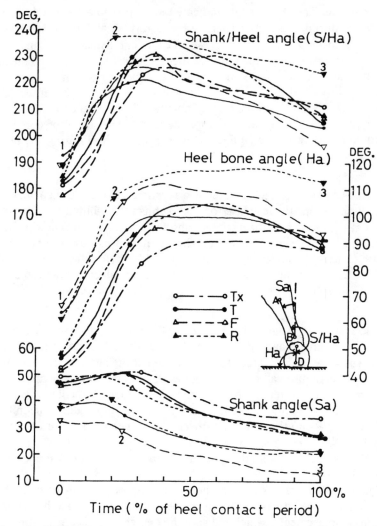

Figure 2—Angular displacement of parameters on the X-Z plane (back view) in the heel contact period of takeoff during trials by Subject A (thick lines) and Subject C (fine lines) employing various spiked shoe models: 1 = Heel contact; 2 = Forefoot contact; 3 = Heel release; Sa = Shank angle between a vertical line through B and the "Achilles tendon," $\overline{AB}$; Ha = Heel bone angle between the "heel bone" $\overline{CD}$, and the horizontal; S/Ha = Shank/Heel bone angle indicates the foot pronation angle between $\overline{AB}$, and $\overline{CD}$.

Foot Deformations Observed with the Parameters

Figure 2 shows the angular displacement of the parameters from the back view in the cases of Subject A and Subject C during the same interval of action.

Shank Angle. The shank angle indicated an approximate inside lean of the body during take-off. There were highly significant differences among the means of the peak values of the angular displacements in each subject corresponding to the levels of their best performances. The mean and the standard deviations of each subject during heel contact period were as follows: Subject A, 50.4 ± 0.45msec, B, 45.3 ± 2.35msec, C, 38.4 ± 4.25msec. The peak of this angular displacement in each trial usually appeared immediately before the moment of forefoot contact, which coincided with the ending of the slip of the heel during the heel contact period.

Heel Bone Angle. The heel bone angle was estimated to be 90° in a normal standing position. In each case for each subject the minimum angle was found at the moment of heel contact and the maximum angle at the flat-footed period. The range of the maximum angle was observed between 119.0° (for the R-model of Subject C) and 90.5° (for the Tx-model of Subject A). The mean of the maximum angles for all cases was 106.2 ± 8.46° (Table 2).

Shank/Heel Bone Angle. The shank/heel bone angle was estimated to be 180° in a normal standing position. In each case, for each subject the angle was close to that value (185.3 ± 4.93°) at the moment of heel contact. The maximum value of the angular displacement usually appeared at or immediately after forefoot contact. The range of maximum values was between 240° and 221° (231.9 ± 5.81°). The maximum value in the trials of the R-model by each subject was larger than those of other models, and in contrast the trials using the T-model and Tx-model tended to be smaller than other trials using the F-model and R-model (see Table 2).

Changes of the shank and the heel bone angles varied widely among subjects. On the other hand, the dispersion of the shank/heel bone angular displacements, which indicate the degree of forced pronation of the ankle joint, decreased with all subjects since the two components of the shank/heel bone angle compensated for each other. Therefore, it may be said that the maximum value determined from the shank/heel bone angular displacement indicates the level of the functional, anatomical limit of the forced pronation of the forced eversion of the ankle joints such as the talocruralis and the subtalaris. In fact, Subject C complained of a slight pain in his deltoid ligament during the take-off when wearing the R-model shoes, and in that situation, the peak value of angular displacement appeared very close to the maximum.

Figure 3 shows the changes in the parameters of the foot/shank and the foot angles from the top view during the heel contact period in the trials by Subject A and Subject C in the same manner as Figure 2.

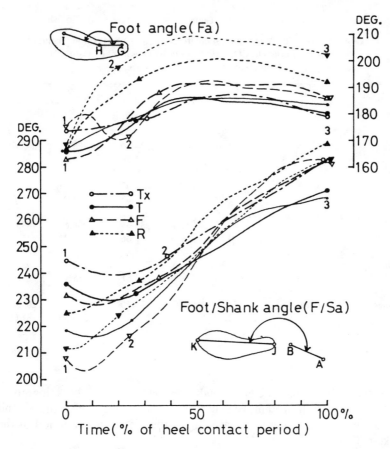

Figure 3—Angular displacement of parameters on the X-Y plane (top view) in the heel contact period of takeoff during trials by Subject A (thick lines) and Subject C (fine lines) employing various spiked shoe models: 1 = Heel contact; 2 = Forefoot contact; 3 = Heel release; F/Sa = Foot/Shank angle indicates the foot abduction angle between the longitudinal axis of the foot, $\overline{JK}$, and the $\overline{AB}$; Fa = Foot angle indicates the forefoot abduction angle between the forefoot, $\overline{HI}$, and the tarsus, $\overline{HG}$.

Foot/Shank Angle. The foot/shank angle indicates the degree of foot abduction of the shank projected on a plane. The values of this variable varied widely during the heel contact period in the trials with each subject, which mainly appeared to be caused by both the inside lean of the shank toward the center of the curved run-up (Shank angle from the back view in Figure 2) and the heel sliding.

The mean angle of this variable during the heel contact period arranged in order was Subject A, Subject B, and Subject C, which coincided with the order of the shank angle and the amount of heel sliding. During the flatfooted period, however, the amount of this variable rapidly increased in all trials for all subjects although the order was

Table 2

**Comparison of the Means and Standard Deviations of Maximum Values
of the Parameters Observed in Each Plane in Each Case
Using Spiked Shoe Models (Degrees)**

Plane		X-Z Plane			Z-Y Plane	
Parameter		Sa	Ha	S-Ha	F-Sa	F
R	$\overline{X}$	44.92	110.33	234.00	288.58	202.67**
	S	3.44	5.48	2.56	9.99	4.55
F	$\overline{X}$	43.50	107.17	231.92	281.33	193.17**
	S	7.68	8.29	5.63	12.07	2.72
T(Tx)	$\overline{X}$	46.81	102.44	230.38	268.67	185.00**
	S	4.60	8.80	7.06	10.22	2.34
Whole	$\overline{X}$	45.25	106.23	231.43	282.35	192.95
	S	5.63	8.46	5.81	15.42	8.13
Range	Max	51.00	118.00	237.70	307.70	209.00
	Min	32.70	90.50	221.30	256.50	181.00

**$(p > 0.01)$

reversed. This rapid increase might result from the abducting force generated by the hinged movement when the forefoot catches the ground as the heel is preparing to release during the amortization period of the take-off.
release during the forefoot-supporting period (not shown in Figure 3). The mean of the peak value was $282.35 \pm 15.42°$. In comparing the peak values, the highest was observed in the trials using the R-model on each subject, following with the F-model and then the T-model (see Table 2).

Foot Angle. The foot angle indicates the angle compounded by the forced abduction of the forefoot to the tarsus and the forced pronation of the whole foot. The rapid increase of the foot angle during the heel contact period might be caused by the changes of the forefoot positioning in the air from the moment of heel contact to the forefoot contact. Therefore, the actual foot deformations causing the forced abduction and the forced pronation should be shown in the displacement of this variable mainly during the flatfooted period. The mean peak value of the trials of the R-model by all subjects was significantly greater than any other of the trials $(202.67 \pm 4.55°)$. In contrast, the mean of the trials by all subjects using the T-model (including the Tx-model) was significantly smaller than in the case of the other trials $(185.00 \pm 2.24°)$.

Table 2 shows the mean of peak values observed on displacement of parameters in each trial by all subjects wearing various spiked shoe models. There were highly significant differences among the means of each trial with respect to the foot angle (P < 0.01). There were no significant differences among the means of each trial on the other variables. It can be said, however, that the mean peak values of all variables tended to be smaller in the trials using the T-model and the Tx-model than in the cases of the other models.

Conclusion

Those tendencies observed indicate that the slanted-shoe sole of the T-model and Tx-model worked to decrease foot deformations such as forced pronation and abduction during the take-off phase of the Fosbury flop, even though there were some complaints from Subject A when wearing the Tx-model which had the largest slant on the sole, and from Subject C who felt somewhat ill-at-ease while wearing the T-model.

These observations suggest that further experimentation should be made to determine the most effective degree of sole tilting or foot padding. Heel sliding might be prevented by wearing adequate spikes and by improving the structure and materials of the shoe soles, especially at the critical point of heel-strike. On the other hand, in actual jump training, coaches and athletes should focus on mastery of correct foot planting and take-off planting and take-off techniques and on the strengthening of specific muscles of the legs and feet for the Fosbury flop.

Acknowledgments

The authors express grateful thanks to Mr. Sakaguchi and Mr. Mimura at the Technical Development Department of the Asics Tiger Company for their kind cooperation.

References

KRAHL, H., and Knebel, K.P. 1978. Medizinische und trainingsmethodische Aspecte der Absprungphase beim Flop (Aspects of medical and training methods in the jumping phase of the flop). Leistungssport 6:501-506.

KUHLOW, A. 1973. A comparative analysis of dynamic takeoff features of Flop and Straddle. In: Cirquilini, C., Venerard, A., and Wartenweiler, J. (eds.) Biomechanics III, pp. 403-408. Karger, Basel.

MURAKI, Y. 1979. Takeoff foot problems complained by Japanese Flop jumpers; questionnaires to the participants of the inter-high school, the inter-collegiate, and the Japanese national championship meets (unpublished data).

SHAPIRO, R. 1978. Direct linear transformation method for three-dimensional cinematography. Res. Quart. 49:197-205.

WALTON, J.S. 1979. Close-range cine-photogrammetry: Another approach to motion analysis. In: Science in Biomechanics Cinematography. pp. 69-97. Academic Publishers.

B.
Gymnastics
and Tumbling

Comparison of the Energy Metabolism and Mechanical Efficiency of Male Gymnasts Aged Nine to Twenty-four years

Kihachi Ishii, Nobuko Ishii and Takeshi Iwai
Nippon College of Health and Physical Education
Tokyo, Japan

A series of free gymnastics as well as calisthenics consisting of individual exercises includes a variety of fundamental movements of the human body, because each exercise in gymnastics is designed to recruit all the skeletal muscles and to arrange their actions systematically based upon a knowledge of segmental movements. It is difficult to express the level of intensity of the exercise by only the patterns of movement and the physiological responses. If this is possible, the exercises will be expressed as only stereotyped ones. Therefore, because in exercises such as sports activities or gymnastics there is a fixed intent, it is necessary to consider the influence of the psychological attitude. This writer assumed the vertical distance moved by one's center of gravity was a measure of physical intensity and using it as a measure determined the amount of effort needed to perform the exercises.

The first purpose of this study was to categorize the fundamental movements into physical patterns and psychological patterns. The second purpose was to determine the mechanical efficiency during the exercises and to compare the changes in the mechanical efficiency during adolescence. There are few studies like this one except the reports by Taylor et al. (1950) and Rode and Shepard (1973).

Methods

The exercises chosen for the experiment are shown in Table 1. They included three exercises for the legs, two for the trunk and one for the arms. In the table, the letters in the right hand column are abbreviations for each exercise. The first large letter signifies the body part and the

Table 1

Description of Gymnastic Movements

Body part	Movement	Stance	Times/ min	Note	Notation
Lower extremities	Half squat	Upright, hands on hips	30		L.h.s.
	Full squat	Upright, hands on hips	30	L.f.s.	
	Jumping	Upright	120		L.j.
Upper extremities	Arms flexion- extension	Upright straddle	20		A.f.e.
Trunk	Forward- backward bend	Upright straddle	10	T.f.e.	
	Lateral bend	Upright straddle	20	Left-right alternately	T.s.b.

small letters following represent the pattern of the individual gymanstic exercise. Oxygen expenditure was measured by standard open circuit techniques.

In the determination of the oxygen expenditure for quiet rest, the subject sat on a chair; however, active 9- to 14-year-old boys could not be expected to sit quietly on the chair, so they were allowed to read comic books. Expired gas was collected three times: 1) for the final five min of the 30-min rest period, 2) during one min of the exercise period, and 3) during 20 min of a recovery period sitting on a chair. Energy expenditure in excess of the resting value was calculated during each exercise and during a 20-min recovery period.

On the other hand, vertical displacement between the highest and the lowest point of the center of gravity during each gymnastic exercise was determined on another day by means of cinematography using the Matsui technique which enables one to calculate the mass center of the subject's body throughout the exercises. The mechanical work was calculated by multiplying the vertical displacement during one repetition of the exercise, by body weight, and the number of repetitions per min. The net mechanical efficiency was calculated as the relationship between the work done and the energy expenditure during the individual exercise over and above that at rest.

All of the exercises were performed under 2° of intensity subjectively perceived as being as light as possible or as heavy as possible, because the intensities of voluntary exercise such as sports activities or gymnastics at higher levels, are determined by the efforts of the individual.

For this study, the subjects were 30 normally healthy boys aged 9, 11,

Table 2

Physical Characteristics of Subjects

Age (yrs)	n	Sex	Height (cm)	Weight (kg)
23.8	5	M	170.1	64.5
2.6			6.1	3.0
17	6	M	173.7	60.2
			4.4	6.0
14	7	M	158.8	48.5
			6.8	8.8
11	6	M	141.8	33.8
			2.5	6.2
9	6	M	128.7	26.2
			4.3	4.1
				Mean
				S.D.

14, 16, and 20 years. Table 2 shows the number of subjects, and the group means and standard deviations of body height and body weight.

Results

The results of the net oxygen requirements were expressed in l/min in all six of the individual gymnastic exercises, because the exercises were performed for one min.

The means and the standard deviations of the oxygen requirements determined are presented in Table 3. The upper part of the table shows the values obtained when the given exercise was performed with only slight effort and in the bottom part of the table, values for the heavy effort are shown. It is clear from the results of the oxygen requirements that large amounts were required to perform the trunk and arm exercises, because the larger muscle groups were utilized. The exercises which showed the highest values were the full squat and jumping. In comparing the intensity of effort, all values under the heavy effort condition were higher than that under the light effort condition except for the full squat for the group of 9-year-old boys.

Table 4 depicts the vertical distance that the center of gravity was displaced in the one repetition of each exercise. It shows higher values under the heavy effort condition than under the light effort condition when the displacements of the center of gravity of the respective exercises were compared. The distance of displacement of the center of gravity, for ex-

Table 3

The Oxygen Requirement of the Individual Gymnastics

O₂ req./min/kg	(ml/min/kg)					light
Age (yrs)	Lhs	Lfs	Lj	Afe	Tfe	Tsb
9	22.5	38.0	35.3	6.9	16.8	12.2
	11.1	9.0	10.2	6.0	10.1	11.7
11	31.2	35.3	32.8	6.5	16.9	13.5
	5.4	5.3	3.3	1.8	5.6	2.4
14	27.6	43.6	39.5	9.4	17.3	12.4
	4.7	7.0	6.7	1.6	3.6	2.1
17	19.4	32.5	36.4	6.4	15.5	15.0
	8.4	7.9	5.4	2.6	8.2	7.7
23.8	24.4	32.5	30.5	4.3	14.3	17.4
	3.5	5.7	2.3	0.7	2.7	4.0

O₂ req./min/kg	(ml/min/kg)					heavy
Age (yrs)	Lhs	Lfs	Lj	Afe	Tfe	Tsb
9	28.2	32.4	42.6	7.2	25.2	21.9
	16.9	11.1	12.5	5.6	10.3	10.8
11	41.8	45.1	44.1	10.3	23.5	23.2
	5.7	5.7	3.5	2.6	5.4	3.4
14	35.7	48.2	49.6	13.4	22.1	21.6
	4.8	5.5	5.1	2.1	4.1	3.7
17	26.0	39.3	40.6	11.7	20.8	29.0
	8.8	8.5	8.0	7.4	8.6	6.4
23.8	38.4	42.4	41.7	12.4	25.7	24.2
	3.7	4.6	6.2	2.8	4.8	0.8

Mean
S.D.

ample, was greatest in the full squat exercise. The near maximum distances were seen when the trunk bent forward and backward alternately, because the exercise involved bending three times in both directions. Therefore, the corresponding values are only about one-third as large as those given in the exercise of lateral trunk bending.

The results of mechanical efficiency in the respective exercise are shown in Table 5, in which the highest values were obtained for jumping and the values for the leg exercises were higher than those for the trunk and arm exercises.

Table 4

The Vertical Displacement of Each Exercise

Displacement of C.G.	(cm)					light
Age (yrs)	Lhs	Lfs	Lj	Afe	Tfe	Tsb
9	20.4	46.4	15.7	4.1	23.2	6.5
	5.8	4.5	3.7	0.8	9.0	4.7
11	25.0	54.3	20.4	5.1	35.7	9.6
	5.7	8.3	4.3	0.8	5.9	4.7
14	27.8	53.4	18.7	6.3	41.5	12.7
	5.1	11.3	3.5	1.0	6.4	6.5
17	26.9	57.3	19.4	5.5	33.8	5.2
	3.9	3.4	2.7	0.6	8.6	1.9
23.8	24.5	49.8	20.7	7.1	39.9	15.0
	4.5	4.9	1.3	0.4	7.7	3.0

Displacement of C.G.	(cm)					heavy
Age (yrs)	Lhs	Lfs	Lj	Afe	Tfe	Tsb
9	31.2	53.8	21.9	5.2	45.3	15.3
	7.7	4.3	4.1	0.8	8.5	4.7
11	33.5	60.8	25.6	5.5	52.0	17.2
	2.6	5.2	2.8	0.7	6.1	5.0
14	37.4	58.9	23.8	6.5	55.9	17.7
	11.4	10.7	2.3	1.5	14.1	6.4
17	35.7	65.9	24.5	6.5	61.0	17.2
	4.9	5.4	2.5	0.7	6.3	7.6
23.8	37.3	60.5	28.6	7.6	72.1	35.8
	5.8	5.8	2.6	0.9	10.0	2.3

Mean
S.D.

Discussion

Comparison Between the Conditions of Perceived Exertion

In the 9-year-old group, the values for displacement in the given exercises were less than the displacements of the other groups in all exercises. It was assumed that the group with the smallest body size would show the smallest values, but the values of the vertical displacement did not change with age. Comparisons with the value of the work done obtained by multiplying the vertical displacement by body weight indicate the work done gradually increased with age under both conditions of per-

Table 5

The Efficiency of Each Exercise

Efficiency Age (yrs)	(%) Lhs	Lfs	Lj	Afe	Tfe	light Tsb
9	16.3	18.4	28.0	10.9	11.2	8.6
	10.5	4.7	12.2	9.5	10.5	7.8
11	11.6	22.4	35.9	8.1	10.7	6.9
	2.3	4.4	8.8	2.4	2.2	3.3
14	14.6	17.7	28.2	6.7	12.1	10.1
	2.7	4.0	8.8	2.1	4.1	5.8
17	26.2	26.4	31.3	17.8	15.2	3.8
	15.0	5.6	7.1	19.4	12.1	1.6
23.8	14.4	22.3	39.3	15.8	13.9	7.5
	1.4	3.1	4.4	2.5	4.3	1.1

Efficiency Age (yrs)	(%) Lhs	Lfs	Lj	Afe	Tfe	heavy Tsb
9	19.4	27.0	30.9	11.9	9.8	8.6
	8.5	11.2	8.1	9.1	3.8	5.3
11	11.6	19.7	33.4	5.4	10.9	7.2
	1.5	3.6	3.7	1.5	2.1	2.2
14	15.2	17.6	27.8	4.7	12.2	7.8
	5.1	3.2	4.2	0.9	6.2	2.5
17	21.8	24.9	35.2	7.4	16.0	5.7
	7.9	5.3	4.4	5.5	5.9	2.0
23.8	14.1	21.8	40.1	6.6	13.6	14.3
	2.4	3.7	6.9	2.4	1.6	1.5

Mean
S.D.

ceived exertion. This means that body weight has a greater influence than vertical displacement on the work done, and therefore, it may be stated that size has less effect on work done than body mass.

For comparison with the condition of the heavy and light exertions, the ratios of the oxygen requirement under both conditions are depicted in Figure 1. All values under the light condition were at the level of 100% as compared to the heavy condition, except the exercises of the full squat for the 9-year-old group. The degree of decrement was less in order of the full squat, the jumping and the half squat, respectively. However, in the arms exercise, the oxygen level under the light condition was 30% for the adult group and gradually increased to near 100% for the youngest

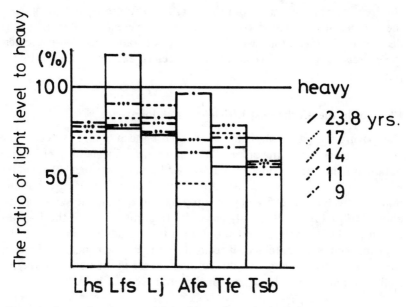

Figure 1—Comparison between the conditions of preceived exertion.

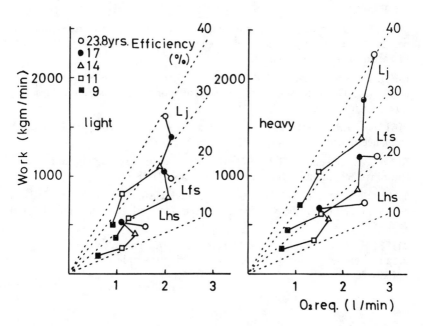

Figure 2—Comparison with the efficiency during adolescence.

children in order of age. From these results, it appears that the greater the number of joints involved in an exercise, the greater is the ratio of the decrement. Namely, that when the number of joints included is greater, the exercise might be affected by the perceived exertion, and not by the physical factors.

Changes in the Mechanical Efficiency in Adolescence

Comparing the values in Table 5, those for the jumping exercise were larger in general than the others. The values of the mechanical efficiency obtained from the jumping exercise showed a range between 30 and 40% which was less than that reported by Thys et al. (1975) Figure 2 shows that the mechanical efficiency in jumping had almost the same values, about 30% from the 9- to 14-year-olds. In addition, from the 14-year-olds it gradually increased to about 40% for the adult group.

From the results obtained, it seemed that the increments of efficiency for the 14-year-olds might be due to the increase in the body mass, especially the muscular volume and/or the improvement in skill.

References

ASMUSSEN, E., and Bonde-Petersen, F. 1974. Apparent efficiency and storage of elastic energy in human muscles during exercise. Acta Physiol. Scand. 92:537-545.

ISHII, K. Funato, K., and Tsumiyama, T. 1980. An analysis of free hand gymnastics (1) — From viewpoint of biomechanics and energetics. Rep. Res. Cent. Phys. Ed. 8:1-11. (in Japanese)

ISHII, K. Tsumiyama, T., and Fukuda, T. 1980. An analysis of free hand gymnastics (2) — Characteristics of the adolescent spurt, Rep. Res. Cent. Phys. Ed., 8:12-20. (in Japanese)

MALINA, R.M. 1974. Adolescent changes in size, build, composition and performance. Human Biol. 46:117-131.

RODE, A., Shephard, R.J. 1973. On the mode of exercise appropriate to a "primitive" community. Int. Z. angew. Physiol. 31:187-196.

TAYLOR, C.M. Bal, M.E. Lamb, M.W., and Mcleao, G. 1950. Mechanical efficiency in cycling of boys seven to fifteen years of age. J. Appl. Physiol. 2:563-570.

THYS, H., Cavagna, G.A., and Margaria, R. 1975. The role played by elasticity in an exercise involving movements of small amplitude. Pflügers Arch. 354:281-286.

Kinematic and Kinetic Analysis
of the Giant Swing on Rings

M.A. Nissinen

Johann Wolfgang Goethe-Universität,
Frankfurt, West Germany

In recent years gymnastics has shown a rapid increase in the development of technical and physical performances. The average age of elite male and female gymnasts has continually decreased, while the intensity and extent of the training has increased. Gymnastic movements have become considerably more difficult with the evaluation and scoring of routines being heavily weighted on the performance of the so-called 'risk movements'. Emphasis on dynamic movements performed with generally larger amplitudes have replaced the earlier emphasis of the static-strength type of movements. Consequently, the number of injuries has increased, thus creating criticism and an attitude towards gymnastics as basically unhealthy and perhaps 'too dangerous' a sport. It seems reasonable, therefore, to assume that there exists a need to develop evaluation methods that will analyze the forces and force patterns as produced during gymnastic performances. The results of these analyses could then be used to design better and safer equipment, and of greatest importance, to improve the technical aspects of gymnastic performances. The purpose of this study was to develop a kinetic and kinematic method of analysis on the rings, and to identify the possible factors that are significant for obtaining a technically correct performance of the backward and forward giant swings on the rings. A special purpose was to look at the feasibility of such a method as a training and coaching tool. The study was designed to investigate reaction forces, displacements of the body center of gravity, and segmental (trunk, arms, legs) velocities and accelerations during performances of giant swings on the rings by male elite, and less skilled gymnasts.

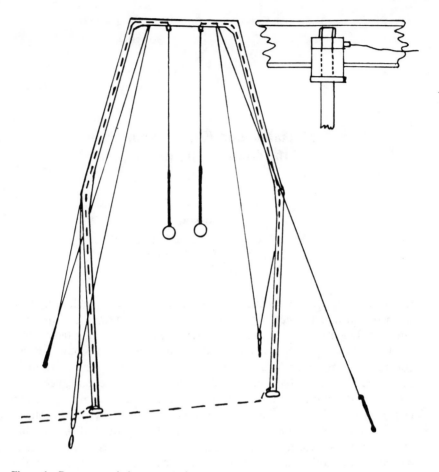

Figure 1 – Dynamometric instrumentation.

Methods

The dynamic instrumentation was designed to allow unrestricted force measurement under competitive, training, and laboratory conditions. The reaction force was measured by two Kistler force transducers (Type 9021) installed inside the metal framework of the rings just above the attachment of the ring cables (see Figure 1). The force signals were recorded and stored on magnetic tape or sent directly into the computer depending on the research situation. The force exerted on each ring was similar in magnitude, therefore the total force was obtained by doubling the value recorded for one ring. A lateral view of these performances was simultaneously filmed with a LOCAM 16 mm camera set to operate at 100 frames/sec.

Every second frame of film during the giant swing was digitized and the resulting displacement time information digitally filtered prior to the calculation of segmental velocities and accelerations. A set of 35 anthropometric measurements was taken from each gymnast. Predictions of individual segmental variables were calculated using a modified Hanavan model. The main investigations were made at two successive German Gymnastic Championships (1979, 1980) and consisted of data from over 60 backward and forward giant swings performed by gymnasts of various ages and skill levels.

Two representative performances are discussed as typical examples in this article.

Results and Discussion

A technically superior performance of the giant swing on the rings is characterized by a straight arm execution and is scored as a very difficult movement or so-called C-movement. A giant swing on the rings with bent arms is scored only as a difficult movement or so-called B-movement. More recently, a well-executed giant has been further characterized by a parallel movement of the arms and rings forward during the downward swing.

Figures 2a and b represent the synchronization of the film and the force-time curve. The criterion for a technically superior (straight arm) giant swing is represented in Figure 2a. Figure 2b represents an inferior (bent arm) performance of the giant swing.

During the downward swing, or first unweighting phase (reaction force < body weight), there was a considerably larger forward displacement of the rings from the body center of gravity by gymnast (a) in order to achieve a higher kinetic energy. As gymnasts (a) and (b) approached a vertical position, the reaction force surpassed that of body weight and increased rapidly. Shortly after the swing of the legs through the vertical position, the maximal peak reaction force was reached. In general, the investigated trials of this study have shown reaction forces as being considerably higher (6.5 to 9.2 times body weight) than so far recorded in the literature (Marhold, 1961; Sale and Judd, 1974). After maximal peak reaction force, the reaction force decreased rapidly, returning to and below that of body weight. In Figure 2a, two smaller peak reaction forces above body weight were recorded. In Figure 2b, these smaller peak reaction forces were not recorded. Therefore, a reaction force equal to that of body weight was not reached until returning once again to a handstand. The investigations of these force fluctuations were then further investigated by film analysis.

The movement of a gymnast on the rings can be shown as a three-segment system (arms, trunk, legs) around a suspension point. Accord-

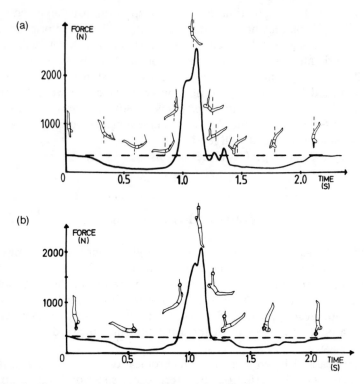

Figure 2 — The synchronization of the film and the force-time curve as measured by one of the two force transducers. The plotted horizontal line represents half of the body weight, F-t curve (a) represents world-class gymnast with straight-arm giant. F-t curve (b) represents club-level gymnast with bent-arm giant.

ing to Djatschkow (1974), every change of the basic swing movement, especially those of the multi-segment system of the human pendulum, is directly related to the accelerating movement of the legs. Therefore, for investigative purposes, the relative movements of the legs with respect to the hips, the trunk with respect to the shoulders, and the arms with respect to the suspension point were of importance. Figures 3a and b represent the time curve of the relative vertical velocities of the trunk, arm, and leg segments of the two gymnasts. Figure 3a shows in the upward swing phase a braking motion of the legs to a complete standstill. During this phase the relative velocity of the arms hardly changed, therefore creating momentarily a new rotating axis around the shoulder joint. The arm contribution began only after hip extension was completed. As mentioned previously, gymnast (a) clearly showed two distinct smaller force peaks in the force-time curves. As seen from Figure 3a, the first peak developed during leg braking and trunk transfer, and the second peak during trunk braking and arm initiation. Comparatively, gymnast (b) failed to show such detailed coordination.

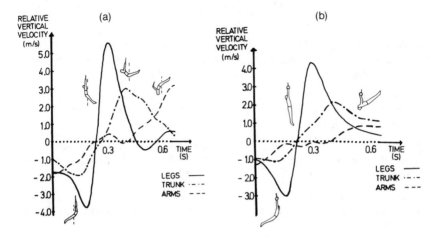

Figure 3 — The time curve of the relative vertical velocities of the trunk, arm, and leg segments of gymnasts (a) and (b).

As an additional aspect to the main investigation, the swing motion of junior gymnasts was studied. A good basic swing motion is an important preparatory movement for all the more difficult movements on the rings, and therefore, an essential prerequisite for a good giant swing. Gymnastic textbooks state that during a forward swing, the gymnast should actively push the rings backward, and then raise the whole body into a horizontal position with respect to the rings. The backward swing is executed similarly except that the rings are actively pushed forward instead of backward.

Our results, however, contradicted the relevant literature and instructional books for the teaching of the giant swing. Our investigation has shown that the arms and shoulders should remain under the rings until hip extension in the upward swing has been fully completed. Therefore, there is no active push of the rings forward or backward. A further investigation showed that these traditional instructions are actually still being used in order to teach junior gymnasts the swing motion, and therefore a majority of these junior gymnasts displayed technically inferior swings, with the exception of only a few.

From this study, the initiation of the following practical coaching instructions could be expected:

1. The initiation of the leg action (hip flexion-extension) is primary to the coordination and execution of the swing.

2. The active movement of the rings by the arms is relatively late in the swing, only after hip flexion-extension has been completed. This delayed arm action allows for a more favorable position of the trunk and legs with respect to the point of support (rings).

References

DJATSCHKOW, W.M. 1974. Die Steuerung und Optimierung des Trainingsprozesses (The guidance and optimization of training processes), pp. 58-68. Berlin, Verlag Bartels & Wernitz KG.

MARHOLD, G. 1961. Über die Belastungsgröben bei Übungen an den Ringen (Concerning the large strain from exercises on the rings). Theorie und Praxis der Körperkultur 5:439-444.

MILLER, D.I., and Morrison, W.E. 1975. Prediction of segmental parameters using the Hanavan human body model. Medicine and Science in Sports 7(3):207-212.

SALE, G.D., and Judd, L.R. 1974. Dynamometric instrumentation of the rings for analysis of gymnastic movements. Medicine and Science in Sports 6(3):209-216.

The Prediction of the Quadruple Backward Somersault on the Horizontal Bar

Hisato Igarashi
University of Oklahoma, Norman, Oklahoma, U.S.A.

The horizontal bar event in men's gymnastics has experienced a rapid improvement with regard to skills performed due to better equipment and the modification of gymnastics rules. The movements on the horizontal bar are mainly based on the swing. Many new routines on the bar have focused on the dismounts. The triple backward somersault (TBS) is, at present, the highest valued skill, as stated in the gymnastics rules pertaining to the horizontal bar event. This skill first appeared in gymnastics competition during the World Gymnastics Championships in 1974. The first successful performance of the TBS opened a new area of movements such as a combination of somersaults with a twist or a pike. During the world championships held at Fort Worth, Texas, in December of 1979, several routines were ended with the successful completion of the TBS.

Advanced techniques of gymnastics skills should be based on scientific evidence so that coaches can assist their students toward their greatest potential. It is important that a safe and successful high-risk movement be achieved with the most efficient techniques and knowledge available. The purpose of this study was to predict the performance of a quadruple backward somersault (QBS) as a dismount from the horizontal bar through analysis of selected biomechanical factors of the single backward somersault (SBS), the double backward somersault (DBS) and the TBS. The biomechanical factors analyzed included flight time, angular displacement of the hip and shoulder at release, location of the shoulder at release, and angular displacement of the hip during flight.

In gymnastics there are similar movement patterns in all of the events. In terms of understanding the nature of a trick, collecting commonalities from the events having similar movement patterns helps to solve problems common to the other events. Previous literature has discussed how a gymnast produced rotational momentum in the air. Cross (1980) described the successful completion of the DBS during floor exercise as completing the first somersault as soon as possible after takeoff, preferably during the ascending phase of flight, and to remain in a tight tuck position for as long as possible until landing. Frederick (1979) studied the TBS in floor exercise and stated that although the center of gravity was parabolic, only one rotation was completed during the ascending phase and the other rotations were finished during the descending phase of flight. Speed of rotation was of paramount importance. Kaneko (1970) discussed factors which were important for the completion of the DBS on the horizontal bar. Body position at release, tuck position in the air, and angles of hips and shoulders were important for successful completion. Mitsukuri et al. (1975) presented a comparison between the TBS and the DBS dismount on the horizontal bar. The shoulder angle of the TBS was larger than that of the DBS yet the TBS showed a rapid decrease in hip angle directly after release from the bar as opposed to a slower decrease in hip angle for the DBS. Mitsukuri and his associates (1975) concluded that whether or not a performer can complete the TBS depends upon how well he can execute the pull into the tuck position. In order to produce angular momentum, the distribution of the mass of the body from the center of rotation at takeoff is critical. By reducing the angles at the hip and knee joints, the degree of sluggishness or inertia is reduced and a gymnast can obtain more spin during flight.

Methods

Three elite male Japanese gymnasts were selected as subjects for the study. They included a Japanese Olympian from the Montreal Olympic Games, 1976, the horizontal bar event champion in the All Japan Championships, 1980, and a member of the 1981 American Cup Championships. All subjects were competent in executing the SBS, the DBS and the TBS.

After a 30-min warm-up, each subject was asked to perform the SBS three times followed by the performance of the DBS and the TBS. Each trial consisted of several giant swings before the dismount was executed. An Actionmaster 200 16 mm camera filmed the subjects at 64 frames/ sec. All trials were analyzed on the PCD motion analyzer to determine angle of release, body segment displacement and the timing of the movement sequence.

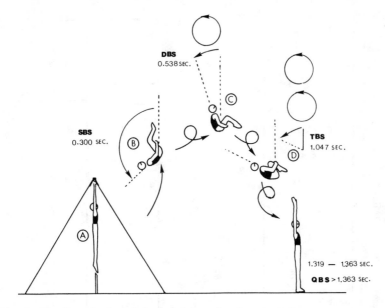

Figure 1—The determination of the completion of each somersault.

Results and Discussion

Flight Time

The mean flight time for each somersault, from release of the bar through landing was calculated. It was observed that it took the subjects 1.319 sec to perform the SBS, 1.325 sec to perform the DBS and 1.363 sec to perform the TBS. Analysis of variance revealed no significant differences among these somersaults with respect to time ($p > .05$). This indicates that the execution of a QBS would have to be performed in only a slightly longer time period.

Hip Angle during Flight

Maximum rotational momentum was determined to occur when the hip angle was the smallest for each somersault. The smallest hip angle in the SBS was 55.89° at .300 sec after the subject left the bar, for the DBS it was 42.67° at .538 sec while that of the TBS was 32.13° at 1.047 sec after release (see Figure 1). The relationship between the time when the smallest hip angle was observed and the total flight time was determined (see Figure 2). It took 22.7% of the total flight time of the SBS to reach the smallest hip angle. In the DBS, it took 40.6% of the total time to reach the smallest hip angle and in the TBS 76.9% of the total time to reach the smallest hip angle. The remainder of the percentage of time for

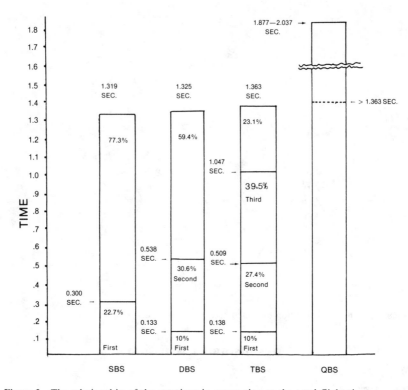

Figure 2—The relationship of the rotations in proportion to the total flight time.

each somersault was used in preparation for landing. The time to execute the first rotation for the DBS was half the time required for the SBS. In the TBS, however, the first rotation was completed in the same amount of time as for the DBS. The second rotation in the DBS and the TBS took approximately the same amount of time. The third rotation in the TBS took the longest amount of time to execute and twice as much time as it took to perform the first two rotations. The time to reach the smallest hip angle was assumed to very closely approximate a geometric sequence. From this geometric sequence assumption an additional .830 to .990 sec would be needed to reach the smallest hip angle for the fourth rotation. The total predicted time of 1.877 sec to 2.037 sec was longer than the possible total flight time.

Shoulder Location at Release

The mean shoulder location at release for the three somersaults were as follows: SBS 84.89°, DBS 75.22°, and TBS 67.38°. Analysis of variance showed a significant difference among these angles ($p < .05$). Assuming a geometric sequence for these angles, the angles of release for a fourth rotation was predicted to fall within the range of 59.9° and 60.3°.

Shoulder Angle at Release

The mean shoulder angle at release for each somersault was as follows: SBS 174°, DBS 162°, and TBS 156°. Analysis of variance revealed that there was a significant difference between the SBS and the other somersaults ($p < .05$) but no significant difference was found between the DBS and the TBS.

Hip Angle at Release

The mean hip angle at release in each somersault was 157° for the SBS, 113° for the DBS and 110° for the TBS. Analysis of variance showed a significant difference between the SBS and the other two somersaults ($p < .05$) but no significant difference between the DBS and the TBS.

Summary

From this study it was determined that the gymnast has only a certain amount of time from release of the bar to landing, and this time appears to be relatively constant regardless of the number of rotations performed. From the geometric sequence assumption used to achieve the smallest hip angles in the performance of the SBS, the DBS, and the TBS, it was predicted that the fourth rotation would take between 0.83 and 0.99 sec; the total time of 1.877 to 2.037 sec predicted to execute the QBS was longer than the possible flight time. The gymnast would need to 1) increase his angular velocity before release, 2) change the angle of release, and/or 3) stay tucked longer to be able to perform the QBS. Based on the data from this study, even with changes in these variables, the QBS would be a most difficult stunt to perform.

Acknowledgments

The author wishes to thank Gail Shierman and Paul Ziert for their assistance in this study.

References

CROSS, T. 1980. Biomechanical Analysis of The Double Backward Somersault. Unpublished Master's Thesis.

FREDERICK, A.B. 1979. A triple back plus a little more. International Gymnastics. 22:64-65.

KANEKO, A. 1970. Instructions for Gymnastics II: Horizontal Bar. Humaido. pp. 241-243.

MITSUKURI, T., Noguchi, Y., and Suzuki, A. 1975. The comparison between double and triple backward somersaults on the horizontal bar. Kenkyu-Bu-Ho. 36:15-23.

Kinematics and Kinetics of the
Backward Somersault Take-off from the Floor

P. Brüggemann
Johann Wolfgang Goethe-Universität,
Frankfurt, West Germany

Only a small number of the many skills in gymnastics have been subjected to biomechanical analysis. A review of the literature indicates that only a few take-off studies have been reported. Payne and Barker (1976) and Feller (1975) analyzed the kinematics and kinetics of the body center of gravity during the backward somersault take-off. Nissinen (1978) investigated the kinematics and kinetics of the running forward somersault. In addition, there are a few kinematic investigations of the forward and backward somersault. The above mentioned works are for the most part descriptive in nature and deal with relatively few subjects.

For a better idea of the structure of a movement, and its influencing factors, a detailed analysis of the total movement of the body and its segments is necessary. In regard to this problem, Hay (1978) has formulated a viable concept. The concept, a mechanical model, describes the investigated movement by single individual segments with respect to time. The essential and most influential model-parameters should be defined by statistical analysis. Based on this concept, this study investigated the take-off of the backward and double-back somersault from the floor.

Procedures

Model

The evaluation criteria of a backward and double backward somersault is maximal height and a sufficient angular momentum for the required rotation (360° or 720°). In this investigation there was a highly significant difference between the mean differences of the investigated groups

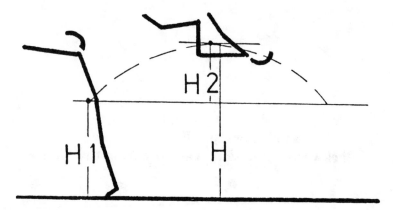

Figure 1—Height of take-off (H1) and height of flight (H2) in the backward somersault.

only for height of the flight, whereas angular momentum showed no significant differences, therefore the model was applied only to the evaluation of the height of the somersault flight. The total height achieved by the body center of gravity during the flight phase of the somersault depends on the height of the body center of gravity at take-off (H1) and the extent to which the body center of gravity is elevated during the flight phase of the somersault (H2) (see Figure 1).

The height of the body center of gravity (CG) at take-off (H1) is determined by the masses of the segments, the locations of the centers of gravity within the segments, the lengths of the segments, and the segmental angles at take-off. The first three factors are anthropometric variables and therefore constant. H2 is determined by the vertical velocity at the instant of take-off. The vertical velocity is determined by the vertical impulse exerted during take-off, the mass of the gymnast and the vertical velocity before the take-off phase. The vertical impulse during take-off may be considered as a sum of the vertical impulses, when the whole impulse is divided into consecutive time intervals. The vertical impulses during these time intervals are determined by the segmental inertial forces, which depend on the angular impulses exerted at the joints. In addition to the vertical impact velocity, the horizontal impact velocity and the angular momentum about the transverse axis of the body's center of gravity, which are produced by preceding movements like the round-off and flick-flack, must also be considered.

Data Collection

Analysis of 40 backward somersaults and 26 double backward somersaults of skilled and highly skilled gymnasts (N = 40) were conducted. The ground reaction forces (F_z and F_x) and the point of application of the forces (a_x) were measured by a Kistler force platform (Type 9261) and

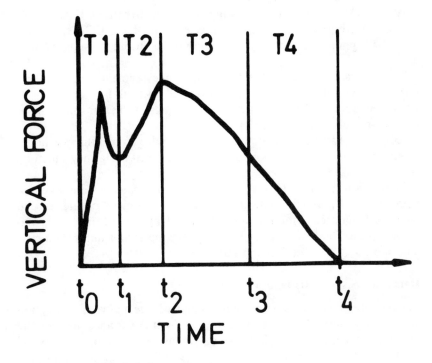

Figure 2 — Time intervals during support phase.

registered on a magnetic tape recorder (Phillips). The force platform was installed level with the gym floor and covered with a special gymnastic mat, therefore enabling measurements under real conditions. The ground reaction forces were integrated, and the vertical and horizontal impulses in the four time intervals were calculated (see Figure 2).

A lateral view of the trials was filmed with a LOCAM 16 mm camera operating at 201 to 209 frames/sec. The camera's lightmarks were set to 10 ms. The optical axis was perpendicular to the plane of motion and the camera distance was 15 m. The films were analyzed with a PCD-digitizing system, recording the coordinates of the segmental endpoints of each frame at take-off using a 14 segmental model. Further analysis was done on a HP 2100 digital computer which included the determination of segmental and total body center of gravity, as well as the body angles, using a five-segment rigid body model consisting of arms, trunk (including head), thighs, lower legs, and feet. It was assumed that the arm and leg movements were symmetrical about the mid-sagittal plane.

After data smoothing with a digital filter, linear and angular velocities, as well as the angular momentum of the whole body about the transverse axis throughout the body center of gravity, were calculated (Hay et al. 1977).

As a check for the data smoothing before and after the vertical (-9.81 m/sec^2) and horizontal (0 m/sec^2) acceleration of the CG before and after the support and the vertical reaction force during the support were used.

These data were then used to compute the segmental vertical and horizontal impulses, the average segmental inertial forces, the joint angular impulses, and the average joint torques in the four intervals of the take-off phase.

As a check for the calculated change of angular momentum the angular impulse during the support phase, as calculated by Ramey's method (1973), was used.

In addition, a set of 35 anthropometric measurements were taken on each analyzed gymnast. The individual mass moment of inertia and segmental mass contribution was calculated by a modified Hanavan-model (Miller and Morrison, 1975). These individual anthropometric data were used for the film analysis.

Data Analysis and Reduction

The model parameters were reduced using variance analysis, correlation analysis and multiple regression analysis. For the variance analysis the variables of skilled and highly skilled gymnasts performing a backward somersault (20 vs 20) and double back (13 vs 13) were compared. For the correlation analysis each parameter of the model, with respect to the subject's weight, was correlated with H2 or with the vertical take-off velocity (Vz_4). For ranking the data during the support intervals, factor analysis was used. Those variables which were seemingly influential on H2 (Vz_4), and which were in accordance with the statistical assumptions, were entered into a multiple stepwise regression analysis. The degree of influence of each parameter of the regression equation was evaluated by changing the mean of one of the variables by a comparable amount using the standard deviation with all other parameters constant. The degree of change in Vz_4 or H2 is the amount of influence of the manipulated variable on the criterion.

Results

Table 1 presents some selected descriptive data. The flight phase of the round-off of the flick-flack is seemingly short (somersault 0.108 sec, and double somersault 0.094 sec). The horizontal velocity of the gymnast's body center of gravity at landing (Vx_0) was approximately 4 m/sec. There was also a relatively large angular impulse about the transverse axis through the body center of gravity. During the take-off as the body center of gravity was changed in an upward direction ($\sim 60°$), the angular momentum (L_0) was drastically reduced ($\sim 50\%$).

Table 1

Summary of Force-time, Angular and Linear Velocity Data

		Mean (N = 40) Somersault	p*	Mean (N = 26) Double somersault	p**
Duration of pre-flight	sec	.108	.015	0.094	.018
Vertical velocity t_o	m/sec	− .600	−	− .420	−
Horizontal velocity t_o	m/sec	3.750	.017	4.120	−
Angular momentum in t_o	kg m²/sec	113.000	.002	128.900	−
Duration of take-off	sec	0.131	−	0.123	−
1. maximum vertical ground reaction force	N	4372.000	.020	4510.000	−
Minimum vertical ground reaction force	N	2839.000	−	3090.000	−
2. maximum vertical ground reaction force	N	6069.000	−	6846.000	−
Maximum horizontal ground reaction force	N	815.000	−	867.000	.041
Minimum horizontal ground reaction force	N	− 1900.000	−	− 2100.000	−
H1	m	1.165	−	1.160	−
H2	m	0.953	.000	1.074	.008
H	m	2.118	.000	2.234	.021
Vertical take-off velocity	m/sec	4.300	.000	4.570	.008
Horizontal take-off velocity	m/sec	2.690	−	2.850	−
Take-off angular momentum	kg m²/sec	55.900	−	64.500	−
Take-off angular velocity	rad/sec	4.130	−	4.850	−

*p indicates the significance of the mean difference between skilled (N = 20) and highly skilled (N = 20) gymnasts (p ≤ 0.05).
**p indicates the significance of the mean difference between skilled (N = 13) and highly skilled (N = 13) gymnasts.

The vertical ground reaction force during the support phase showed a typical force curve pattern with two maximal and one relatively minimal peak. The impact forces were especially reduced by the mats placed over the force platform. The beginning of the horizontal ground reaction force curve showed a very short interval of propulsion followed by a longer braking period.

The contributions of the average vertical segmental inertial forces on the take-off phases are presented in Figure 3. The trunk, because of its large mass, was responsible for the majority of the total impulse. The inertial force-time curve of the arms was of major interest because this motion is often emphasized by coaches. Especially in the time interval T4 the average inertial force of the arms was negative, not positive, and had a large influence on the acceleration of the whole body.

The height of the body center of gravity at take-off (H1) showed no significant differences between the groups (see Table 1), whereas the

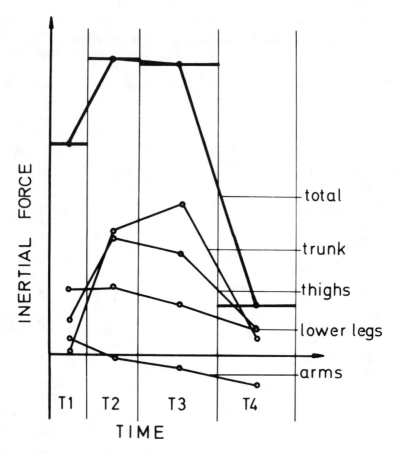

Figure 3 — Contribution of the average total and segmental vertical inertial forces.

height of the center of gravity during the flight (H2) differed significantly between the skill level groups. After take-off, the angular momentum showed very little variance between the groups for both the backward and double backward somersault. Therefore, the statistical analysis used only H2 and the measurement data taken from the backward somersault (N = 40). The parameters, Vx_o and L_o, have already been shown to have influential input for the performance. This result was supported by the correlation coefficients of Vx_o (r = 0.42; p ≤ 0.01) and L_o (r = 0.44; p ≤ 0.01) with H2.

With help from the factor analysis, the movements of the body segments during the individual support intervals were investigated. In T1 the following three factors were extracted: trunk, arm, and lower extremity movement. At the moment of impact, the movements of the feet, thighs, and lower limbs were dependent on each other. In T2 and T3 once again, the above three factors (trunk, arms, and lower extremity movements) were extracted, and proven to be statistically independent of each other. However, during the support phase, intervals T2 and T3, as in T1, the movement of the feet, lower legs, and thighs were not statistically independent. During the interval T4, the factor analysis extracted four factors (arms, trunk, lower legs, and feet movements) as independent factors.

The weight of the gymnast was then partialled out and all of the parameters from the model were correlated with H2. Although the average acceleration of the body center of gravity in T1 indicated no significant correlation with H2, a highly significant correlation ($p \leq 0.01$) was found between the average acceleration and H2, for the time intervals, T2, T3, and T4. In addition to the previously mentioned factor analysis, independent factors found during the intervals T2, T3, and T4 were correlated with H2. The input parameters, Vx_0 and L_0, as well as the significant ($p \leq 0.01$) correlates of H2 during T2, T3, and T4, were further entered into a multiple stepwise regression analysis. Thereby, H2 or Vz_4, was described as a function of the approach velocity of the center of gravity of Vx_0, of the angular momentum of the pre-flight L_0, of the average vertical inertial force of the arms during T2 ($\bar{a}m_{z4T2}$), and of the average vertical inertial force of the feet during T3 and T4 ($\bar{a}m_{z1T3}$, $\bar{a}m_{z1T4}$) with an accuracy of 0.19 m/sec.

The vertical inertial force of the feet was taken as an indicator for the torque produced about the ankle joints.

Table 2 shows the variation on take-off velocity of the CG (ΔVz_4), when each parameter in the regression equation was changed by one standard deviation. The table presents the expected change of the take-off velocity with a 95% probability and includes the minimal $\Delta\bar{V}z_{4(MIN)}$ and the maximal $\Delta\bar{V}z_{4(MAX)}$ calculated values.

It could be concluded that when a skilled gymnast performed a backward somersault after a round-off or a flick-flack, that the most influencing factors seemed to be the horizontal velocity of CG before the support, the angular momentum about the transverse axis through CG be-

Table 2

Relative Influence of the Selected Parameters on the Vertical Take-off Velocity

	$\Delta Vz_{4(min)}$ m/sec	$\Delta Vz_{4(max)}$ m/sec
Vx_0	0.12	0.17
L_0	0.10	0.19
ma_{z4T2}	0.02	0.09
ma_{z1T3}	0.09	0.19
ma_{z1T4}	0.01	0.25

fore the support, the inertial force of the feet in the final take-off intervals (T3 and T4), and the arm action in interval T2.

In conclusion, from a practical point of view, it can be said that the linear velocity and the angular momentum before take-off is extremely influential in performing a good somersault backward. Therefore, it is important that the gymnast prior to the somersault perform an excellent round-off or flick-flack.

For future investigations, the modification of joint torques and further testing for validity of the procedure are recommended.

References

FELLER, I. 1975. Absprünge rückwärts in Kunstturnen (Backward somersaults in gymnastics). Diplomarbeit am Laboratorium für Biomechanik der ETH, Zürich.

HAY, J.G. 1978. The identification and ordering of the technical factors limiting performance. A paper presented to the XXI World Congress in Sport Medicine, Brasilia.

HAY, J.G., Wilson, B.D., Dapena, J., and Woodworth, G.O. 1977. A computational technique to determine the angular momentum of a human body. Journal of Biomechanics 10:269-277.

KNIGTH, S.A., Wilson, B.D, and Hay, J.G. 1978. Biomechanical determinants of success in performing a front somersault. International Gymnast 18:54-56.

MILLER, D., and East, D. 1975. Kinematic and kinetic correlates of vertical jumping in women. In: P.V. Komi (ed.), Biomechanics V-B, pp. 65-72. University Park Press, Baltimore.

MILLER, D., and Morrison, W.E. 1975. Prediction of segmental parameters using the HANAVAN human body model. Medicine and Science in Sports 7(3):207-212.

NISSINEN, M.A., 1978. Kinematic and kinetic analysis of the hurdle and the support phases of a running forward somersault. M.S. Thesis, University of Washington.

PAYNE, A.H., and Barker, P. 1976. Comparison of the take-off forces in the flic-flac and the back somersault in gymnastics. In: P.V. Komi (ed.), Biomechanics V-B, pp. 314-321. University Park Press, Baltimore.

RAMEY, M.R. 1973. Significance of angular momentum in long jump. Research Quarterly of the AAHPER 44:489-497.

Swinging as a Way of Increasing the Mechanical Energy in Gymnastic Maneuvers

W.L. Bauer

Universität Bremen, Bremen, West Germany

This research demonstrates that the solution of the equation of the pendulum of variable length can be used for constructing congruent trajectories for different gymnastic maneuvers in the r, ϕ-plane when considering the displacement of the center of gravity.

The purpose of the study was to show relationships between different gymnastic maneuvers based on common basic principles of mechanics and to provide graphic instructions supporting the teaching and learning process.

Model

A gymnast is swinging in the inverted hang as shown in Figure 1a. For the sake of mathematical simplification he may be regarded as a particle whose distance (r) from the suspension of the rings is L_1 in the bent body inverted hand and L_2 in the straight body inverted hang. Although in reality the changeover from straight to bent position is smooth it may be assumed as instantaneous for the same reason. Having this in mind, the motion of the gymnast may be described by the differential equation of a suspended pendulum with variable length r.

$$\frac{d}{dt} (mr^2 \dot{\phi}) + mgrsin\phi = 0 \qquad (1)$$

During the gymnast's maneuver, r is a function of the angular displacement (ϕ) and velocity ($\dot{\phi}$). Thus we have

$$r = r(\phi,\dot{\phi}) \qquad (2)$$

The basic idea of any further analysis is to consider the length of the pendulum as a piecewise constant. Thus the solution

$$r(\phi,\dot\phi) = \frac{L_1 + L_2}{2} - \frac{L_1 - L_2}{2} \, \text{sgn}\phi \, \text{sgn}\dot\phi \qquad (3)$$

given by Magnus (1976) can be used to describe the arrowed path of Figure 1a even when friction and damping is involved. The change of energy after each semicycle is

$$\Delta E_n = mg\,(L_1 - L_2)\,(1 - \cos\phi_{o(n+1)}) + \tfrac{1}{2} m L_1^2 v_{1n}^2 \left(\frac{L_1^2}{L_2^2} - 1\right) - E_{fn} \qquad (4)$$

E_{fn} considers the work done by friction and damping forces. Applying the energy equation and introducing

$$E_{on} = mgL_1 \,(1 - \cos\phi_{on}) \qquad (5)$$

$$E_{fn} = E_{1fn} + E_{2fn} \qquad (6)$$

we have

$$\tfrac{1}{2}\, m L_1^2 v_{1n}^2 = E_{on} - E_{1fn} \qquad (7)$$

$$\tfrac{1}{2}\, m L_2^2 v_{2n}^2 = mgL_2 \,(1 - \cos\phi_{o(n+1)}) + E_{2fn} \qquad (8)$$

Conservation of angular momentum gives

$$L_2^2 v_{2n} = L_1^2 v_{1n} \qquad (9)$$

By substituting and rearranging Equations (4) to (9) one can obtain the change of energy after each semicycle as

$$\Delta E_n = (c^3 - 1)E_{on} - \Delta E_{fn} \qquad (10)$$

where $c = \dfrac{L_1}{L_2}$ and $\Delta E_{fn} = c(c^2 E_{1fn} + E_{2fn})$

By calculating the total energy we obtain after n semicycles

$$E_{o(n+1)} = c^{3n} \left(E_{o1} - \sum_1^n \frac{\Delta E_{fn}}{c^{3n}}\right) \qquad (11)$$

Assuming negligible friction and damping ($\Delta E_{fn} = O$) the total energy after n semicycles is

$$E_{o(n+1)} = c^{3n}\, E_{o1} \qquad (12)$$

In reality the system behaves in three different ways, depending on the friction and the damping forces and wrong control activities of the gymnast. It can increase or decrease its energy from initial conditions ending up in limit cycles or come to a rest.

Periodic Motion

When periodic motion is established the mechanical energy of the system remains constant. (ΔE_n = O). Assuming $E_{1fn} = E_{2fn} = E_{fn}/2$ we have

$$E_{fn} = 2 \frac{c^3 - 1}{c(c^2 + 1)} E_{on} \tag{13}$$

Conditions for a stable limit cycle are

$$\Delta E_{n+1} > O \text{ for } E_{o(n+1)} < E_{on}$$
$$\Delta E_{n+1} < O \text{ for } E_{o(n+1)} > E_{on} \tag{14}$$

By setting $c = c_{max}$ Equation (13) gives the maximum admissible energy dissipation which can be compensated for by the gymnast.

Active Damping

Even when there is no friction or passive damping involved, the gymnast can decrease the mechanical energy through control actions by reversing his movements according to Figure 1d. The arrowed path can be described by changing the first minus sign into a plus sign in Equation (3).

Applying the previous principle of calculations accordingly, the change of mechanical energy is obtained as

$$\Delta E_n = (\frac{1}{c^3} - 1) E_{on} - \Delta E_{fn}^\dagger \tag{15}$$

and the total energy after n semicycles

$$E_{o(n+1)} = \frac{1}{c^{3n}} (E_{o1} - \sum_1^n c^{3n} \Delta E_{fn}^\dagger) \tag{16}$$

with $\Delta E_{fn}^\dagger = \frac{1}{c}(\frac{1}{c^2} E_{1fn} + E_{2fn})$. The change in mechanical energy is always negative. Thus, the system comes to rest in any case. There are only a few stunts in gymnastic routines which apply this principle.

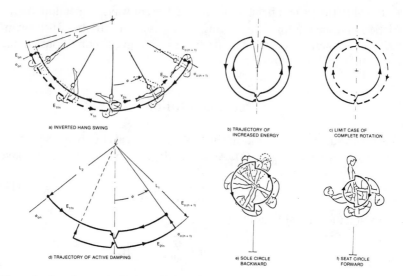

a) INVERTED HANG SWING

b) TRAJECTORY OF
INCREASED ENERGY

c) LIMIT CASE OF
COMPLETE ROTATION

d) TRAJECTORY OF ACTIVE DAMPING

e) SOLE CIRCLE
BACKWARD

f) SEAT CIRCLE
FORWARD

Figure 1 — Rotation as the limit case of a swing movement.

Applications

Complete Rotations

Assuming a particle having mass (m) is connected to a rigid bar of variable length whose mass is negligible, by increasing the energy of the system, the two upward paths (see Figure 1b on top of the circles) coincide and either of the trajectories of Figure 1c will occur.

This trajectory configuration describes the basic principle of movement for a number of gymnastic stunts where complete rotations are performed , e.g., a giant swing backward (see Figure 2a), sole circle backward (see Figure 1e), giant swing forward (see Figure 2d), seat circle forward (see Figure 1f), on the horizontal bar. Similar exercises can be found on the uneven parallel bars. On the parallel bars backward and forward rolls fit this principle. Since these exercises include complete rotations and may be performed periodically Equation (13) can be applied.

Assuming negligible angular velocity ($\dot{\phi}_{on} = \dot{\phi}_{o(n+1)} \sim o$) on top of the trajectory the mechanical energy of the system is

$$E_{no} = 2mgL_1 \tag{17}$$

By setting $c \simeq 1 + \dfrac{\Delta L}{L_1}$, which describes the gymnast's ability to change the distance of his or her center of gravity relative to the point of support, one obtains the maximum permissible energy loss per revolution through friction and damping as

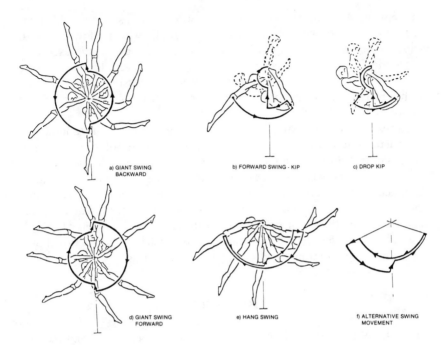

Figure 2—Relationships to non-periodic movements.

$$E_{fn} \simeq 6mg\Delta L_{max} \tag{18}$$

Swing Movements

Since the anatomy of the human body does not always allow one to do symmetrical movements as shown in Figure 1a, trajectories of Figure 2e are used to overcome friction, e.g., when swinging during the hang on the flying rings, doing the cast, the flank cut-away, and the hang swing on the horizontal bar. Trajectories of Figure 2f can also be found in exercises on the flying rings.

Other exercises can be described by constructing trajectories of sectors with piecewise constant radii. This was done in Figure 2c for a drop kip and in Figure 2b for a kip starting with a forward swing.

Conclusions

The movement principle of exercises with complete rotations, where the mechanical energy is increased from within by displacements of the center of gravity, have been interpreted as a special case of the swing movement. Movement principles were presented as trajectories consisting of sectors with piecewise constant radii, which allowed one to show the common principles involved in different gymnastic exercises. Since the changeovers of the trajectories from one radius to another were considered to be instantaneous, the maximum permissible energy dissipation through friction and damping for periodic stunts could be calculated approximately.

In reality the changeovers of the trajectories cannot be instantaneous. The determination of the real trajectory leads to the solution of an optimal control problem similar to that described by Bauer (1980).

Practical experience, however, reveals that the instantaneous changeovers coincide with the gymnast's points of decision where different muscle groups are activated from flexion to extension or vice versa. Thus, according to the theory of perceptual motor learning of Ungerer (1977) it should be sufficient to use the idealized trajectories for instructional purposes in order to initiate and complete a learning process.

References

BAUER, W.L. 1980. Mathematical modelling and optimization and their influence on sport movements — possibilities and limitations. Paper presented at International Symposium on Biomechanics of Sport. Deutsche Sporthochschule Köln, Institut für Biomechanik, Dec. 4-6.

MAGNUS, K. 1976. Schwingungen. (3.Auflage), B.G. Teubner Verlagsgesellschaft mbH, Stuttgart.

UNGERER, D. 1977. Zur Theorie des sensomotorischen Lernens (The Theory of Motor Learning). (3.Aufl.), Verlag Karl Hofman, Schorndorf bei Stuttgart.

C.
Aquatic
Sports

The Dynamic Features of the Canadian Canoe Paddle

Hideo Oka
Osaka Kyoiku University Highschool, Ikeda, Japan

Tsutomu Okamoto
Kansai Medical School, Hirakata, Japan

Hiroshi Nakagawa
Osaka University of Economics, Osaka, Japan

Noriyoshi Yamashita and Minayori Kumamoto
Kyoto University, Kyoto, Japan

Masatada Yoshizawa
Fukui University, Fukui, Japan

Yasuhiko Tokuhara
Teikoku Women's University, Moriguchi, Japan

Fujio Hashimoto
Osaka Electro-Communication University, Osaka, Japan

In racing Canadian canoe and kayak, their length, beam, and weight are regulated by the international rules, and it requires a high level of technique to build them. Almost all Canadian canoes and kayaks used in the world championship regattas or Olympic events were made in Denmark, and thus there is little room for selection on the part of the canoeists in every country of the world. On the other hand, there is no international regulation for the paddles, and the paddles have been mostly designed by the experimental estimation of the champion canoeists rather than on a scientific basis. The present experiments were carried out to investigate the dynamic features of the various kinds of Canadian canoe paddles in a water tank.

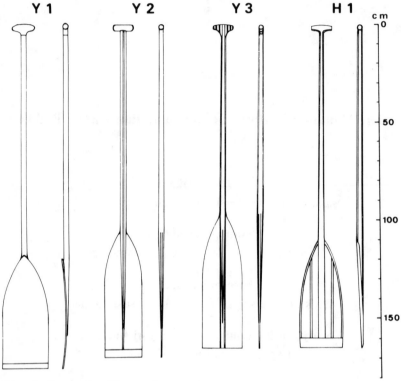

Figure 1—Shapes of paddles tested.

Method

Four kinds of Canadian canoe paddles used in the experiments were different in length, shape, and areas as shown in Figure 1 and Table 1.

Hardness of Shaft

Experimental set-up illustrated in Figure 2 show how the paddle was held horizontally at the grip (A) by wooden blocks attached to the board, the supporting block (B) was set up 80 cm from the grip, and the blade was faced upward. The various loads (W) were applied to a point 20 cm from B. A thin steel wire (C) of sufficient length was attached to the top of the blade along the shaft, and the centimeter scale (S) to measure the vertical deflection at the point 80 cm from the weighing point disregarding the differences in the paddle lengths.

Water Tank Experiments

The paddles were held at the grip by wooden blocks attached to the iron

Table 1

Description of the Paddle Tested

Paddle	Length (cm)	Weight (g)	Blade Length (cm)	Blade Width (Top) (cm)	Blade Area (cm²)	Remarks
Y1	174	1080.0	56	24.0	1117	Concave face, Danish
Y2	170	673.0	63	19.5	1108	Straight blade, Japanese
Y3	164	1087.5	67	20.5	1233	Straight blade, Rumanian
H1	164	942.0	54	22.0	976	3.5° backward leaned face, Hungarian

frame built over the water tank, and various loads (50 kg, 40 kg, 20 kg, and 10 kg) were applied at a point 80 cm down from the grip by way of a chain. Strain developed on the paddle was recorded via a strain gauge attached to the shaft. Angular shift of the paddle centering around the grip was recorded by a video camera (60 frames/sec) or filmed by a 16 mm movie camera (48 frames/sec). The water level of the tank was kept at the level where the blade did not come out during the angular shift of the paddle. The paddle loaded with a weight was hooked at the starting position and was checked to be certain no strain developed on the shaft. After sudden release of the hook, the strain curve of the shaft and the angular shift of the paddle were recorded from the starting positions, 10° backward from the vertical line or at the vertical line, to just after passing the vertical line or about 10° after passing the vertical line. The strain curve of the paddle and a signal on the video tape or the frame of the film were simultaneously recorded on the recording paper.

Results

Hardness of Shaft

The relation between stress and weight loaded on each paddle was linear as is shown in Figure 2. There were quite large differences in the hardnesses of the shafts of the paddles; that is, the deflection of Y2 was about 2.5 times and Y1 and H1, about 1.5 times that of Y3, the hardest paddle.

Angular Shift of Paddle

From 16 mm movie film, degrees of angular deflections of the shaft centering around the grip in each frame were measured and plotted against the frame number as was shown in Figure 3.

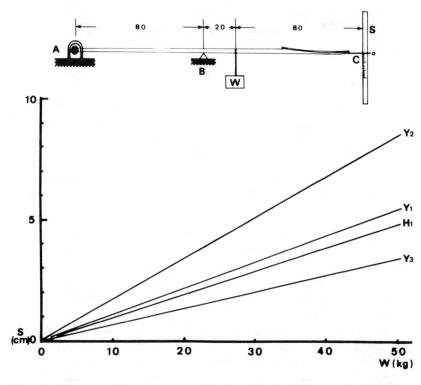

Figure 2 – Experimental device to measure deflection of the shaft (upper). Relation between deflection and weight loaded (lower).

When the paddles were pulled with a load of 40 kg, significant differences were observed in the angular deflections of each paddle. That is, the softest Y2 showed the fastest angular deflection, while the hardest Y3, the slowest, and Y1 and H1 were in between Y2 and Y3. In the blade area, the largest Y3 was about 10% larger than Y1 and Y2, where Y1 was nearly equal to Y2. H1 was about 15% smaller than Y1 and Y2. Therefore, it was suggested that the significant differences in the angular deflections were due to hardness of shaft rather than blade area. Whereas, in the case of 10 kg load, the hardness of the paddle did not show any consistency in the angular deflection of the paddle.

Strain Curve of Shaft

Figure 4 shows the typical strain curves of the paddles tested loaded with the various weights. In this figure, the time required for the strain curve to reach its peak might illustrate responsiveness of paddles to pulling strength. The harder the shaft, the shorter the time in the cases of 20 kg and 40 kg. However, in Y1, Y2, and H1, a 50 kg load did not induce any

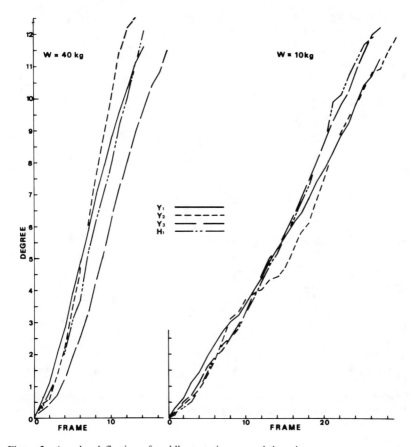

Figure 3 — Angular deflection of paddle centering around the grip.

shortening of the time as in the times at 40 kg and 20 kg loads; in other words, Y1, Y2, and H1 did not respond to a heavier load if it was increased to more than 40 kg. Only the hardest Y3 responded to the load of 50 kg.

The calibration of the strain of each paddle against a certain external load applied to the shaft was adjusted to a unit value as shown in Figure 4. Therefore, the peak values of the strain curves showed the degrees of water resistance developed by the blades.

The highest peak value of Y3 and Y1 might be due to the facts that the blade of Y1 was deepest in the water and the Y3 had the largest area. The Y2 which had almost the same area as Y1, but was shorter than Y1, showed middle peak value. H1 with the shortest shaft and the smallest blade area showed the smallest peak value.

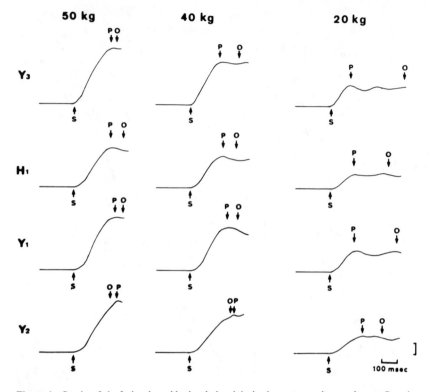

Figure 4—Strain of shaft developed by loaded weight in the water tank experiment. S: point where paddle started to move, P: peak point of strain curve, O: point where shaft just passed vertical line.

Discussion

The results obtained in the present experiments strongly suggest that the canoeist who has a high pulling force should use a paddle of sufficiently hard shaft; otherwise a weak paddle cannot respond to develop a large pulling force. The hardness of shaft predominantly affected the responsiveness of the paddles to pulling force rather than any other characteristics of the paddles, shape, length, and area.

Referring to the electromyographic studies of the kayak (Yoshio et al., 1974), correlation between the muscular activities in the trunk and upper extremities and the strain curve of the paddle shaft was examined during Canadian canoe paddling. When the canoeist did not perform the proper arm pulling motion to develop sufficient force, the strain curve of the paddle shaft showed an oscillatory curve depressed from a normal sustained curve (Kumamoto et al., 1980). In the water tank experiments, when the weight load was reduced, the harder the shaft, the greater the oscillation in the strain curve of the shaft (Kumamoto et al., 1980).

Therefore, the canoeist should select the hardness of the shaft according to the pulling muscular force.

Effects of paddle shape on canoeing performance are now under investigation, but it is presumed that shape has much to do with paddling style, whereas hardness of shaft is closely responsive to the muscular force.

References

KUMAMOTO, M., et al. 1980. Paddle and paddling style of Canadian canoe. Bulletin of Medical and Scientific Research of Sports 4(2): 67-76.

YOSHIO, H., et al. 1974. The electromyographic study of Kayak paddling in the paddling tank. Research Journal of Physical Education 18(4):191-198.

Analysis of Rowing Movements with Radiotelemetry

T. Ishiko and S. Katamoto
Juntendo University, Japan

T. Maeshima
Senshu University, Japan

Movement of a boat is achieved by the propelling force of the boat against the water's resistance. This force is generated by the muscular force each rower exerts at the grip of the oar and is the resultant force generated at the oarlocks. Rowing performance, therefore, is influenced by the magnitude, frequency, style, and conformity of each rowing force in addition to environmental factors such as wind and waves. In this experiment repetitive rowing forces of rowers obtained during actual rowing performances with radiotelemetry were analyzed.

Methods

The subjects were 11 first class Japanese rowers. The age and physical characteristics of the subjects are listed in Table 1. The rowing forces of the rowers were transmitted telemetrically from the strain gauge glued on the inboard side of the test oar to the shore, where the recordings were made. The strain on the oar was calibrated to determine the rowing force by hanging different weights on the grip (Ishiko, 1968).

In order to avoid the variations in rowing style due to the position of the seat, the subject sat in turn on a designated seat (No. 4 or 5) of an eight-oared shell and rowed using the test oar with maximal exertion. The other seven rowers rowed in union with the subject. The rowing forces were recorded continuously from the beginning of rowing performance through 120 strokes, which covered approximately 1,000 m distance.

From the strain curve of oar in Figure 1 peak force, impulse, and dura-

Table 1

Physical Characteristics of Subjects

Number	Age years	Stature cm	Body Weight kg
11	22.2 ± 2.3	181.3 ± 1.8	77.9 ± 6.6

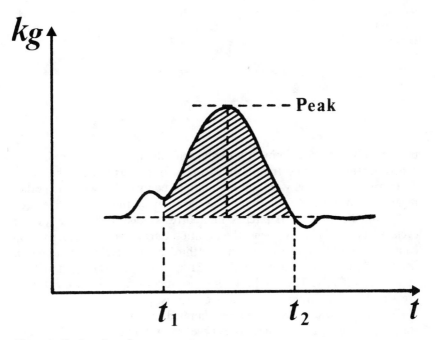

Figure 1 — Explanation of measurements.

tion during one stroke were obtained as the peak value, hatched area, and time interval between t1 and t2, respectively. Mean force was calculated as the ratio of impulse to duration.

Results

Frequency, duration, impulse, peak and mean forces were measured during strokes 21 to 30, 51 to 60 and 81 to 90. They were considered as representative values of 120 strokes. The first and last stages of the 120 strokes were deleted since they exhibited considerable variability. The results obtained are shown in Table 2. The frequency of the strokes was ap-

Table 2
Results of Measurements during Rowing

Strokes	21-30	51-60	81-90
Frequency (pitches/min)	34.8 ± 1.13	34.8 ± 1.66	35.0 ± 0.94
Duration (msec)	628.0 ± 32.00	640.0 ± 39.00	643.0 ± 38.00
Impulse (kg • sec)	30.7 ± 5.50	30.9 ± 6.20	30.1 ± 6.40
Peak Force (kg)	80.0 ± 16.90	78.3 ± 19.00	75.2 ± 19.70
Mean Force (kg)	49.6 ± 10.80	48.4 ± 11.00	46.8 ± 12.30

M ± SD

proximately 35 pitches/min. Although each subject rowed in turn in each experimental bout, the frequency of the strokes was almost the same. The results, therefore, were considered to be comparable with each other regardless of rowing frequency. The duration of the strokes averaged 0.63 to 0.64 sec. Since 35 pitches/min corresponded to 1.71 sec for one cycle of the rowing movement, the duration during which the oar was in contact with the water was about one third of one cycle. The impulse, peak and mean forces averaged 30 to 31 kg • sec, 75 to 80 kg, and 46 to 49 kg, respectively. These values were almost constant with a slightly decreasing tendency following repetitive strokes.

Figure 2 illustrates rowing curves compiled for five strokes at the three stages of rowing performance. The curves showed marked individual differences, indicating the lack of conformity in the rowing style among the subjects. Although the form for each rower seemed fairly consistent, differences were detected when the five curves were compared. The magnitude of the impulse and forces showed considerable differences among subjects.

Subjects 1 and 3 rowed powerfully, but the former was lacking in consistency in each rowing style while the latter decreased in impulse and force with advancing strokes. Those who took much time for each stroke lacked force (Subjects 5 and 10).

Discussion

Rowing performance cannot provide information on the evaluation of each subject on board. The present study was aimed at analyzing actual

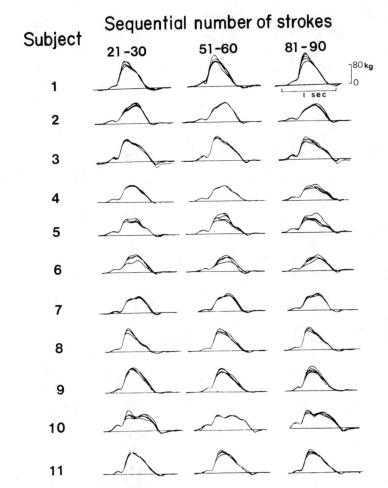

Figure 2 — Rowing curves of subjects compiled for five strokes.

rowing force in each subject using telemetry. Asami et al. (1978) studied the rowing force exerted by first class Japanese oarsmen using a rowing tank and reported that the total impulses and peak force calculated by the same procedure as the present study were 79330 N • sec (37.0 kg • sec/stroke) and 82.11 kg, respectively, when the subjects rowed for 6 min with a frequency of 35 pitches/min. These values were comparable to those obtained in the present study. Therefore, rowing tank experiments should provide a good simulation of actual rowing.

The impulse had a higher correlation with peak and mean forces than duration (see Figure 3), which indicated that impulse was influenced more by force. This finding was derived from the fact that the variation in duration was small.

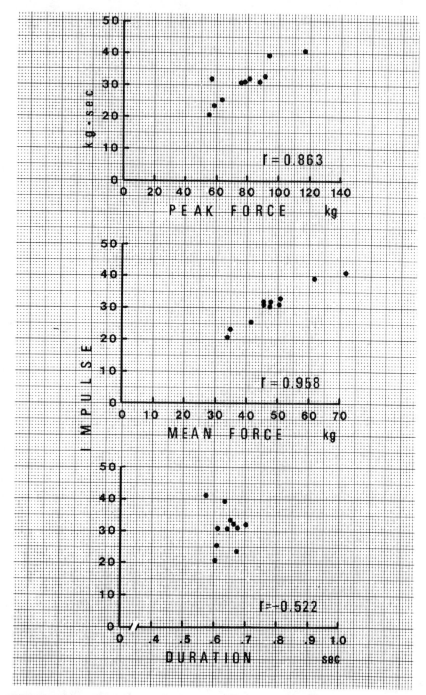

Figure 3—Relationships between impulse and parameters of peak force, mean force and duration.

The marked individual differences in rowing curves (see Figure 2) have been reported by Ishiko (1971) and Schneider et al. (1978). From this finding it was felt that some effort should be taken toward the improvement in rowing styles for better rowing performances. However, Schneider et al. (1978) reported that two members of a crew of a four-oared shell could change their rowing curves but the other two failed to change after a special effort was taken in distance training. This finding would suggest that a correct and uniform rowing style should be established at an earlier stage of the rowing career.

In conclusion, analysis of rowing curves can provide useful information to the rowers and coaches for better rowing performance. Rowing performance would be expected to be improved by reforming the rowing style and increasing the rowing force of rowers.

References

ASAMI, T., Adachi, N., Yamamoto, K., Ikuta, K., and Takahashi, K. 1978. Biomechanical analysis of rowing skill. In: E. Asmussen and K. Jorgensen (eds.), Biomechanics IV-B, pp. 109-114. University Park Press, Baltimore.

ISHIKO, T. 1968. Application of telemetry to sports activities. In: J. Wartenweiler and E. Jokl (eds.), Biomechanics I, pp. 138-146. Karger, Basel/New York.

ISHIKO, T. 1971. Biomechanics of rowing. In: J. Vredenbregt and J. Wartenweiler (eds.), Biomechanics II, pp. 249-252. Basel.

SCHNEIDER, E., Angst, F., and Brandt, J.D. 1978. Biomechanics in rowing. In: E. Asmussen and K. Jorgensen (eds.), Biomechanics IV-B, pp. 115-119. University Park Press, Baltimore.

Springboard Reaction Torque Patterns during Nontwisting Dive Take-offs

Doris I. Miller
University of Washington, Seattle, Washington, U.S.A.

If the performance is to be successful, a springboard diver must build up angular momentum of appropriate magnitude and direction during the take-off which immediately precedes the flight. During this period, the only external force having a torque with respect to the diver's center of gravity (CG) and, therefore, the only force with the potential for changing the diver's angular momentum is the springboard reaction. Because this force also plays an important role in establishing or modifying the diver's linear momentum, the effects of its horizontal (R_x) and vertical (R_y) components upon translation must be considered in conjunction with their influence upon rotation.

If the line of action of R_y passes in front of the CG, it promotes backward (or reverse) somersaulting angular momentum and, if behind the CG, forward (or inward) somersaulting angular momentum (see Table 1). In both cases, it increases the diver's upward momentum. If R_x promotes somersaulting in the correct direction for dives from the forward and backward groups, it reduces horizontal momentum away from the board. By contrast, in the reverse and inward groups, considered the more dangerous by many divers, if R_x is consistent with producing the desired rotation, it also increases the diver's momentum away from the board (Miller, 1980).

The question being addressed in the present study, therefore, centered upon the way in which the springboard reaction force meets the differing rotational demands of the four nontwisting groups by producing an appropriate angular impulse during take-off while, at the same time, furnishing an adequate linear impulse to meet the common requirements for height and safe distance in the flight.

Table 1

Characteristics of Nontwisting Springboard Dives

Diving Group	Type of Approach	Take-off Orientation	Somersault Direction
1. Forward	Running	Forward	Forward
2. Backward	Standing	Backward	Backward
3. Reverse	Running	Forward	Backward
4. Inward	Standing	Backward	Forward

Procedures

Through the cooperation of the U.S. Diving Association, 16 mm films of 3 m performances of competitors in the 1979 Fort Lauderdale International Diving Competition were obtained for analysis. These films were taken by T.M. McLaughlin of Auburn University during practice sessions. A Locam camera with 25 mm lens set 8 m back from the major plane of motion at about board height was used to photograph the dives. The field of view included the final approach step, hurdle, take-off, flight and entry. The frame rate of 101.3 fps was determined from marks exposed on the side of the film by an internal LED flashing at 100 Hz.

From these films, the take-offs of multiple somersault nontwisting dives of six male competitors representing the United States (3), Mexico (1), Great Britain (1), and Canada (1) were selected for analysis. In total, four forward 3½s, four backward 2½s, five reverse 2½s and three inward 2½s were investigated. All dives were performed in the tuck position. The beginning of take-off was clearly defined by the simultaneous two foot contact with the board in landing from the hurdle in the forward 3½s and reverse 2½s. In the backward and inward 2½s, it was considered to coincide with the initiation of final depression of the springboard. In all four diving groups, take-off was completed as the diver's feet left the board.

For the purpose of analysis, the divers were treated as seven-link systems (feet, legs, thighs, trunk, head-neck, upper arms, and forearms-hands). Bilateral symmetry was assumed and movement outside the primary plane of action was considered negligible. Each frame ($\Delta t = 0.01$ sec) of the take-off and at least 10 frames preceding and 10 following were digitized using an Altek microprocessor system to obtain coordinates of the body segment endpoints, diving board tip, and background references. Prior to differentiation, the position data were smoothed with a second order Butterworth digital filter with cut-offs for individual coordinate-time histories ranging from 4 to 7 Hz in accordance with an analysis of the residuals and their first derivatives (Jack-

Table 2

**Average Horizontal Velocities (m/sec) at Critical Points
during Multiple Somersault Dive Take-offs[a]**

		Depression			Recoil	
Dive	N	Initial	Min.	Maximum Depression	Max.	Final
Forward 3½ Tuck	4	0.75 (0.10)	0.64 (0.07)	0.87 (0.10)	1.01 (0.15)	0.79 (0.13)
Back 2½ Tuck	4	0.15 (0.04)	0.07 (0.10)	0.55 (0.12)	1.09 (0.22)	0.97 (0.20)
Reverse 2½ Tuck	5	0.72 (0.05)	0.29 (0.06)	0.36 (0.09)	0.83 (0.27)	0.80 (0.26)
Inward 2½ Tuck	3	0.11 (0.08)	−0.28 (0.04)	0.15 (0.01)	0.69 (0.05)	0.68 (0.05)

[a]Standard deviations are given in parentheses.

son, 1979). Segment masses and CG locations were predicted from the data of Dempster (1955) and Dempster and Gaughran (1967). Magnitudes of R_x and R_y were estimated as follows: $R_x = ma_x$ and $R_y = ma_y + W$. In order to calculate the springboard torque with respect to the CG, the springboard reaction force was assumed to act through the metatarsal-phalangeal (MP) joint during the take-off.

Results and Discussion

Previous research (Miller, 1980) had indicated greater interindividual variability in the springboard reaction torque patterns of half somersaults from a particular diving group than in multiple somersaults which required closer to maximum effort. Consequently, the selection of the 2½s and 3½s for analysis in the present study reflected an effort to obtain torque patterns which characterized the unique requirements of the four diving groups with a minimum of contamination from style and technique influences. Because of the commonality in the direction of rotation, there were similarities between the take-offs for the reverse and inward 2½s and likewise between the forward 3½s and backward 2½s.

In both inward and reverse 2½s, there was a substantial reduction in horizontal velocity during board depression (see Table 2). This was necessary to compensate for the increase in horizontal velocity during recoil when R_x was directed toward the tip of the board and was responsible for promoting most of the required inward (see Figure 1) and re-

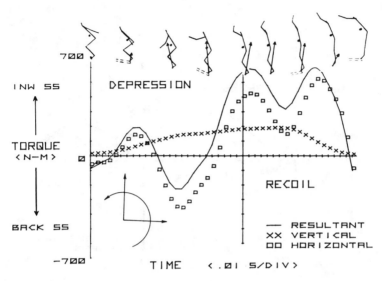

Figure 1 — Torque produced by the resultant springboard reaction force and its horizontal and vertical components with respect to the center of gravity during the take-off for an inward 2½ somersault tuck. The magnitude and direction of the springboard reaction at designated points are indicated by the length and orientation of the arrows on the stick figures. The reaction force shown at maximum springboard depression (vertical line separating depression and recoil) is four times the diver's body weight (687 N).

verse (see Figure 2) somersaulting angular momentum. In world class divers, however, during recoil *both* R_x and R_y provided torque consistent with the rotational direction of the dive. For R_y to make such a contribution, the diver's CG had to be closer to the fulcrum than the MP joint which was considered the point of application of the reaction force. For these divers, during recoil, the CG was between 8 and 2 cm closer to the fulcrum in the inward 2½s (N = 3) and between 6 and 1 cm closer in the reverse 2½s (N = 4). In contrast, with less skilled (but still national caliber) divers including one individual in the present investigation performing a reverse 2½, two top male and female Canadian intercollegiate divers executing reverse 2½s tuck and 1½s layout, respectively (Miller, 1974) and two female Canadian national team members doing inward 2½s tuck and a reverse 2½ tuck and 1½ layout (Miller, 1980), R_y either did not contribute to or retarded the build up of inward and reverse somersaulting angular momentum because the CG was directly over or further from the fulcrum than the MP joint during recoil.

The horizontal velocity pattern during forward 3½s and back 2½s showed little initial decrease but rather increased to a maximum value slightly preceding or coinciding with final contact with the board (see Table 2). This pattern reflected an R_x directed toward the board tip during most of the take-off and hence retarding the build up of angular

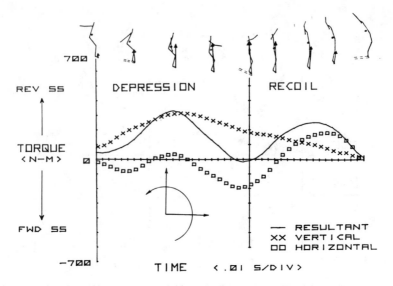

Figure 2 — Springboard reaction torque with respect to the center of gravity during the take-off for a reverse 2½ somersault tuck. The diver shown had a greater contribution to reverse somersaulting angular momentum from the vertical component than did the other divers analyzed. The reaction force arrow shown at maximum springboard depression is 5.2 times the diver's body weight (575 N).

momentum appropriate for the dive. In all four foward 3½s, however, R_x reversed its direction shortly before final contact largely due to the acceleration of the upper extremities. The torque provided by this fulcrum-directed force thus contributed to increasing the diver's forward somersaulting angular momentum just preceding the flight. These findings were in agreement with those reported previously (Miller, 1980). The torque patterns of R_x with respect to the CG in the back 2½s, however, were inconsistent and no firm conclusions could be drawn on the basis of these data. In both the forward 3½s and back 2½s studied, R_y was primarily responsible for building up the required angular momentum. Its torque was consistent with the somersault direction not only during the entire recoil period but also during the final half of the depression which preceded it.

   The analysis indicated that the key to differentiating the performance characteristics of dives among the four groups lies in the horizontal springboard reaction force. The magnitude and direction of R_x establish in a large part the diver's horizontal velocity during the flight and contribute to or retard the build up of angular momentum required for the dive. They also influence the horizontal position of the diver's CG and thereby determine the direction of the torque produced by R_y. Since R_x reflects the horizontal acceleration of the segments, each contributing in proportion to its mass, the next step in this type of analysis is to investigate in detail the segment acceleration patterns in the four diving group take-offs.

References

DEMPSTER, W.T. 1955. Space Requirements of the Seated Operator. WADC TR 55-159, Wright-Patterson Air Force Base, Ohio.

DEMPSTER, W.T., and Gaughran, G.L.R. 1967. Properties of body segments based on size and weight. Am. J. Anat. 120:33-54.

JACKSON, K.M. 1979. Fitting of mathematical functions to biomechanical data. IEEE Trans. Biomed. Eng. 26(2):122-124.

MILLER, D.I. 1974. A comparative analysis of the take-off employed in springboard dives from the forward and reverse groups. In: R.C. Nelson and C.A. Morehouse (eds.), Biomechanics IV, pp. 223-228. University Park Press, Baltimore, MD.

MILLER, D.I. 1981. Body segment contributions of female athletes to translational and rotational requirements of non-twisting springboard dive take-offs. In: J. Borms, M. Hebbelinck and A. Venerando (eds.), The Female Athlete, pp. 206-215, Basel.

Effects of EMG-biofeedback Training
on Swimming

Masatada Yoshizawa
Fukui University, Bunkyo, Japan

Tsutomu Okamoto
Kansai Medical School, Osaka, Japan

Minayori Kumamoto
Kyoto University, Kyoto, Japan

Biofeedback technique is known as an effective method for the acquisition of self-control of physical movements. Therapeutic effects of EMG-biofeedback have been reported by many investigators since the mid-1960's, but only a few EMG-biofeedback applications have been reported using sports activities. In this study, the effects of EMG-biofeedback training on swimming were determined.

Procedures

Subjects employed in this EMG-biofeedback experiment were seven top-level Japanese swimmers including participants in the Third International Championships held in Berlin and the 21st Olympic games held in Montreal. There were three breaststroke swimmers (BR), two butterfly stroke swimmers (BF), and two back stroke swimmers (BA).

Referring to Tokuyama et al. (1976) and Yoshizawa et al. (1976, 1978), from the muscles acting during the arm pull in swimming, the following four muscles were selected for electromyographic (EMG) tests: deltoideus pars spinata (Ds), pectoralis major pars abdominalis (Pa), latissimus dorsi (Ld), and flexor carpi radialis (Fc).

Electrical discharges from the muscles tested were transmitted by a four channel telemetric system and recorded by a multichannel electroencephalograph. At the same time, the discharges were converted into audible signals, and the signals were given to the subjects through water-proof earphones. The subjects were requested to improve the audible patterns of the discharges of the muscles during the arm pull motion in accordance with the desirable EMG patterns obtained from the Olympic winners of one BR, one BF, and one BA. The feedback information was different to some extent from subject to subject.

Results

Representative results of the female BR subject are shown in Figure 1, where Record I was obtained before EMG-biofeedback training, Record II after the subject received the first EMG-biofeedback training, and Record III after the second EMG-biofeedback training. In Record I, the Ds, one of the prime movers of the shoulder extension, showed less discharge in the initial phase of the arm pull than in the middle, whereas the Ds of the Olympic gold medalist showed the strong discharge throughout almost the entire arm pull phase. The discharge of the Ld from the middle to the end of the arm pull was about the same as the one obtained from the Olympic gold medalist (Yoshizawa et al., 1978). The marked discharge of the Fc tended to decrease or disappear at the middle of the arm pull for about 100 msec which was long enough to induce complete muscular relaxation in the Fc. Whereas, the Fc of the Olympic gold medalist showed a strong continuous discharge throughout the arm pull phase. Thus, the BR subject did not perform the strong arm pull in the first half of the arm pull phase and did not catch enough water throughout the arm pull phase. Therefore, the biofeedback treatments were planned as follows: the BR subject was requested to make the audible signal from the Ds as long as possible at the initial stage of the arm pull phase, and to maintain a strong audible signal from the Fc throughout this phase.

After such biofeedback training, as was shown in Record II, the subject was able to produce the strong discharge of the Ds at the initial stage of the arm pull phase and to continue the discharge of the Fc, even though it was weaker than the one in the Record I. The discharges of the Ds and the Pa in Record II, showed no overlapping between them. The Ds and the Pa are the principal shoulder extensors and have lateral and medial components, respectively (Okamoto et al., 1966). Sequential activities from the Ds to the Pa account for the arm pull motion of the breaststroke where the arm is pulled laterally, then medially. In order to get the stronger propulsive force along the body axis, however, the longer overlapping of the Ds and the Pa is desirable as was suggested from the EMG pattern of the Olympic gold medalist. Therefore, the second biofeedback treatment applied was to make the subject continue the

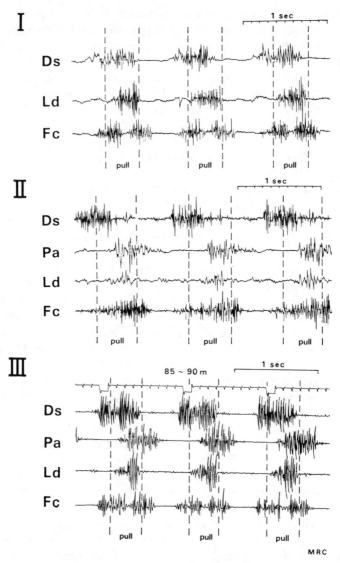

Figure 1 — EMG recordings of a breaststroke subject. Record I was obtained before EMG-biofeedback training was given; Record II after the subject received the first EMG-biofeedback training; and Record III after the second EMG-biofeedback training.

strong audible signal from the Ds up until the end of the arm pull phase.

As was shown in Record III, after such biofeedback training the subject was able to produce the strong discharge of the Ds in the latter half of the arm pull phase which overlapped with the Pa. Thus, the discharges of the Ds and Pa were much improved, but the discharge of the Fc returned to the two-burst pattern as seen in Record I.

In the EMGs of the male BF subject before the biofeedback training, the discharge of the Fc tended to decrease or disappear while the Ds showed strong discharge. This indicated that, since the palm was not able to catch enough water, the shoulder extension force produced by the Ds did not contribute to the propulsion of the body. The biofeedback treatment applied to this subject was to make the subject maintain the strong audible signal from the Fc throughout the arm pull phase as was seen in the Olympic gold medalist of BF.

After such biofeedback training, the strong discharge of the Fc overlapped well with the discharges of the Pa and the Ds. However, there was still some intermission in the discharge of the Fc at the time when the discharge of the Pa switched to the discharge of the Ds during many strokes.

In the EMGs of the female BA, before the biofeedback training, the discharge of the Ds disappeared earlier than the desirable pattern, and the discharge of the Fc appeared only during the initial ⅓ of the arm pull phase. The first biofeedback training, where the subject was requested to prolong the audible signal from the Ds, resulted in a little prolongation of the Ds. The second biofeedback training, in which the subject was requested to prolong the audible signal from the Fc, resulted in only fractions of the discharges of the Fc at the latter half of the arm pull phase.

Discussion

In the present experiments, only a short period of training of about 30 min was necessary to obtain sustained discharge patterns in both the Fc and Ds for all subjects during the arm pull period. According to Koyama et al. (1976), in English consonant pronunciation, the normal subject showed much improvement after several hours of training referring to the electromyographic information from the muscles around the mouth. The fact that the improvement of the muscular activity by EMG-biofeedback training was achieved in such a short period of time may be because the subjects were the top adult swimmers while the patient needed fairly long term EMG-biofeedback training (Kameyama et al., 1981). It took a long time, however, for even the top swimmer of BR to establish a fixed pattern, as is seen in Record III. Even though the EMG-biofeedback training succeeded in improving the Fc activity and the overlapping of the Ds and Pa activities, this did not lead to improvement in the swimming records during or just after the training. However, one to three months later their swimming speeds improved.

References

KAMEYAMA, O., Oka, H., Hashimoto, F., and Kumamoto, M. 1981. Electromyographic study of the ankle joint muscles in normal and pathological gaits. In: A. Morecki (ed.), Biomechanics. VII-A, pp. 50-54. University Park Press, Baltimore.

KOYAMA, S., Okamoto, T., Yoshizawa, M., and Kumamoto, M. 1976. An electromyographic study on training to pronounce English consonants unfamiliar to Japanese. J. Human Erogol. 5:51-60.

OKAMOTO, T., Takagi, K., and Kumamoto, M. 1966. Electromyographic study of extension of the upper extremity. Japan J. Phys. Fitness 15:37-42.

TOKUYAMA, H., Okamoto, T., and Kumamoto, M. 1976. Electromyographic study of swimming in infants and children. In: P.V. Komi (ed.), Biomechanics. V-B, pp. 215-221. University Park Press, Baltimore.

YOSHIZAWA, M., Tokuyama, H., Okamoto, T., and Kumamoto, M. 1976. Electromyographic study of the breaststroke. In: P.V. Komi (ed.), Biomechanics. V-B, pp. 222-229. University Park Press, Baltimore.

YOSHIZAWA, M., Okamoto, T., Kumamoto, M., Tokuyama, H., and Oka, H. 1978. Electromyographic study of two styles in the breaststroke as performed by top swimmers. In: E. Asmussen and K. Jørgensen (eds.), Biomechanics. VI-B, pp. 126-131. University Park Press, Baltimore.

Factors Influencing Stroke Mechanics and Speed in Swimming the Butterfly

U. Persyn, H. Vervaecke, and D. Verhetsel

Katholieke Universiteit Leuven, Heverlee, Belgium

It is important that the coach be able to observe particularities or deviations in a swimming stroke and to know whether or not these faults are mechanical or are due to the influence of factors such as body structure, flexibility, strength, buoyancy, stamina, or endurance. Further, the coach must know which factors can be altered by training. This investigation was limited to the importance of technical factors in the butterfly stroke. These are concerned with the movement as well as the body structure, strength and flexibility. The relationship between movement and other factors was also studied. On this basis an individual profile was developed, which was used to evaluate an elite Belgian swimmer.

An appropriate swimming stroke involves a compromise between: 1) an optimal continuity of propelling impulses generated by the limbs (and in the butterfly stroke presumably also by the trunk itself), and 2) a minimum of resistance caused by the limbs and by the trunk, in conjunction with an optimal body balance. Propulsion is developed by drag and/or lift forces applied by the surfaces of the hands and the feet (Barthels and Adrian, 1975). The analysis of the propulsion factor, in principle, requires the knowledge of the speed of these surfaces along the movement path, which is described in relation to still water, and of their angle of incidence to this path (Schleihauf, 1979) (see Figure 1).

In a stroke cycle, not only are forward forces developed but, in addition, lateral (balance disturbing) and backward (retarding) forces are also generated. In our opinion, these disturbances of balance and of continuity of displacement are corrected chiefly by the kicking actions (Persyn, 1969). Thus, the kick must not only place the trunk in a position of minimal resistance, allowing an effective arm action and easy breathing (balance), but must also propel the body during phases that the arms cannot (continuity).

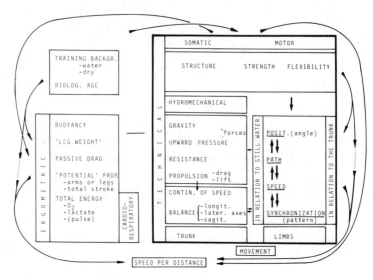

Figure 1—Relationship of factors assumed to be relevant for speed (per distance). → = some investigated relations.

Further, it is clear that these movement patterns and hydromechanical implications depend on body structure, strength, and flexibility, such as body shape, limb length, surface and shape of hands and feet, shoulder strength, and ankle flexibility (Cureton, 1975).

Hydromechanical Interpretations from Observation and Movement Analysis of Butterfly Specialists

If measurements of positions, movement paths, and speeds can be made on film of: 1) the limbs and the head in relation to the trunk and to still water and 2) the trunk in relation to the horizontal, hydromechanical interpretations of not only propulsion principles but of balance mechanisms can be made for the different phases of a stroke cycle. In previous film analyses, a repartition of the cycle of the movement patterns of top butterfly specialists into limb, head, and trunk phases, showed that a homogeneous synchronization of their movements as well as specific flexibility exists in the shoulders, the chest, the lumbar region, and even in the knees (see Figure 2A). From the observation of this particular synchronization and flexibility and of the formation of bubbles by the hands, feet, and more importantly by the trunk movements, some hydromechanical hypotheses had been derived with top swimmers and can be specified as follows per arm phase (Persyn et al., 1975):

1. Downward press phase: As a result of the first downward kick

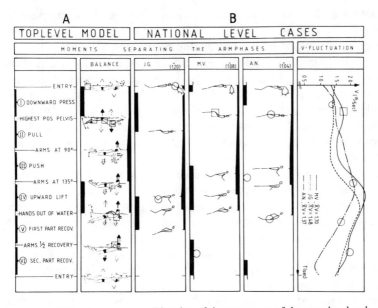

$\mu\mu$ = Waterline; $>>$ = Direction of the movement of the arm, leg, head, . . .; VV = Pressure of the water on . . .; ———— = Legs; – – – – = Arms; ········· = Head; wwww = Chest; ∧, ∧ = Resultant of . . .; ▲, ▼ = Preponderant resultant of . . .; □ = Pronounced flexibility; ○ = Irregularities in the speed fluctuation curve resulting from movement deviations;

= Hands enter the water.

= Highest position of the pelvis.

= Arms at 90°.

= Arms at 135°.

= Hands leave the water.

= Hands enter the water.

= Kick;   = Path of the movement of the wrist. The hand positions at 45°, 90°, and 135° are indicated; V-FLUCT = Speed fluctuation of a fixed point on the hip during one cycle.

Figure 2A—Moments separating the arm phases in the butterfly stroke of top level champions and interpretations of balance mechanisms. **B**—Three cases of national level swimmers, with the resulting curves of speed fluctuation.

followed by a short pause in the hip flexion (and, in addition, of the downward head position), the thorax and the arms are pressed bottomwards. By anteverting the pelvis during the kick, the legs do not finish too deeply. It is further presumed that by extending the chest and the shoulder girdle, the upward water pressure on the elevated upper limbs compensates for the downward pressure on the back, thus avoiding an exaggerated sinking of the thorax and lifting of the pelvis.

Due to the undulation of the entire body, one could assume that: a) no important up and downward displacements of the center of gravity occur during this phase and b) a decrease of water resistance results in the bubble sector along the back, when the body dives into the water.

2. Pull phase: The most important body acceleration caused by the arms begins at this time because the legs return to a position in line with the trunk.

3. Push phase: The legs are raised further and flexed, the head is lifted, and the chest is arched, causing frontal drag forces resulting in a couple action on the trunk and thus allowing the head to rise in preparation for breathing. This upward trunk rotation is supported by a pushing action of the arms.

4. Upward lift phase: As a result of the second downward kick, the acceleration of the body is prolonged and the pelvis remains high, notwithstanding the upward displacement of the arms.

5. First part of recovery: Due to the second raise of the legs and to the continuation of the arching of the chest, another couple action allows the support of both the arms and the head out of the water. (In order to minimize the upward displacement of the center of gravity and the frontal resistance, most swimmers breathe every two cycles.)

6. Second part of recovery: In order to prepare the body undulation, the head dives in before the arms.

Compared to the homogeneous patterns of top level swimmers, very heterogeneous styles were found during film studies of Belgian national level competitive crawl swimmers (Persyn et al., 1979). In Figure 2B the moments separating the butterfly arm phases are drawn for three cases with the corresponding fluctuations of speed: a) Subject J.G. shows a long glide in the downward press phase, which results in a prolongation of the low speed section in the curve. b) Subject M.V. places the first kick too early during the recovery. This causes an irregularity in the curve. In spite of the absence of the kick during the downward press phase of the arms, however, an acceleration is seen, which can only be explained by the pronounced undulation of the body. c) Subject A.N. does not demonstrate trunk undulation and thus could not produce any significant acceleration at the moment when the pelvis reached its highest position, following the phase of the first kick. She did not use a second kick. This resulted in a slow acceleration preceding the upward arm lift phase and a steep deceleration during the recovery phase, as a result of the deep body position.

Selection of Relevant Technical Measurements

Due to the variation in movements within competitors at the national level, this population can be considered as an interesting group in order to detect the relevancy of movement parameters to swimming speed.

In order to obtain the statistical information necessary to answer a request of the Belgian Olympic Committee to evaluate the Olympic selection, 150 national level swimmers were measured, tested, and filmed. The investigated factors are shown in Figure 1. All the measurements, tests, and films took place in an evaluation circuit, which lasted three hours for 25 swimmers.

In this specific report only the technical factors will be considered. The somatic, strength, and flexibility data were obtained by a set of measurements, assumed to be relevant for swimming performance from the previous observations of top level champions (see Figure 2A). This procedure has been discussed elsewhere (Vervaecke and Persyn, 1981). Information on swimming performances was collected from official ranking lists and from the basic speed tests during the filming sessions in the circuit.

The films were taken underwater from a side view (16 mm, battery-powered, Bolex camera at 24 frames/sec) and from a front view (super 8 mm Canon battery-powered camera at 18 frames/sec) and analyzed first in a simple practical manner. The angles of the body segments were measured in relation to the trunk and to the horizontal at each moment separating the phases of the limbs, the head, and the trunk. Information on parts of the paths, described by hands and feet in relation to the body and to still water, was obtained by measuring the movement amplitudes in the moments separating the phases as well as the minimal and maximal amplitudes in the entire cycle. In order to obtain the individual synchronization all these data were set on the same time base.

The entire set of heterogeneous data in relation to the investigated factors (see Figure 1) was processed by SPSS (Nie et al., 1970) and BMDP (Dixon et al., 1979) computer programs. In addition to the general statistic analysis until now Pearson product-moment and partial correlations (adjusted for the effect of age) were calculated.

The Construction and Use of a Profile Card

From the obtained statistical data, a profile card for butterfly for 16-year-old males is shown in Figure 3. With the use of a microprocessor, this profile can now be printed for the 25 swimmers during the testing and filming day.

The columns include the following information: In columns C and E movement parameters are presented (which can be observed during the

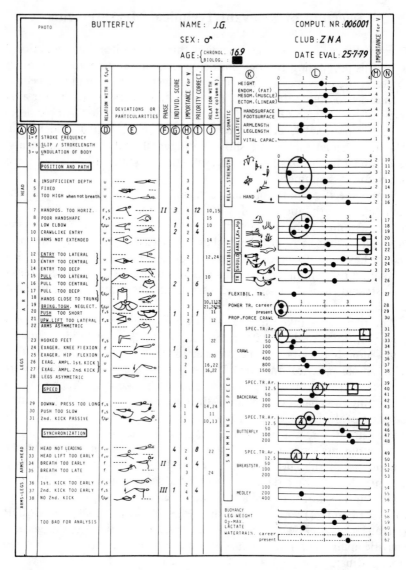

Figure 3—Profile card used to rank relevant movement, somatic, strength, flexibility, training background, swimming speed, and ergometric parameters in the Belgian population of competitive butterfly male swimmers of 16 years old. • = Individual score; ○ = Poor score; □ = Excellent result.

film projection by specifically trained coaches). First, the central criteria for speed and balance are set: stroke frequency (f), slip in relation to stroke length (s), and body undulation (u). Next, more detailed deviations and particularities are given with specifications of positions, paths, speeds, and synchronization. Relationships between the first three criteria and the following specifications are presented when correlations of at least .33 (5% level of significance) were found. These significant correlations are indicated in column D. The importance of the style parameters for swimming speed from 10 m up to 200 m is indicated on a 4-point scale in column H. (1 = a correlation from .26 to .32; 2 = .33 to .45; 3 = .46 to .54; 4 = > .54).

In columns K and L (N, 1-26) somatic, strength, and flexibility data, which are relevant for speed in one of the four strokes, are drawn with specifications. Using the same 4-point scale, the importance of each factor for butterfly performance is indicated in column M. In the horizontal scales (L) the individual scores are printed (in relation to the results of the investigated elite swimmers of the particular age and sex). A score of 3, for example, means that 25% of this population obtains a higher score. A shortcoming is indicated by a circle on the score.

In column J, the relationships between the movement parameters and somatic and motor factors are specified when correlations of at least .33 were found.

When the coaches and the swimmers have discussed the individual somatic, strength, and flexibility data in relation to sprint speed (12.5, 50 and 100 m; in 12.5 m for arms (A), legs (L), and total stroke (T), the film is observed and noticeable style deviations and particularities are measured and marked in column G. This is done, once again, on a 4-point scale for individual score. The scale was constructed from the range of the results obtained in the investigated population. The phase is specified in column F (1-6). Simultaneously, with the training advice for weaknesses, the priorities in style correction for sprint speed can be proposed by a score in column I (first priority is 4 [for importance score] × 4 [for individual score] = 16).

The present performance has to be considered in relation to the: 1) biological (skeletal) age (information only available to the medical doctor) and 2) training background (as compared to top level swimmers). The training background is obtained by oral questioning; water training is specified in total stroke (N, 61-62) and pulling with arms alone (N, 31, 39, 44, 49) and dry land training in flexibility and power training (N, 27-29).

If poor performances are seen in longer events in relation to the sprint, this could, of course, not only be caused by some of the preceding (somatic, motor, and style) parameters but complementary by: 1) the relation between gravity and upward pressure forces (column N, 58) and 2) a low level of maximum total energy use (derived from gas- and blood analysis) (column N, 59 to 60) (Daly et al., 1980). This is, of course, essentially a crawl stroke problem.

One of the three cases in Figure 2 (J.G.), evaluated from the pure movement point of view, will now be briefly discussed using his profile (see Figure 3). This discussion will start with deviations from the three criteria (B, 1 to 3):

1. The stroke frequency is too slow (column B, 1). This results from a long gliding, downward press phase (column B, 29), which frequently is combined with a very effective kick (column N, 45), resulting from extremely flexible ankles (column N, 20, 22) and exaggerated knee flexion (column B, 24).

2. There is too much slip in relation to the stroke length (column B, 2). This can be explained by a too horizontal hand position (column B, 7), a low elbow position (column B, 9) and a small pull (column B, 16). These deviations, however, could be due to shortcomings in strength (column N, 10, 12, 14), which further could be explained by a neglect of power training (column N, 28, 29) as well as of pulling during water training (column N, 44).

3. The body undulation must be more pronounced (column B, 3). The initial impulse is not given by diving the head into the water (column B, 32) and the shoulder extension is limited (column N, 25).

Discussion

It is expected that information necessary to develop the profile presented in this article is a basis for coaches' training. Of course, a technical (and ergometric) diagnosis is of little help if the concrete individual advice for style correction and improvement in shortcomings is not extensively explained and demonstrated in the pool and in the weight room.

Acknowledgment

This article has been elaborated with the help of D. Daly, X. Thevelein, and L. Van Tilborgh.

References

BARTHELS, K.M., and Adrian, M.J. 1975. Three dimensional spatial hand patterns of skilled butterfly swimmers. In: J. Clarys and L. Lewillie (eds.), Swimming II, pp. 154-160. University Park Press, Baltimore.

CURETON, T.K. 1975. Factors governing success in competitive swimming: a brief review of related studies. In: J. Clarys and L. Lewillie (eds.), Swimming II, pp. 9-39. University Park Press, Baltimore.

DALY, D., Thevelein, X., and Persyn, U. 1980. Effectiveness and efficiency of crawl stroke swimming: a case study. Paper presented at the 4th Intern. Council of Physical Fitness Research, Leuven. (in press)

DIXON, W.J., Brown, M.B., Engleton, D., Frane, J., and Jennrich, R. 1979. Biomedical computer programs. Univ. of California Press, Berkeley.

NIE, N.H., Hull, C.H., Jenkins, J.G., Steinbrenner, K., and Bent, D.H. 1970. Statistical Package for the Social Sciences, McGraw-Hill, New York.

PERSYN, U. 1969. Hydrodynamische gegevens die aan de basis liggen van de zwemtechnieken. (Hydrodynamical data at the basis of swimming techniques). Sport, XII nr. 2 (46):119-123.

PERSYN, U., De Maeyer, J., and Vervaecke, H. 1975. Investigation of hydrodynamic determinants of competitive swimming strokes. In: J. Clarys and L. Lewillie (eds.), Swimming II, pp. 214-222. University Park Press, Baltimore.

PERSYN, U.J.J., Hoeven, R.G.G., and Daly, D.J. 1979. Evaluation center for competitive swimmers. In: J. Terauds and E. Bedingfield (eds.), Swimming III, pp. 182-195. University Park Press, Baltimore.

SCHLEIHAUF, R.E. 1979. A hydrodynamic analysis of swimming propulsion. In: J. Terauds and E. Bedingfield (eds.), Swimming III, pp. 70-117. University Park Press, Baltimore.

VERVAECKE, H., and Persyn, U. 1981. Some differences between men and women in various factors which determine swimming performances. Medicine and Sport XV, Karger, Basel. (in press)

Physiological Energy Consumption during Swimming, Related to Added Drag

G. Takahashi, T. Nomura, and A. Yoshida
University of Tsukuba, Ibaraki-ken, Japan

M. Miyashita
University of Tokyo, Tokyo, Japan

While swimming at constant velocity, the swimmer must exert a propulsive force equal to the water resistance. Therefore, numerous studies have reported on water resistance in the static position in relation to velocity, body size, etc. (Karpovich and Millman, 1944; Alley, 1952; Clary et al., 1974; Miyashita and Tsunoda, 1978).

Since the swimmer must overcome water resistance, the top swimmer is required to have a large capacity for energy expenditure to maintain a high velocity. The energy expenditure during swimming related to velocity, stroke, level of skill, etc., and also the aerobic work capacities of the top swimmers have been studied extensively (Klissouras, 1968; Holmér, 1972, 1979; Pendergast et al., 1980).

Competitive swimming has developed rapidly, partly because of the marked increase in the volume of training work (Holmér, 1979). In addition, various devices for training may have contributed to the development of world records. Recently a swimming training device to increase drag force which is called a 'drag suit' has been developed.

The purpose of the present study was to investigate the differences in water resistance and energy expenditure during swimming when using a normal suit and a drag suit.

Methods

The subjects were four male and five female college swimmers, ranging from 18 to 23 years. They were not well trained at the time of the experi-

Table 1

Anthropometric Data and Swimming Performances of Subjects

Subj	Sex		Age (yr)	Ht (cm)	Wt (kg)	BAS (m^2)	BMR (ml/min)	Best record (min:sec)
AM	M		22.0	180.0	72.0	1.93	290.1	1:06 FC
KH	M		22.0	184.0	78.0	2.02	306.1	0:58 FC
NT	M		22.0	174.0	72.0	1.87	283.4	0:58 FC
YH	M		22.0	164.0	61.0	1.68	253.4	1:03 FC
		M	22.0	175.5	70.8	1.87	283.3	
		SD	0.0	8.7	7.1	0.15	22.0	
FY	F		19.0	160.0	57.0	1.59	222.0	1:18 BR
SY	F		18.0	157.0	55.0	1.55	214.4	1:20 BR
TZ	F		18.0	160.0	55.0	1.58	221.4	1:00 FC
MH	F		18.0	166.0	60.0	1.68	235.9	2:28 BA*
FC	F		23.0	164.0	61.0	1.67	228.6	1:10 FC
		M	19.2	161.3	57.3	1.61	229.1	
		SD	2.2	3.7	2.5	0.06	15.0	

FC = front crawl, BR = breast stroke, BA = back stroke.
200 m back stroke. Others are all 100 m swim times.

ment, but were highly skilled swimmers with an average of seven years experience in swimming. The swimming experiments were performed in a specially designed swim flume at the University of Tsukuba which was very similar to the flume described by Åstrand and Englesson (1972). Water is circulated in a deep loop by a motor-driven propeller. The water velocity in the swim flume can be varied from 0 to 2.5 m/sec in intervals of 0.01 m/sec. The dimensions of the measuring section are 5.5 m length, 2.0 m width, and 1.2 m depth. The water temperature was kept from 25 to 27°C in this experiment. Each subject was asked to swim at several speeds using the crawl stroke. The subjects swam twice at the same speed for 3 to 4 min, once while wearing a drag suit originally designed by Councilman, and the other time wearing a normal suit.

Oxygen uptake ($\dot{V}O_2$) was determined using the Douglas bag technique. The volume of expired gas was measured in a calibrated dry gasometer and samples were analyzed for O_2 and CO_2 content using a Respiratory Masspectrometer (Perkin-Elmer Co. U.S.A.).

According to the pretests, each subject showed a submaximal 'steady state' of $\dot{V}O_2$ at a relatively low velocity of from 0.4 m/sec to 1.0 m/sec for the males and 0.4 m/sec to 0.8 m/sec for the females. Therefore, energy expenditure for swimming at low velocity was determined from $\dot{V}O_2$ during swimming and resting ($\dot{V}O_2$).

Energy expenditures for swimming at a higher velocity than 1.0 m/sec for males and 0.8 m/sec for females was calculated from the oxygen consumption during swimming and 30 min of rest after swimming and resting $\dot{V}O_2$.

One end of a cord (about 2 m in length) was attached to a load cell; the other end was held in the hands of the swimmer lying in the water. The load cell (measuring range: 0 to 50 kg) was fixed at the edge of the flume ahead of the swimmer. The tension of the cord was recorded by a pen writing oscillograph through the load cell, which was calibrated with a known weight before and the after measurements.

The drag in the normal suit and the drag suit was measured at several velocities (range: 0.2 to 2.0 m/sec).

Results and Discussion

Anthropometric data and swimming records of the subjects are shown in Table 1. Drag increased exponentially with water velocity in both conditions (drag suit and normal suit).

Regression equations calculated by the method of least squares between drag force and water velocity were as follows:

$$Dn = 3.134\ v^{1.805} \qquad \text{(male)} \qquad (1)$$

$$Dn = 2.456\ v^{1.695} \qquad \text{(female)} \qquad (2)$$

$$Dd = 3.724\ v^{1.805} \qquad \text{(male)} \qquad (3)$$

$$Dd = 3.075\ v^{1.788} \qquad \text{(female)} \qquad (4)$$

where Dn is the drag in kg while wearing the normal suit, and Dd is the drag in kg while wearing the drag suit, and v is the water velocity in m/sec. These results were very similar to the values of previous reports (Karpovich and Millman, 1944; Clarys et al., 1974; Miyashita and Tsunoda, 1978).

At any speed, the drag of the male swimmers was greater than that of the female swimmers. The reason why males had greater body drag was mainly because of the sinking of the legs which is the result of heavy legs (Clary et al., 1974; Holmér, 1974) and greater body dimensions (Clary et al., 1974). The clear difference in drag between wearing the drag suit and the normal suit appeared above the velocity of 1.4 m/sec in both groups (see Figure 1). The delta drag (Dd-Dn) increased with velocity, and was

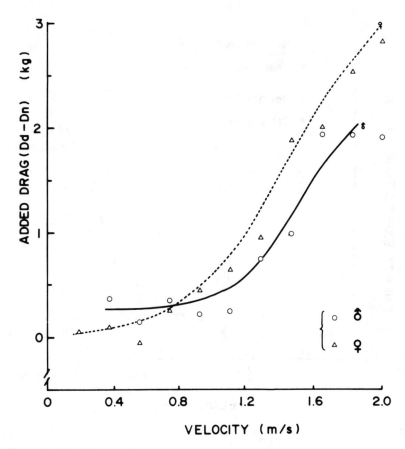

Figure 1 — Added drag in relation to water velocity of swimming.

2.0 kg for males and 2.5 kg for females at a velocity of 2.0 m/sec.

Energy expenditure increased gradually up to the swimming velocity of 1.0 m/sec and thereafter it increased rapidly with an increase in velocity (see Figure 2). Energy expenditure while wearing the drag suit was greater than for the normal suit at all speeds for both male and female subjects. The delta energy expenditure between the drag suit and the normal suit definitely increased above the velocity of 1.0 m/sec. At the velocity of 1.0 m/sec, delta energy expenditure was 9.2% (0.24 1/min) for males and 17% (0.52 1/min) for females, respectively. It increased up to 37.5% (3.3 1/min) at a maximal velocity of 1.4 m/sec for males and 56.0% (2.9 1/min) at 1.2 m/sec for females.

These results showed that the drag suit increased drag and energy expenditure at higher velocities. Therefore, the drag suit provides a certain amount of mechanical and physical stress during competitive swimming training. At the maximal velocity (1.4 m/sec for males and 1.2 m/sec for

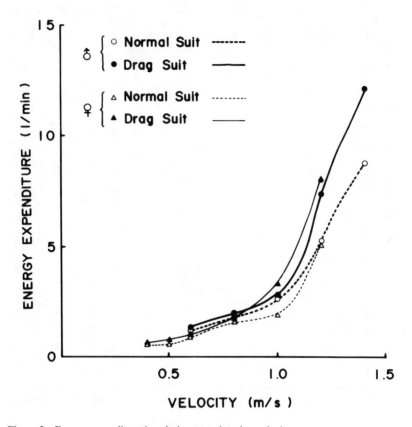

Figure 2 — Energy expenditure in relation to swimming velocity.

females) at which swimmers could swim for 3 to 4 min, drag with the drag suit was greater (15.3% for males, 22.9% for females) than when swimming using the normal suit, and energy expenditures during swimming with the drag suit were more (37.5% for males, 56.0% for females) than those with the normal suit. These results indicated that at higher velocities swimming efficiency decreased when swimming with the drag suit compared with swimming using the normal suit.

References

ÅSTRAND, P.-O., and Englesson. 1972. A swimming flume. J. Appl. Physiol. 33:514.

ADRIAN, J.J., Singh, M., and Karpovich, P.V. 1966. Energy cost of leg kick, arm stroke and whole crawl stroke. J. Appl. Physiol. 21:1763-1776.

ALLEY, L.E. 1952. An analysis of water resistance and propulsion in swimming the crawl stroke. Res. Quar. 253-270.

CLARYS, J.P., Jiskoot, J., Risken, H., and Brovwer, P.J. 1974. Total resistance in water and its relation to body form. In: R.C. Nelson and C.A. Morehouse (eds.), Biomechanics IV, pp. 187-196. University Park Press, Baltimore.

DI PRAMPERO, P.E., Pendergast, D.R., Wilson, D.W., and Rennie, W. 1974. Energetics of swimming in man. J. Appl. Physiol. 37:1-5.

HOLMÉR, I. 1972. Oxygen uptake during swimming in man. J. Appl. Physiol. 30:502-509.

HOLMÉR, I. 1974. Propulsive efficiency of breast stroke and freestyle swimming. Eur. J. Appl. Physiol. 33:95-103.

HOLMÉR, I., Lundin, A., and Eriksson, B.O. 1974. Maximum oxygen uptake during swimming and running by elite swimmers. J. Appl. Physiol. 36:711-714.

HOLMÉR, I. 1979. Physiology of swimming man. In: R.S. Hutton and D.I. Miller (eds.), Exercise and Sport Sciences Reviews. Vol. 7. The Franklin Institute Press, Philadelphia.

KARPOVICH, P.V., and Millman, N. 1944. Energy expenditure in swimming. Res. Quar. 140-144.

KLISSOURAS, V. 1968. Energy metabolism in swimming the dolphin stroke. Arbeitsphysiologie 25:142-150.

MILLER, D.I. 1975. Biomechanics of swimming. In: J.H. Wilmore and J.F. Koegh (eds.), Exercise and Sport Sciences Reviews. Vol. 7. The Franklin Institute Press, Philadelphia.

MIYASHITA, M., and Tsunoda, R. 1978. Water resistance in relation to body size. In: B. Eriksson and B. Furberg (eds.), Swimming Medicine IV, pp. 395-401. University Park Press, Baltimore.

PENDERGAST, D.R., di Prampero, P.E., Craig, A.B., Jr., and Rennie, D.W. 1980. Metabolic adaptations to swimming. Exercise bioenergetics and gas exchange. Biomedical Press, Elsevier/North-Holland.

D.
Other
Sports

An Evaluation of the Diagonal Stride Technique in Cross-Country Skiing

A. Dal Monte, S. Fucci, L.M. Leonardi, and V. Trozzi
Institute of Sports Medicine of C.O.N.I., Rome, Italy

The factors which determine cross-country skiing performances are numerous, and depend on the composition and type of material from which the equipment is made and from some of the characteristics of the skier. These characteristics can be classified as individual endurance capacity and perfection of the skills of skiing.

This study describes the plan and results of an experiment undertaken in order to arrive at a technical analysis of the diagonal stride in cross country skiing on the flat.

Dynamometric System

The dynamometric instruments used enabled one to record the acceleration of the skis and of the athlete's body, and the force applied by the arms to the ski poles.

The transducer placed on the foot plate of the ski was made in such a way as to be sensitive only to the displacements in the same direction as the skier's walk, while the transducers placed on the poles were sensitive only to the force components on the pole parallel to the major axis of the pole.

The data obtained were transmitted telemetrically to a mobile laboratory, where they were recorded on magnetic tape for later analysis. Simultaneously, a cine camera filmed the athlete during the test at a speed of 64 frames/sec.

The beginning and end of the athlete's path over a prescribed distance, activated an electronic flash and produced a trigger impulse in order to allow the synchronization of the biomechanical signals.

Dynamometric Analyses

To analyze the dynamometric curve and interpret it in terms of the phases of the step, first the frames of film of the start and finish of each phase were identified. They corresponded to the characteristic points of the action and therefore determined the limits of the diagram.

Nine points were identified for the legs, and are labeled by the letters A to I, and seven points for the arms, indicated by Q to Z.

Kinegrams of the Legs

With regard to the legs, which are more complex from the point of view of analysis, the following phases were identified: a) pull, b) load, c) push-off, d) hanging phase, e) leg recovery phase, f) body weight transfer, and g) glide.

Each one of these phases presents a characteristic and identifiable curve on the diagram. The diagram, therefore, makes it possible to identify each phase of the step once the typical diagram is found.

The characteristic traces of the curve are recognized thus:

1. I-A: pull of leading leg.
2. A-B: load of leading leg.
3. B-D: push-off (where C represents the point of maximum push and D the point of maximum leg extension).
4. D-E: take-off and leg suspension.
5. E-G: recovery of leading ski (F recovery of the leg).
6. G-H: body weight transfer from one ski to the other. In this trace, Point G represents the point of impact of the ski with the ground, and Point H the instant in which the transfer of the body weight is completed.
7. H-I: glide.

Kinegram of the Arms

After the kinegram of the arms is available, one needs to determine where to insert the action of the force applied by the arms on the kinegram. To accomplish this, the characteristic points of the action of the arms are determined after which these points are made to correspond with those of the legs, on the basis of the sub-divisions made by the computer on the graphs and in correspondence to the photographs.

In the graph of the arms, seven points were identified and indicated by the letters Q to Z.

The action phases are: (see Figure 1)

1. Q-R: preparation and pole implantation.
2. R-S: pull phase (start).
3. S-T: pull phase (max).

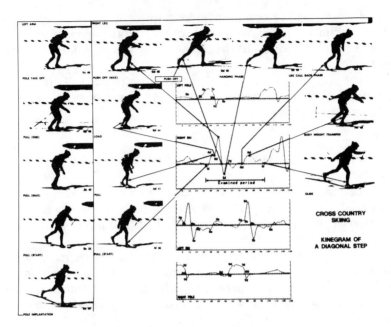

Figure 1—Kinegram of a diagonal step.

4. T-U: pull phase (end).
5. U-V: pole take-off.
6. V-Z: hanging phase.
7. Z-Q: forward movement of the legs and pole.

Noting the phases which comprise the push of the arms it is possible to see, following their graphs, how the arm action is more or less effective depending on the configuration of the graph.

In fact, in order for the push to be most effective, essentially it should approach the following pattern:

1. Reduction of the distance of Q-R which represents the preparation phase and pole plant. On the best push, this distance is almost zero or Q nearly coincides with R.

2. The pull phase, R-S, is almost perpendicular which signifies that in a short time the maximal pushing force is attained.

3. The S-T phase, which represents the maximum push, should be extended for a longer period of time.

Complete Analysis of the Kinegrams for Legs and Arms

The complete analysis of the four limbs allows the evaluation of the coordination of the action of the extremities in the propulsion of the cross-country skier.

From the preceding considerations, in order for the movement to be most efficient, the following conditions must be satisfied:

1. Analyzing the graph of one arm with that of the opposite leg, there should be correspondence of the characteristic points, S-T for the arm, and I-B for the opposite leg.

2. Analyzing the two legs, the characteristic phases A-C (load and push of a leg) should correspond with G-H (change of body weight to other leg) (see Figure 1).

Technical Conclusions

From the analysis of the graphs of the skiers under observation, it was ascertained that some of the characteristic elements of running existed in the diagonal stride; for example the pull, load, and push-off are easily identified on the graph of the legs.

More specifically, it could be observed that the salient characteristic of this step is represented by the push-off phase and by the corresponding weight change, from one ski to the other (characteristic Point H). In fact, depending on the instant at which such correspondence occurs, one can essentially verify the following three events:

1. The body weight transfer comes immediately after the moment in which the push-off reaches its maximum efficiency (best condition).

2. The body weight transfer occurs at the moment of maximum push-off or with slight delay. On the graphs representing this situation one could, above all, ascertain for the part of the curve representing the real push-off, a greatly reduced graphical excursion, while the portion between the maximum push-off and recovery is accentuated.

3. The body weight transfer precedes the moment of maximum push-off slightly or even simultaneously with the load of the push-off limb. The long distance skier, in this situation, can be compared to a subject who lets himself be carried by a scooter on which he rests with one leg, while the other leg effects the push-off.

This case provides evidence more so than the preceding ones, of the inefficiency of the push-off as the load will no longer be able to produce efficient muscle stretching, related to the real push-off.

It is clear that the best skiing technique is that illustrated by first class skiers. The leading leg alone, however, supports the body weight and the push-off. In order to adopt such a technique, one needs to have sufficient strength of the legs and balance to permit skiing on only one leg. Second and third class skiing, therefore, may be caused by either a lack of strength or by a lack of balance.

Concerning the action of the arms, as previously noted, one can essentially observe that the pole plant must not be preceded by a long suspension phase of the same pole held in front of the body. In other words, one needs to start the work of the arms as soon as possible (within the limits of a coordinated movement), and therefore reach in a short time the highest force values that the arms are able to achieve, trying at the same time to maintain them as long as possible.

Apart from what has been formerly expressed, it has also been possible to see that in some subjects, the pole take-off from the snow is more rapid than with others. In a particular way in those subjects in which the maximum push-off stroke (S-T) is prolonged, the Point V nearly coincides with Point U (final push-off). The considerations up to now have yielded the ideal diagram for skiing. It is clear that the actual graphs made of the athletes are far from ideal in relation to the style adopted by the athletes, considering their capabilities and peculiarities.

Aerodynamic Investigation of Arm Position during the Flight Phase in Ski Jumping

Kazuhiko Watanabe
University of Hiroshima, Hiroshima, Japan

The effect of ski jumping form (flight posture) on performance has been scientifically investigated. Straumann (1926) pointed out the advantage of the forward leaning posture. Tani and Iuchi (1971) suggested the upper arms should be extended backward for achieving a greater flight distance. Recently, some questions have been raised by coaches concerning the arm position during the flight phase, the reason being that the suggested best arm position of extending backward and close to the body is sometimes difficult even for top level jumpers. With the exception of the investigation of postural disturbance in jumping, from a neurophysiological standpoint, there are few reports about the effect of the arm position (abduction) on the ski jumping performance. This experiment has been directed toward this problem.

Wind Tunnel

The 3 m wind tunnel at the Institute of Space and Aeronautical Science, University of Tokyo, was used. The model was supported by flunnion axis and wires in the wind tunnel, and measurement of the drag and lift components were made at a wind velocity of 25 m/sec. The drag area, S_D, and left area, S_L, were determined.

Model

The model which was used in this experiment was the same as the model used by Tani and Iuchi (1971) and furthermore refined to include the shoulders and wrists. The model was made of wood (1.73 m in height, 48

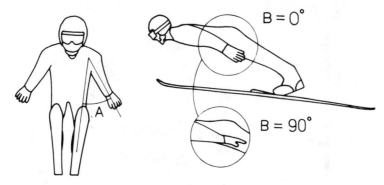

Figure 1—The model was refined to include movement at the shoulders (abduction) and wrists (rotation).

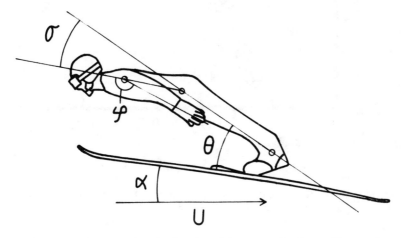

Figure 2—α = ski angle against the direction of the stream; θ = forward lean angle; σ = bent angle; φ = arm angle.

cm in shoulder width) and equipped with helmet, goggles, pants, gloves, boots, and skis. The arms of this model were positioned against the body from 0° (close to the body) to 90° continuously (see Figure 1A). This model also enabled changes in arm position at the shoulder axis with rotation of 360° in the sagittal plane. The position of the hands were also movable through 360° of rotation at the wrist (see Figure 1B).

Postural Conditions

As indicated in Figure 2, α, the ski angle against the direction of the stream, was changed at 0°, 5°, 15°, 20°, 25°, and 30°, while θ, the forward lean angle, was fixed at 20°. σ, the bent angle, was fixed at 20° and

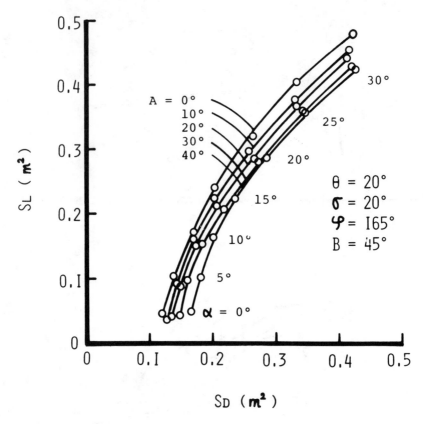

Figure 3 — S_L = the lift area; S_L/S_D = the lift to drag ratio. The increase in the lift to drag ratio is essential to improved jumping performance.

take off, x°; the arm angle was also fixed at 165°.

The abduction of the arms from the trunk (angle A), was changed as follows: 10°, 20°, 30°, and 40°. The hand (palm) position (angle B) was selected as parallel to the stream (0°), (45°), and at a right angle against the stream (90°). Therefore, the positions were changed in (α), (A), and (B), respectively.

Result and Discussion

Figure 3 indicates the values of the S_L, S_D as a function of (α), and (A, B). In the flight phase of ski jumping, the 'lift to drag ratio' (S_L/S_D), is an important indicator, since increasing this value is associated with improved performance. From Figure 3 it can be seen that for all positions of the arm (0°, 10°, 20°, 30°, 40°), the lift to drag ratio increases in relation to the increment of (α). Also, the change in the arm position affects

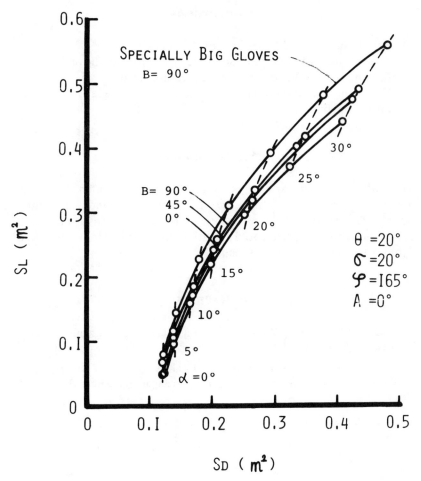

Figure 4—The hand (palm) position influences the lift to drag ratio with the best position being at a right angle to the air stream.

the lift to drag ratio. The curve for A = 40°, shifts upward to the left in comparison with A = 0°. This means that a greater ratio has been obtained. These results clearly show the merit of having the arms positioned close to the body and also shows the disadvantage of extending the arms away from the body for all conditions of the ski angle (α).

In Figure 4, the effect of the hand (palm) position against the stream is shown. This result clearly shows that the best hand (palm) position is at a right angle (90°) against the direction of the wind stream.

Conclusions

The results support the following conclusions:

1. The position of the arm close to the body provides the optimum lift to drag ratio (S_L/S_D) and is therefore recommended.

2. The hand (palm) should be positioned at a right angle against the air stream for the best performance.

Acknowledgments

The author wishes to express his thanks to Prof. A. Azuma, Mr. Iuchi, and Mr. I. Watanabe, all from the Institute of Space and Aeronautical Science, University of Tokyo, for their useful advice and assistance in the completion of this experiment. The author also thanks Mr. M. Akashi of Jyohsai University for his assistance.

References

STRAUMANN, R. 1926. Vom Skiweitsprung und seiner Mechanik. (The mechanics of ski jumping.) Jahrbuch des Schweiz Ski Verbandes, pp. 34-64.

TANI, I., and Iuchi, M. 1971. Flight-mechanical investigation of ski jumping. In: K. Kinoshita (ed.), Scientific Study of Skiing in Japan, pp. 33-52. Hitachi Ltd., Tokyo.

New Effects of Release Bindings for Alpine Skiing

Hans Ekström
Linköping Institute of Technology, Linköping, Sweden

Ulf Bengtzelius
Chalmers Institute of Technology, Gothenburg, Sweden

Stig Ottosson
Sweden

The first ski binding for Alpine skiing was invented in 1850 by Søndre Norheim from Telemark in Norway. This binding made it possible to turn with parallel skis, which is one of the conditions for Alpine skiing. The first release bindings were introduced during the 1940s with single point releasing (toe-side releasing) and were subsequently developed to two point releasing (toe-side, heel-up). In line with this development the next generation of bindings will employ three-point fixing with three release directions (toe-side, heel-up, toe-up) (see Figure 1).

Release bindings for Alpine skiing have three principal functions:

1. Retention. The binding should constitute a stable connection between the boot and the ski. It should transmit steering forces without play or delay.

2. Release. In a skiing situation with an inherent risk of injury (fracture, sprain, dislocation, etc.), when the energy applied to the ski-binding-boot-skier system is too high, compared with the tibia strength, the binding should free the skier from the ski. The release should take place semistatically for impacts, torsional, and bending moments on the tibia.

3. Shock absorption and damping. The ski and the binding should be able to absorb a sufficient amount of energy from short duration impacts so that inadvertent release of the binding does not occur. The binding should preferably be able to discriminate between very short, harmless

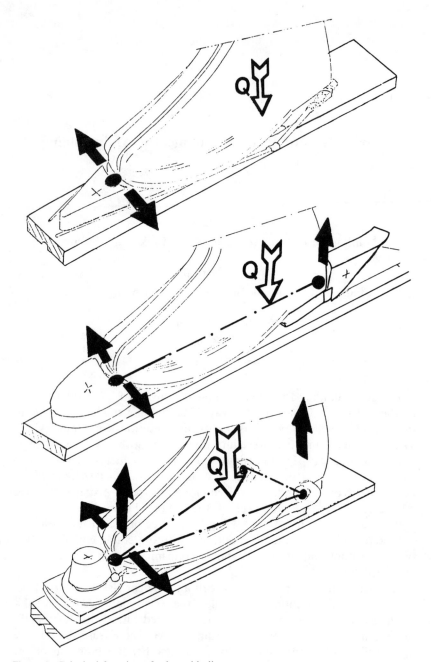

Figure 1 – Principal function of release bindings.

impacts and forces of longer duration, e.g., slow torsional moments. In other words, some inertia in the system is of advantage so that most of the impacts encountered during skiing can be absorbed in the mechanics and friction of the binding.

Procedure

A requirement specification for a good release binding could be formulated as follows:

1. The binding should release in as many directions as possible.

2. The release characteristics should be independent of the ski bending, friction, contamination, temperature, and ice and require practically no maintenance.

3. The binding should vary for semistatic and dynamic force according to accepted biomechanical principles.

4. The binding should consist of a minimum number of parts to be mounted on the ski and the boot sole, if possible without changes to the boot or the sole.

5. It should be easy and comfortable to step into the binding, both initially and again after a release on the slopes.

6. Mounting, adjustment, maintenance, and repair should be easy to perform by any skier.

Today's norms for Alpine release bindings are based only on simple technical requirements on semistatic loadings at right angles to the ski with toe-side, heel-up releasing. Oblique loading directions and combinations of bending and torsional forces must be given greater consideration, however, on the basis of the actual loadings applied by the skier to the binding and the strength of the leg.

In composite loadings of the bending and torsional moment type it is especially important that the release forces be reduced, when they deviate from the normal, in order to avoid resultant forces which exceed the breaking strength of the tibia (Doré and Riedl, 1978).

According to Young and Loo (1978) concerning fall types for various types of skiers and Johnson et al. (1978) concerning reported knee injuries, high loadings occur especially on the knee during downhill skiing, when the skier is moving quickly and is suddenly forced into a short uphill slope. If the binding does not permit release for toe-up loading there is a high risk of, among other things, rupture of the ligaments in the knee joint and/or tibia fractures.

With the increased use of toe-side and heel-up release equipment, the frequency of leg fractures has decreased. The accident data reported (Eriksson, 1976; Johnson et al., 1978) shows that the frequency of knee injuries is large, which indicates a need for more release directions than those permitted by today's bindings. With the introduction of a further

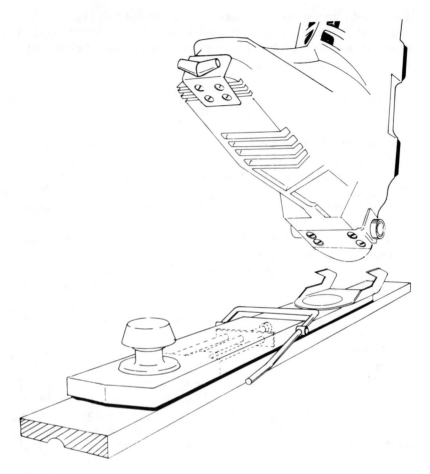

Figure 2—The BOX release binding.

toe-up mechanism the possibility of overloading of the knee ligaments will be reduced.

Results

The BOX-binding is built in one complete section consisting of box, nose wheel, guides, spring system, adjusting unit, rear fixing, and ski-stopper. The ski-boot is anchored to the binding at three points, where the contact surfaces in action are reduced to a minimum. There is no direct contact between ski-boot, sole, and the ski. The bending and torsion moments between ski-boot and ski are transmitted via the three points of attachment in two different axes of motion. The BOX-binding releases in three directions; toe-side, heel-up and toe-up (see Figure 2).

In the basic mechanics of the BOX-binding, the fixing and release mechanism is built up on the spring characteristics and the geometry of the nose plate, nose wheel and rear fixing claws with regard to diameter and angles (see Figure 3). The BOX-binding has few moving parts which depend on the spring force. The demands on elasticity of a ski binding vary between skiers of different ability. In modern ski bindings there are very few possibilities of varying the elasticity without changing the other characteristics of the binding. The elasticity of the BOX-binding can be changed according to the skier's wish by altering the length and the angle of the toe plate and the stem diameter and concial diameter of the front wheel, spring system, and locking angles in the rear fixing plate.

Static and dynamic testing were performed in a material testing system, type MTS 16 Mp, with the binding mounted in a special jig together with the boot sole, including angle bracket and heel spindle. The force was transmitted to the bindings via hydraulic cylinders with an error of $\pm 1\%$ of the recorded force and displacement. The movement of the boot in the binding was recorded with an inductive transducer built into the movable piston of the test equipment.

The binding was loaded in three directions, toe-side, heel-up, and toe-up, at $90°$ and $60°$ to the ski. Comparative tests were also performed on other makes of binding with regard to semistatic release characteristics. The dynamic testing and the force and movement as functions of time were registered using an ABEEM Ultralette 5651 UV recorder.

The proper function of conventional two-part toe-heel bindings can be jeopardized by the flexibility of the ski. When the ski is bent upwards, the toe and the heel parts come closer to each other and lock the boot harder to the ski. Many manufacturers have tried to solve this problem by making the heel part movable along the ski. However, this approach complicates the design since it requires at least one more spring.

The release characteristics for several two part-bindings depend on the ski's bending and loading direction. Comparative measurements show that higher release forces are obtained if the loading direction differs from the normal direction for two-part bindings. The BOX-binding is mounted on the ski in such a way that the bending of the ski cannot change the release characteristics of the binding. The binding does not impede the movements of the ski during skiing, which means a better grip on ice and uneven surfaces.

The pressure distribution between ski and snow in skiing will vary between the two binding principles because of the ski-boot stiffness in the section between the front and rear fixing points in the two-part binding. This gives local increases in pressure between ski and snow at the front and rear fixing points and thus affects the friction between ski and snow. With the BOX-binding the ski is permitted total flexibility under the binding. The ski thus follows the snow, which leads to a more even

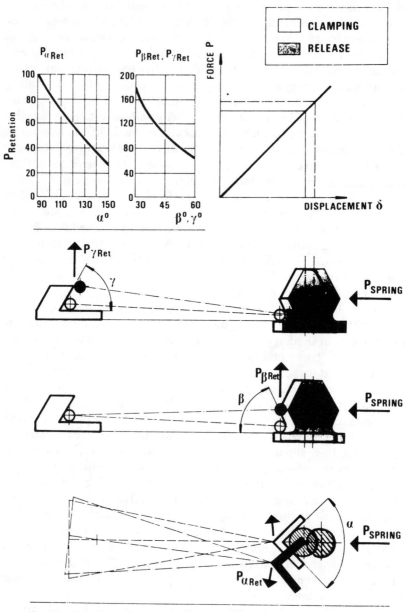

Retention force is in % of P_{SPRING}

Figure 3 – Basic mechanics of the BOX-binding.

pressure distribution of the ski upon the snow, accompanied by lower friction and greater reliability compared to the two-part binding.

Comparative glide tests have been carried out on a two-part binding and the BOX-binding fitted to Fisher Racing Cut skis and used at speeds of about 120 km/h. Twenty races were performed during a period of two days with alternating types of binding. The BOX-binding gave a faster result in all the races and a speed difference of $> 0.5\%$ was measured. At the same time observations were made of the bending oscillations in the skis fitted with two-part bindings, these bending oscillations also being indicated by dirt streaks on the gliding surfaces of the skis. No sensation of bending oscillations from the ski through the BOX-binding into the leg could be detected, nor were there any dirt streaks on the gliding surfaces of the skis.

Summary

1. By fixing the ski boot to the ski at three well-defined points in the BOX-binding it is possible to ski downhill using smaller fixing and release forces compared to two part bindings.

2. Since the binding is attached to the ski with two movable fixings the ski has total freedom for bending upon loading also under the ski boot. This permits lower friction between ski and snow compared to the two-part binding and makes a ski fitted with the BOX-binding faster.

3. The binding permits toe-up release, which means a smaller risk of overloading primarily the knee and lower leg.

4. The spring force on the nose plate in the V-track in the ski's longitudinal direction gives the binding faster retention than in today's two-part bindings, since the spring force in the binding seeks to center the ski boot to a neutral position in both the vertical and lateral directions.

5. Since the binding can be removed from the ski by opening a simple lock it can be protected during transport. It also provides higher reliability compared to two-part bindings.

References

DORÉ, R., and Riedl, M. 1978. Mechanical and biomechanical aspects of ski bindings. In: E. Asmussen and K. Jorgensen (eds.), Biomechanics VI-B, pp. 77-82. University Park Press, Baltimore.

EKSTRÖM, H. 1980. Biomechanical research applied to skiing. Dissertation, Linköping University, Linköping, Sweden.

ELLISON, A. 1977. Skiing injuries. Clin. Symp. 29:2-40.

ERIKSSON, E. 1976. Ski injuries in Sweden; A one year survey. Orth. Clin. North Am. 7:285-291.

JOHNSON, R.J., et al. 1978. Knee injuries in skiing. In: J.M. Figueras (ed.), Skiing Safety II, Int. series sport sci., Vol. 5, pp. 15-17. University Park Press, Baltimore.

MOTE, C.D., Jr., et al. 1973. Remarks on the dynamic performance of ski release bindings. In: J.L. Bleustein (ed.), Mechanics and sports, pp. 251-266. ASME Corp, New York.

MOTE, C.D., Jr., and Hull, M.L. 1976. Fundamental considerations in ski binding analysis. Orth. Clin. North Am. 7:75-94.

YOUNG, L.R., and Loo, D. 1978. Skier falls and injuries: Video tape and survey study of mechanics. In: J.M. Figueras (ed.), Skiing Safety II, Int. series sport sci., Vol. 5, pp. 36-45. University Park Press, Baltimore.

Kinetic Analysis of Judo Technique

Masataka Tezuka
Meiji University, Tokyo, Japan

Sandy Funk, Michael Purcell, and Marlene J. Adrian
Washington State University, Pullman, Washington, U.S.A.

Data collected over the last 40 years indicate that injury is common among participants in the sport of judo. Judo injuries in Japan from 1938 to 1942 included: 769 contusions, 643 sprains, 233 fractures, and 90 dislocations (Nakata and Shirata, 1943). In 1971, Hirata surveyed 666 judo students in 24 Japanese universities. In summary, he found 9 subacute inflammations, 22 lacerations, 151 bruises, 20 sprains, 5 lower leg fractures, 5 upper limb fractures, and 1 death. In an Amateur Athletic Union judo study in the United States, injuries were cited as follows: 21 fractures, 27 dislocations, 4 concussions, 2 torn knee cartilages, and 7 major sprains. Eighteen percent of these injuries were classified as being related to improper falling technique, 18 percent to throwing surfaces, and 18 percent to aborted throws. Since many of the injuries were identified as resulting from a lack of proper technique an apparent need exists to identify the patterns of an effective technique. Qualitative investigations of judo technique have been conducted (Miura et al., 1970; Takhashi et al., 1971), which correlated electrogoniometric data with electromyographic (EMG) data during progressive static positioning. Dynamic competitive situations were not investigated. Furthermore, these studies provided no quantitative values.

In judo, quantitative analyses of dynamic sports under competitive conditions are difficult to conduct. When using EMG, or the electrogoniometer, a performer is restricted by the electrical wires or the placement of measuring devices. If photography is used in the analysis, the performer has an area confinement due to camera focal view and angle. With force plates, the performance is limited by the surface and area of the platform.

There are added difficulties with the quantitative investigation of judoists during a throw because tori (the thrower) cannot be separated from uke (the one thrown). Any evaluative technique, therefore, must attempt to isolate tori from the two-party system. Also, within the activity of judo, uke is always resisting tori. One then views a relative technique rather than some absolute movement pattern.

The purpose of this research was to investigate the use of combined cinematography and dynamography in the determination of: 1) consistency of performance of skilled judoists, and 2) differences among throwing techniques.

Procedure

Skilled judoists (a 6th and a 3rd Dan) performed ten trials each of taiotoshi (body drop) and harai-goshi (sweeping loin) while standing on a 0.91 m × 0.6 m force platform (see Figure 1). The force platform used in this study was an adaptation of the one described by Cooper and Smith (1977). The platform was constructed of two flat sheets of aluminum 3/8 cm thick and embedded in a 12 cm wooden container. The strain gauge circuits allowed for the amplification of three directions: vertical (Z), horizontal-anterior posterior (X_1, X_2), and horizontal-medial lateral (Y). The analog output of each amplified signal was recorded on a galvanometric type oscillograph (Honeywell 1508A) throughout each performance. Calibration of the force output was achieved by static loading with known weights and static pulling on a cable tensiometer connected to the platform. A male weighing 823 N and having a year's experience in judo served as uke for all baseline trials. The landing mat for the uke was a standard gymnastics landing mat the same height as the platform. Simultaneously, two high speed movie cameras filmed side and front views of the performances. Force platform records and film were synchronized by means of a triggered light in the view finder of both cameras and recorded on the oscillograph.

Results and Discussion

Taiotoshi was analyzed with respect to three phases: preparatory, step-in, and flight of uke. The force-time patterns of the vertical forces were similar for both judoists and all trials of each throw (see Figure 2). The heavier, higher ranked tori (MT, 213 lbs, or 948 N), however, preceded the throwing phase with an unweighting impulse equal to the positive impulse of the throwing phase followed by an unweighting during the flight phase. The lighter tori (MP, 185 lbs, or 823 N) executed the throw 0.1 sec faster and without the initial unweighting impulse. During the step-in

Figure 1 — Contourgrams.

phase, there was an increase in vertical force until the uke's foot left the ground (the take-off point), at which time the vertical force decreased, intersected the baseline (body weight of tori plus uke), and continued to decrease to about 50% of the baseline value at landing. Variations of the horizontal force-time patterns existed for both tori indicating that these were the compensatory movements in response to the uke.

The magnitude of forces in the Z direction (vertical) were always greater for the heavier tori than for the lighter tori. In addition, the magnitudes generally were greater in the Y (medial-lateral) direction and about the same in the X (anterior-posterior) direction. These differences in medial-lateral (Y) and anterior-posterior (X) force curves indicated that the heavier tori (213 lbs, or 947 N) pivoted a half turn to the left and flexed the legs more than did the lighter weight tori (185 lbs, or 823 N).

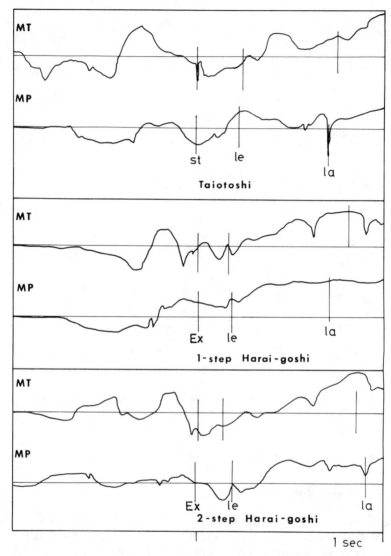

Figure 2—Vertical forces (z). st = End of preparatory phase, step in with right leg; Ex = Tori's right leg applies force to Uke; le = Leaving the ground by Uke, start of the flight phase; la = End of the flight of Uke. Base lines equal body weight of Uke plus Tori.

Both tori executed the one-step harai-goshi with near identical X, Y, and Z forces and patterns (see Figure 3). For both tori, the medial-lateral force dominated prior to the sweep. During the sweep there was an unequal push-pull causing a greater medial-anterior force.

The two-step harai-goshi showed exceptional intra-tori consistency. It appeared that the movement was initiated in the medial-anterior direc-

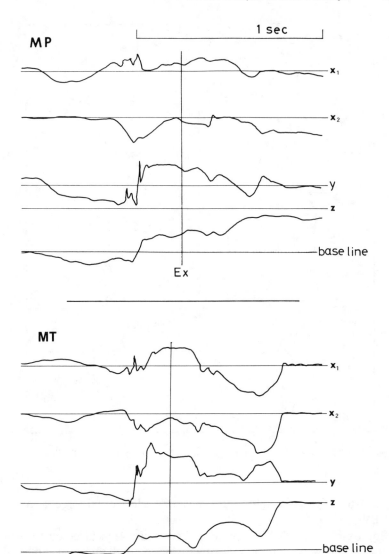

Figure 3—1-step Harai-goshi. Ex = Tori's right leg applies force to Uke. Base lines equal body weight of Uke plus Tori.

tion followed by a dominant vertical force. Differences in timing of maximum and minimum forces appeared to be related to variations in magnitudes of force. Again, the heavier tori showed greater magnitudes of force. The two-step style harai-goshi showed greater magnitude of forces than did the one-step.

Average vertical forces for the tori performances on all throws were

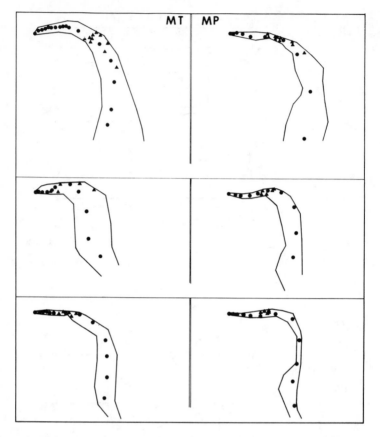

Figure 4—Flight pattern point-plot ranges. ▲ = Take-off point; • = Typical trial.

1.2 to 1.6 BW (body weight of uke plus tori). Horizontal forces averaged 0.4 BW.

Point-plots of uke's flight patterns (see Figure 4) yielded comparisons of throw, throwers, and the style of a throw. As both taiotoshi and harai-goshi are forward throws where the uke's feet arc over his head, it is not surprising that the patterns were similar. However, the effect of tori's leg sweeping backward can be seen as a "bump" in Figure 4, just after the take-off point in harai-goshi (particularly of MP). The greater twisting motions of MT implied by the force platform data are confirmed in the high density of points seen early in the plots. The tighter range of all points and take-off points in MP's throws reflect greater consistency in all techniques. Likewise, both throwers demonstrated greater consistency in the two-step as opposed to the one-step harai-goshi.

Results indicate that different styles of performance may be utilized by different skilled judoists while successfully performing the same throwing technique. Flight patterns of the uke, obtained from the film, provided evidence for the interpretation of force-time patterns. These flights also correlated with the initial position of the tori. When the tori was unbalanced (weight mostly on one leg), the flight was shorter.

Forces did not appear to be high compared to collision-type sports. It could not be ascertained which style was mechanically more efficient, more effective, or safer. Differences in weight and size may prevent the identification of any one pattern.

References

COOPER, J.M., and Smith, S.L. 1977. Design of Force Platform Utilized at Indiana University. Newsletter No. 3, Force Platform Group, International Society of Biomechanics, pp. 6-17.

MIURA, S., et al. 1970. An Electrogoniometric Study of a Judo Throwing Technique (Uchimata-Inner Thigh). Judo 41(10):51-59.

NAKATA and Shirata. 1943. Judo, pp. 54-59.

TAKHASHI, K., et al. 1971. An Electromyographic Study of a Judo Throwing Technique (Osotogari-Major Outer Reaping). Judo 42(12):55-61.

A Kinematic Study of the Trunk Rotation during a Gyaku-Zuki using Tilted-plane Cinematography

B. Van Gheluwe and H. Van Schandevijl
Vrije Universiteit Brussel, Brussel, Belgium

Among karate specialists, there is no agreement on the matter of how karateka rotate their trunks during the execution of a Gyaku-Zuki, one of the basic karate punches. Some (Habersetzer, 1976) believe that the trunk is rotated as a solid block, while others (Häsler, 1970; Myasaki, personal communication) argue that the rotation starts at the hips and moves up to the shoulder, creating in this manner a torsion of the trunk. Therefore, it was decided to investigate the rotation of the spine, using an appropriate cinematographical technique.

Experimental Procedures

In order to emphasize the movement of the spine, special sticks were attached with suction cups at five different vertebral locations: C7, T6, T10, L3, and S1 (see Figure 1). T6 and T10 locations were approximate since palpation of these thoracic vertebrae is quite difficult. Maximal contact was obtained by gluing the base of the suction cups to the skin. The sticks themselves were regular pipe-cleaners, fixed firmly on the suction cups. Pipe-cleaners, when stiffened with glue, proved to be light and rigid enough to withstand bending during maximal acceleration of the trunk.

Because the movement of the sticks during a Gyaku-Zuki did not deviate more than 7° from the horizontal plane, considering this movement as essentially flat was found to be a fair approximation (error of less than 1%). Therefore, one single Photosonics 16 mm film camera was used to record the movement. The film speed was set at 200 frames/sec. However, it was not possible to place the camera in an overhead position, looking directly down vertically on the subjects, as the natural lordosis of

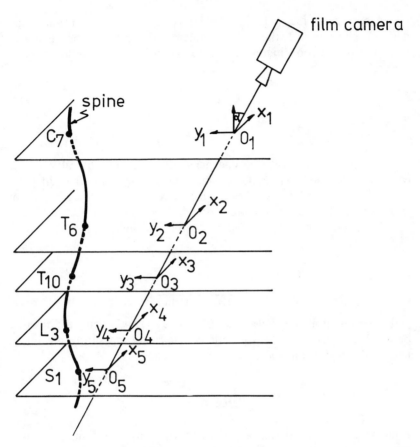

Figure 1—Geometric configuration of the five vertebral locations of the spine and the film camera.

the karateka would block the view of the lowest sticks in the pictures. Therefore, the camera was placed behind the subject looking down at his back from a tilted position (see Figure 1): actually the amount of tilt was 39°. The real movement of a stick was reconstructed from the film recordings calculating the coordinates of the base of the stick and of a second point near the end of it. Using these coordinates the orientation (and thus the angular displacement) of the stick was easily calculated for any arbitrary moment during the Gyaku-Zuki.

Formulas for Tilted-plane Cinematography

Because of the camera tilt, the conventional formulas for calculating the life size coordinates with two dimensional cinematography would not apply; a correction factor had to be introduced. If the coordinates are defined as in Figure 1, the real coordinates (X and Y) can be calculated from the image coordinates (x and y) according to the following formulas:[1]

$$X = mx \, C_x \tag{1}$$

$$Y = my \, C_y \tag{2}$$

where m is the magnification coefficient (from image to real size). As each stick has its own plane of movement, there are five different magnification coefficients. They are calculated using the image length of measuring-rods placed in the plane of action of each stick. C_x and C_y are the correction factors mentioned and can be calculated as

$$C_x = (1 - \frac{y}{y - f \, \cot g\alpha}) \tag{3}$$

$$C_y = (\frac{f}{f \cos\alpha - y \, \sin\alpha}) \tag{4}$$

with α representing the angle of tilt (see Figure 1) and f being the distance between lens and film. f is calculated indirectly measuring the image length of a measuring rod at a known distance.

Results

Eight karate experts (from second to fourth dan) were filmed and analyzed. Because of the limited space, only the results of one subject (Tsuchi, 4th dan and captain of the karate team of the University of Teicho, Tokyo, in 1980), considered as being the most representative, are displayed. Figure 2, showing the planar movement of each stick assigned to one of the five different vertebral locations of the spine, illustrates the translational and rotational motion of the vertebrae. The second (T6), third (T10), and fourth (L3) sticks are subjected to skin shifts near the end of the movement. Therefore, the representation of the corresponding vertebral movement in Figure 2 must be considered as unreliable for that time period. This is also true for the curves representing the corresponding angular data past the vertical dash line in Figure 3. The numbers accompanying the sticks relate to the corresponding pictures of the film.

In order to demonstrate the rotational nature of the vertebral movement in more detail, Figure 3 displays the angular displacement (full line), the angular velocity (dash line) and the angular acceleration (dotted line) for each of the five sticks. The curves representing these variables were obtained using smoothing techniques with Quintic Splines. The er-

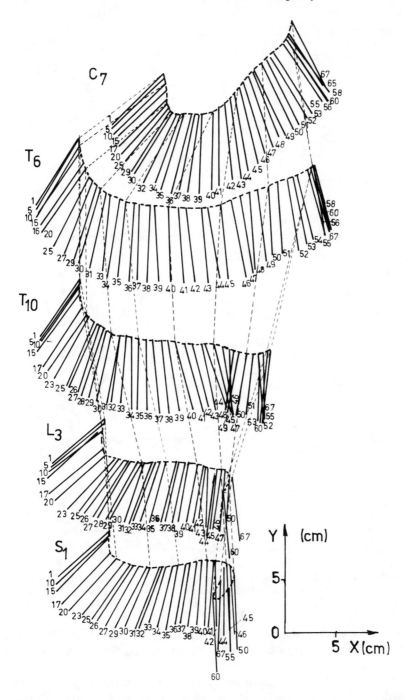

Figure 2 — Planar movement of the sticks corresponding to each of the five vertebral locations of the spine during a Gyaku-Zuki.

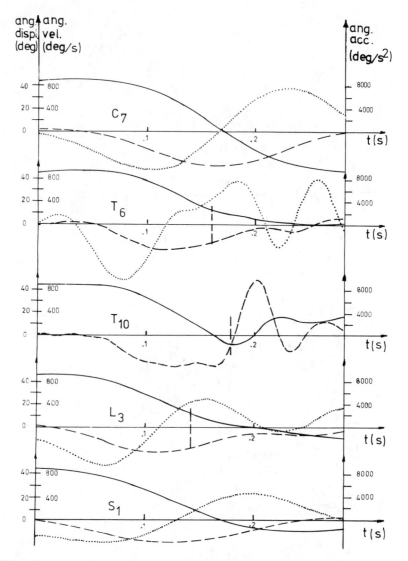

Figure 3—Angular displacement (—), angular velocity (---) and angular acceleration (...) of the sticks corresponding to each of the five vertebral locations of the spine during a Gyaku-Zuki.

ror on the angular displacement is estimated to be less than 1°, on the angular velocity less than 10°/sec and on the angular acceleration less than 100°/sec².

As perfect alignment of the sticks on the back of the subject was very hard to achieve, especially for T6 and T10 (due to skin shifts), the different sticks do not start with identical orientations, although no torsion of the trunk was observed at the start of the Gyaku-Zuki. The curves for angular displacement in Figure 3, however, are corrected for this mis-alignment in order to allow for relative comparisons of the orientation of different sticks.

Discussion

Examining the motion of all the sticks simultaneously in Figure 2, one may compare the movement of the spine to the movement of a long curtain, set in motion by someone pulling at the lowest end of it in order to get the top of the curtain gliding over the rails in the desired direction. Indeed the spine in Figure 2 starts with a marked scoliosis with thoracic concavity to the right and also ends up with a marked scoliosis, but with thoracic concavity to the left. The thoracic vertebrae are moving thereby steadily to the right, while L3 and S1 are curling back at the end. All this indicates that the hips are the first, prior to the shoulders, to start and to finish their lateral translation.

As for the rotational movement, the same observation can be made looking at the curves for the angular displacement and velocity of C7 and S1 in Figure 3. The torsion of the trunk created by the asynchronous rotation of the hips and shoulders can be evaluated comparing the angular displacement curves of C7 and S1. The hip rotation is leading with maximal 10° (negative torsion) during about ¾ of the execution time (± .25 sec) of the Gyaku-Zuki. Thereafter the shoulders are taking over, however, creating a positive torsion up to 25° at the end of the movement.

Conclusion

As the trunk of a karateka during a Gyaku-Zuki is subjected to torsion and scoliosis (with the concavity going from right to left), it can hardly be considered as moving like a solid block. On the contrary, the results in this study show a lateral and rotational deformation of the spine starting at the hips and moving up to the shoulder region.

Footnote

1. Because of the lack of space, it is impossible to demonstrate the mathematics behind the formulas.

References

HABERSETZER, R. 1976. Karate-Do. Amphora, Paris.

HASLER, R. 1970. Ich lerne Karate (Learning karate). Fackelverlag, Stuttgart.

Foil Target Impact Forces during the Fencing Lunge

Anne K. Klinger

Clatsop Community College, Astoria, Oregon, U.S.A.

Marlene J. Adrian

Washington State University, Pullman, Washington, U.S.A.

Fencing equipment and apparel have been designed empirically. The same protective equipment is used in all three weapons, and to protect all target areas. Furthermore, no differentiation has been made between the kinds of protection used with electric and non-electric weapons, although electric weapons tend to be heavier and less flexible than the non-electric weapons.

Part of the problem in devising adequate protective apparel for fencers is that the magnitude of the forces exerted on the human body during fencing is not known. Therefore, the purpose of this investigation was to determine the magnitude of these forces during the fencing lunge.

Methods and Procedures

Specially designed non-electric and electric foils were used to impact wood, padded, boney, and fleshy targets to determine the magnitude of forces exerted on the target during the lunge.

Subjects were 20 ranked and unranked members of fencing clubs in the United States. Each subject practiced the standardized test situation and then performed four or more lunges at the presented targets.

Each subject assumed the engarde position with the heel of the lead foot on a make-break electrical switch pad (see Figure 1). The foil arm was extended toward the target. When ready, the subject executed the lunge as forcefully (quickly) as possible. A second make-break switch mat was placed at an appropriate distance for the lunge. The trial was considered to be valid if the tip of the foil struck the target and the lead

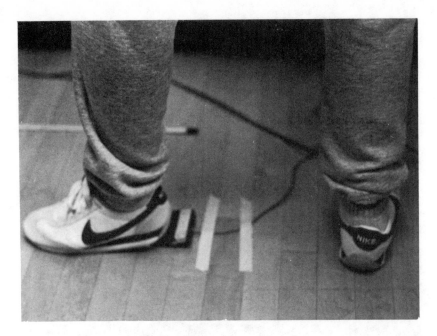

Figure 1—Lead foot of fencer on switch.

Figure 2—Lead foot on landing mat after lunge.

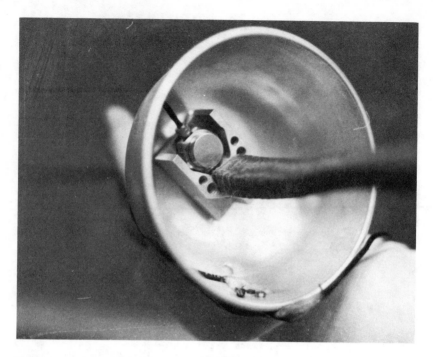

Figure 3—Accelerometer inside modified bell guard of foil.

Table 1

**Mean Speed, Mean Peak Forces, Target Condition,
and Foot Strike in the Fencing Lunge**

Condition	Mean speed (m/sec)	Mean peak forces (g's)		
		F	T	OK
Padded	2.30	26.16	40.20	31.75
Wood	2.35	49.20	78.80	82.00
Sternum	2.37	19.00	25.60	40.36
Pectoralis	2.45	32.00	22.20	26.67

foot impacted the switch mat on the lunge (see Figure 2). The following targets were used in random order: padded target, wood target, fleshy part of human torso (pectoralis), and boney part of human target (sternum). The more skilled subjects executed the trials with modified non-electric and electric foils, and the lesser skilled fencers performed with only the modified non-electric foils.

The two make-break switch mats were connected to a chronoscope (Dekan Timer), which recorded the duration of the lunge as measured by movement of the lead foot. The velocity of the lunge was calculated by dividing the horizontal distance travelled by the lead foot by the duration of the lunge.

The bell guards of the foils used in the tests were modified in size and shape in order to house an accelerometer (see Figure 3). This accelerometer was calibrated in g's and connected to a storage oscilloscope. At foil target impact, tracing was obtained on the oscilloscope. Peak forces, duration of the force, and general pattern of the impact, and the three conditions of lead foot contact were recorded. The three conditions of the lead foot contact were: 1) lead foot contacted before target was hit (F), 2) approximately simultaneous contact of foot and target (T), and 3) target contacted before foot contacts pad (OK).

Results and Discussion

The speed of the lunges ranged from 1.5 m/sec to 4 m/sec with the average speed being 2.41 m/sec. There were no differences in the speed and distance lunged due to target conditions. The most skilled fencers achieved the greatest speed, and therefore were able to impart the greatest g forces to the targets. Each individual subject showed greatest impact force in the wood condition.

As can be seen from Table 1, the smallest mean peak forces were recorded when the foot struck the floor before the foil struck the target. This, of course, is due to the fact that less of the body's mass was driving the foil into the target, and the acceleration of the foil had been sharply decreased.

When the lead foot landed at the same time the target was struck, the peak forces were larger in three of the four conditions. Only in the case of the fleshy target (pectoralis) was the peak force less than the peak force in the foot first condition.

It can also be seen from this table that the mean peak forces in all conditions except the padded condition are greatest in the target strike before foot landing situation (column T). This is probably due to the fact that in striking the target before the foot lands, more of the mass of the body is driven into the target, which means more force must be absorbed by the target. Also, the blade itself continues to accelerate into the target in this condition, while in the foot strike first the blade does not continue to accelerate. Since highly skilled fencers strike the target before foot strike, this means that skilled fencers must absorb more force when they are touched than non-skilled fencers.

In the padded and pectoralis condition, it can be seen that the mean peak forces are greatest when the target is struck at the same time the foot lands (column T). This suggests that in these conditions some of the force is being absorbed by the padding and by the pectoralis muscle, while in the wood and sternum conditions this is not the case.

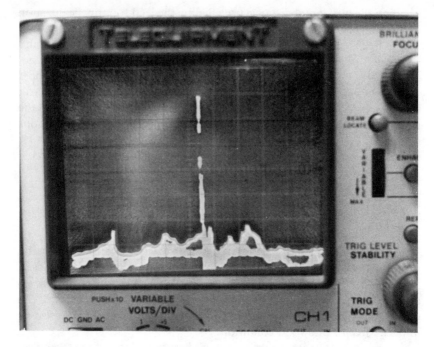

Figure 4 — Force time tracing on oscilloscope during electric foil: wood target impact.

It should also be remembered that the fencing foil blade is flexible. Therefore, in the target strike first condition, some of the force of the strike will be absorbed in the bending of the blade (Klinger, 1978).

Only the wood and padded targets were struck with the electric foil. The peak forces in the padded electric condition were 160 g's, and in the wood electric condition 200 g's.

The spike obtained on the oscillograph for the peak force in the electric wood condition can be clearly seen in Figure 4. It is much larger than the spike in Figure 5, which represents the wood non-electric condition, or the small spike in Figure 6, which represents the padded non-electric condition.

The electric padded peak forces were almost four times as large as the non-electric padded forces, and the wood electric were over twice as large as the non-electric wood forces. Part of the reason was the stiffer, heavier blade, but part also was the skill level of the fencers involved. Only skilled subjects used the electric foils when striking the target, and their skill permitted them to use their masses to produce force more effectively.

The recording from the oscilloscope is divided into 50 millisecond divisions, which allows us to examine the duration of the impacts. As can be seen in the figure depicting the wood condition, there is a large peak im-

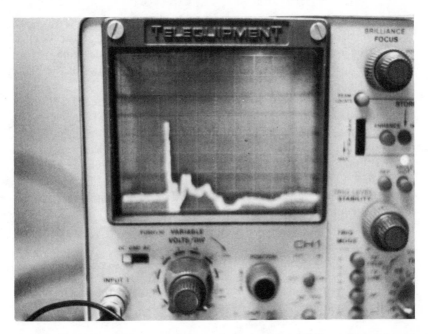

Figure 5 — Force time tracing on oscilloscope during non-electric foil: wood target impact.

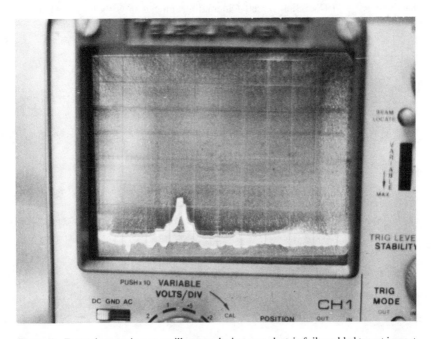

Figure 6 — Force time tracing on oscilloscope during non-electric foil: padded target impact.

pact, but over a small period of time, while in the padded condition, the peak is much less, but the duration of the impact is longer. Other oscillations in the graph may be due to differences in technique, time of foot strike, the deceleration of the blade, and the characteristics of the blade itself.

Calculations of the actual impact forces can be estimated if foil-target contact occurred prior to foot strike. Since F = ma, and body weight = ma, and mass = BW/g, and acceleration = gG, therefore F = BWG.

Positive but not significant relationships existed between speed and length of lunge (r = .48985), and length of lunge and height of subject (r = .36183), and length of lunge and leg length (r = .63152).

It is hoped that this investigation can aid in determining the basis for protective equipment development and evaluation.

Acknowledgment

The authors thank Elton Huff for the fabrication of the accelerometer utilized in this study.

Reference

KLINGER, A. 1978. High Speed Cinematographic Analysis of the Foil Blade during Target Impact. Unpublished paper, Washington State University.

Distribution of Cycling-induced Saddle Stresses

Martha Jack
Washington State University, Pullman, Washington, U.S.A.

The need to prevent saddle soreness and/or to alleviate its discomforts was the reason for this investigation. The purpose was to determine under dynamic conditions the distribution of cycling-induced saddle stresses on the female human body from three different bicycle transducer mounts congruent to three saddle shapes (Avocet, Brooks, and Cinelli) at 13 locations orthogonal to the saddle surface.

Procedures

The three saddle shapes were used to construct similarly shaped transducer mounts. Each transducer mount contained 13 bearing plates located in similar arrays and a fixed saddle clip such that the transducer mount remained parallel to the ground when mounted on the seat post. The force measurement locations were lettered and numbered for identification. Female subjects (n = 14) were tested while pedaling a track bicycle mounted on rollers in the laboratory. While they pedaled at 90 revolutions/min, the reactive forces in three planes orthogonal to the saddle surface were measured at one of 13 locations on each of three different transducer mounts. The track bicycle was adjusted for each rider so that the angle of her back was standardized with respect to the ground.

Three orthogonal dynamic force components were measured simultaneously using a piezoelectric crystal as a transducer. The crank position was monitored with a reed switch at the same time forces were measured. Strain gauges mounted on the seat post measured the total static compressive force imparted by each subject. In addition to the foregoing quantitative data, visual qualitative data were obtained simultaneously at each location in two planes using a video tape monitoring system.

Results

Stresses were computed from the force data [Equations (1), (2)] and categorized as static stress or change in stress [Equations (3), (4)].

$$\sigma_z = \frac{F_z}{A} \tag{1}$$

$$\tau_{xy} = \sqrt{(\tau_x)^2 + (\tau_y)^2} \tag{2}$$

$$\Delta\sigma_z = \max (\sigma_{z,max} - \sigma_{z,min}) \tag{3}$$

$$\Delta\tau_{xy} = \max \sqrt{(\tau_{xy,max} - \tau_{xy})_1{}^2 + (\tau_{xy,max} - \tau_{xy})_2{}^2} \tag{4}$$

where σ_z = normal stress in the z direction (kilopascal), F_z = normal force in the z direction (newton), A = area of applied force, $0.000707 \, m^2$ (square meter), τ_{xy} = shear stress in the xy plane (kilopascal), ($\sigma_{z,max} - \sigma_{z,min}$) = normal stress range from maximum value (kilopascal), and ($\tau_{xy,max} - \tau_{xy}$) = shear stress range from maximum value (kilopascal). Stress change represented the dynamic data.

Static normal stress (σ_z) (see Figure 1) and normal stress change ($\Delta\sigma_z$) (see Figure 2) in excess of each mean for the mode of subjects (≤ 7) occurred at two locations toward the back of the transducer mounts (Locations 7 and 9) and in one area toward the nose of each transducer mount (Locations 2 and 3). These static normal stresses were postulated to lie beneath the ischial and pubic rami elliptic arch bony structure on all three transducer mounts. Static normal stress ($\overline{X}$ = 13.9 kPa) and normal stress change ($\overline{X}$ = 13.3 kPa) less than each mean for the majority of subjects (> 7) occurred at the transducer mount nose, the transducer mount sides, and across the transducer mount cantle (Locations 1, 4, 5, 6, 8, 10, 11, 12, and 13), which was a little used area for support. These stresses were postulated to lie beneath the tissue of and/or dorsal to the perineal area, lateral to the ischia, and ventral to the pubic symphysis.

Static shear stress (τ_{xy}) (see Figure 3) and shear stress change ($\Delta\tau_{xy}$) (see Figure 4) in excess of each mean for the mode of subjects (≤ 7) occurred at two locations toward the rear of the transducer mount (Locations 7 and 9); these locations were postulated to lie beneath the dorsal ischial rami. Static shear stress ($\overline{X}$ = 10.2 kPa) and shear stress change ($\overline{X}$ = 10.4 kPa) in excess of each mean for the mode of subjects (≤ 7) occurred at two locations toward the rear of the transducer mount on the Brooks (Locations 7 and 9); these locations were postulated to lie beneath the dorsal ischial rami. Static shear stress ($\overline{X}$ = 10.2 kPa) and shear stress change ($\overline{X}$ = 10.4 kPa) in excess of the mean for the mode of subjects (≤ 7) occurred toward the transducer mount nose on the Avocet (Locations 1 and 2) and Cinelli (Locations 1, 2, and 3) transducer mounts; these locations were postulated to lie ventral to and beneath the pubic symphysis.

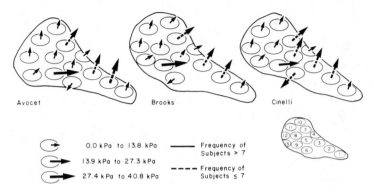

Figure 1 — Static normal stress during cycling.

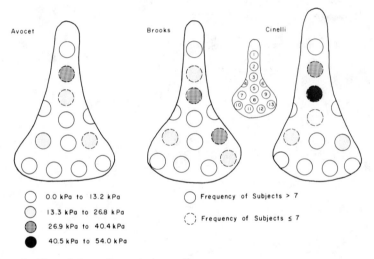

Figure 2 — Normal stress change during cycling.

Static shear stress and shear stress change below each mean for the majority of subjects (> 7) generally occurred toward the transducer mount nose on the Brooks transducer mount (Locations 1, 2, and 3), at the transducer mount middle (Locations 4, 5, and 6), and across the transducer mount cantle (Locations 8, 10, 11, 12, and 13) on all three transducer mounts. These stresses were postulated to lie beneath the tissues of and/or dorsal to the perineal area, ventral to the ischia, and ventral to the pubic symphysis.

The magnitude and direction of reactive stresses were not equally distributed over these three different transducer mounts during a crank rotation. Stress was distributed asymmetrically about an axis drawn longitudinally down the center of each transducer mount. This asym-

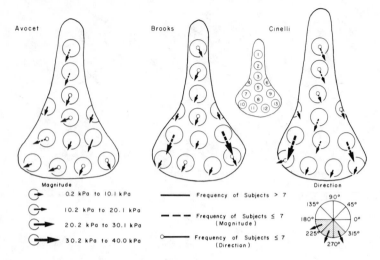

Figure 3—Static shear stress during cycling.

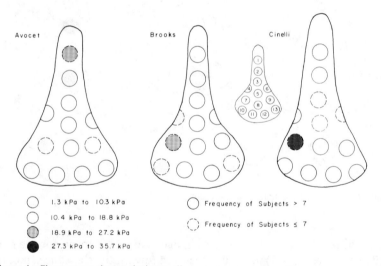

Figure 4—Shear stress change during cycling.

metry of stress was introduced partially in the fabrication of the transducer mounts. There was no trend for stress on either the right or left side.

It was determined that a mean of 47.6 ± 8.4% of the total body weight was applied to the seat post across the transducer mount. The variability of this measurement was due to the deviation in the angle of the back of the rider.

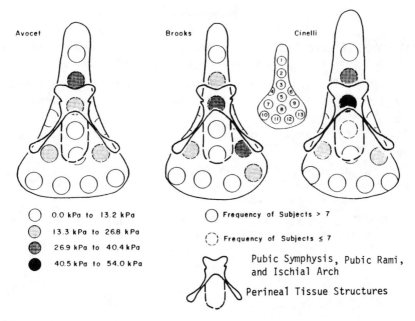

Figure 5 — Normal stress change during cycling with the mean pelvic body structures superimposed.

Discussion

A model of the mean pelvic size of these 14 subjects was generated (see Figure 5). This size was determined from the mean anthropometric measurements. Since the highest normal stress and normal stress changes occurred at Locations 1, 2, 7, and 9; Locations 2, 3, 7, and 9; or Locations 3, 5, 7, and 9, the generated mean pelvic size was positioned on the transducer mounts. As reflected by these patterns, shorter pelvises were positioned over Locations 3 and 5; longer pelvises, over Locations 1 and 2. Wider and narrower pelvises were positioned over some part of Locations 7 and 9.

The distribution of stresses was also dependent upon pelvic sizes as they were positioned on each transducer mount. For these pelvic sizes and riding positions and limited to these transducer mounts, of the three shapes investigated, the Avocet transducer mount was the most comfortable and the Cinelli the least comfortable for most of the subjects. This was substantiated by comments from the subjects, by visual qualitative data, and by the quantitative data, which indicated stresses of lower magnitude on the Avocet transducer mount and of higher magnitude on the Brooks and Cinelli transducer mounts for the majority of the subjects.

References

CHOW, W.W., and Odell, E.I. 1978. Deformations and stresses in soft body tissues of a sitting person. J. Biomechanical Engineering 100(2):79-87.

EXTON-SMITH, A.N., and Sherwin, R.W. 1961. The prevention of pressure sores, significance of spontaneous bodily movements. The Lancet November 18(ii):1124-1126.

JACK, M. 1981. Distribution of Cycling-Induced Saddle Stresses. Ph.D. Dissertation. Washington State University.

KOSIAK, M., Kubicek, W.G., Olson, M., Danz, J.N., and Kottke, F.J. 1958. Evaluation of pressure as a factor in the production of ischial ulcers. Arch. Phys. Medicine and Rehab. 39(10):623-629.

SODEN, P.D., and Adeyefa, B.A. 1979. Forces applied to a bicycle during normal cycling. J. Biomechanics 12(7):527-541.

Development of a Motor Skill using the Golf Swing from the Viewpoint of the Regulation of Muscle Activity

Hidetaro Shibayama and Hiroshi Ebashi
Physical Fitness Research Institute, Tokyo, Japan

Regarding the special pattern of any kind of voluntary motion, the subcortical brain center is considered to be related to its formation, and it is well known that the motion gradually becomes nearly a reflexive one, if the pattern is memorized in the center. For the stabilization of the motion pattern associated with the development of motor skill, the coordination of the tonus of all the muscles participating in that motion and the augmentation of the kinetic features of the motion seem to be indispensable. The present study was conducted for the purpose of investigating the features of voluntary motion, using the golf swing as an example. Analyses of the motion pattern and observations of the development of motor skill accompanied by changes in muscle activity were performed.

Procedure

Observation by Means of High Speed Cinematography

The cinematography of the golf swing (tee shot) was made outdoors with a camera (Photosonic 1P type) at a distance of 12 m from the subject, with the lens at a height of 1.2 m, and the speed at 400 frames/sec. The lens used was an F2.2, 50 mm, and the film was Kodak 4X (ASA 500). Analyses of pictures were made on a film analyzer.

Observation of Muscle Activity

Electromyograms were taken by the surface electrode method from the M. biceps brachii (right and left), M. triceps brachii (right and left), M. deltoideus (right and left), M. latissimus dorsi (right and left), M. teres

Table 1

Physical Characteristics of Subjects

Subjects	Sex	Age	Body height (cm)	Body weight (kg)	HDCP
Professional golfer					
S. Kanai	M	37	170.0	68.5	—
F. Ishii	M	37	173.0	75.0	—
T. Tsuchiyama	M	29	163.0	66.0	—
Amateur golfer					
H. Shimada	M	44	158.0	63.0	4
A. Furukawa	M	44	170.0	68.5	7
T. Ando	M	28	174.0	67.0	18
Y. Goto	M	36	162.0	59.5	26
M. Shimoyama	F	43	160.0	50.0	36

major (right), M. rectus femoris (right and left), M. sartorius (right and left), M. biceps femoris (right and left), M. gastrocnemius (right and left) and M. tibialis anterior (right and left). Electromyography was performed in the laboratory. The electrodes used were concave silver discs, 10 mm in diameter, which were fixed by adhesive plaster on the middle of each muscle along the direction of muscle fibers.

Simultaneously with the electromyography, an electrocardiogram using the sternal lead and respiration curves using the thermistor method were recorded by a polygraph.

Golf Club and Ball

The golf club had a black shaft, weighed 336 g, and was 112.7 cm long. The ball weighed 45.1 g (small size—English style).

The Phases of Motion of the Golf Swing

Analyses of the films were completed in 10 phases of swing motion around the impact point. On the other hand, the electromyography was made throughout the entire course of swing from take-away to follow through.

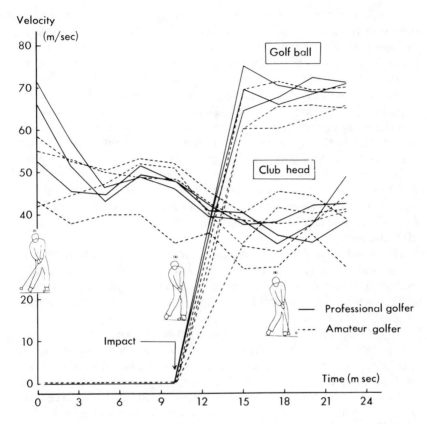

Figure 1 — Changes in velocity of the club head and golf ball around the moment of impact.

Subjects

The subjects were three professional golfers of the first class in Japan and five amateurs (including a woman and two men with single digit handicaps). Physical characteristics of the subjects are shown in Table 1.

Results and Discussion

Kinetic Features of the Golf Swing

In the analyses of motion patterns of the golf swing, attention was especially paid to the regularity of the motion in the phases around the impact point. In Figure 1 are shown changes in the velocity of the club head and the ball around the moment of impact. The solid lines indicate the professional golfers and the broken lines represent the amateurs. There was a tendency for the velocity of the club head to increase gradually from the top of the backswing through the impact and decrease thereafter toward the finish. Regarding the velocity of the club head on the average during the same phase for professional golfers and amateurs, it was 51.76 m/sec for the former and 48.86 m/sec in the latter just before impact, 43.63 m/sec in the former and 44.80 m/sec in the latter at the moment of impact; 40.53 m/sec in the former and 36.87 m/sec in the latter just after the impact, respectively. It can be noted that the velocities before and after the impact were significantly larger in the professional golfers than in the amateurs, athough the velocity at the moment of impact was not significantly different between the two groups.

The initial velocity of the ball after the impact on the average was 67.65 m/sec in the professional golfers and 62.86 m/sec in the amateurs, namely, the former was 4.79 m/sec faster than the latter. The fact agrees with the differences in the changes in club head velocities before and after the impact.

The movement of the hypothetical center of gravity of the golfer with the club was obtained by Matsui's (1972) method. There are remarkable individual differences in the patterns of movement in the professional golfers as well as in the amateurs, although they show a general tendency of moving toward the left.

The kinetic features of the golf swing at the moment of impact in each person calculated by the conventional formulae are presented in Table 2. The impulsive force at impact on the average was 249.1 Kg • m/sec in the professional golfers while it was 231.2 Kg • m/sec in the male amateurs and 106.4 Kg • m/sec in the female amateur. In addition, the force at the grip of the club was 35.28 Kg • m/sec² in the professional golfers, 27.83 Kg • m/sec² in the male amateurs and 10.54 Kg • m/sec² in the female amateur. The work done at the impact was estimated at 30.61 Kg • m in the professional golfers, at 20.90 Kg • m in the male amateurs and at 13.14 Kg • m in the female amateur.

Regarding the movement of the center of gravity of the club, it was noted that its velocity was fairly constant in the phases around the impact. However, as the swing motion comprised a rotating motion around the shoulder joint, it is presumed that a centripetal force toward the center of rotation works during the rotatory motion. In this respect the centripetal force as on the arm was estimated. It was 74.58 Kg • m/sec² just before the impact and 45.00 Kg • m/sec² just after the impact in the professional golfers, while it was 68.83 Kg • m/sec² just before the impact and 38.55 Kg • m/sec² just after the impact in the male amateurs. No significant difference in the changes was found between the former and the latter. Such changes around the impact point seem to suggest that there is a certain amount of kinetic energy given to the ball through the club. In the case of the female amateur, the centripetal force at impact and the changes around the impact point were remarkably smaller than in the above-mentioned two groups.

Table 2

Kinetic Features of Swing Motion in the Moment of Impact

Subjects	Sex	Velocity of club head m/sec	Initial velocity of golf ball m/sec	Impulsive force at the impact kg m/sec	Force on the club grip at the impact kg m/sec²	Work done at the impact kg m	Centripetal force given to the upper arm at the impact kg m/sec²		
							Before	After	B-A
Professional golfer									
S. Kanai	M	52.013	68.654	252.8	31.44	34.825	83.257	60.187	23.070
T. Tsuchiyama	M	51.117	66.275	244.0	34.45	30.440	61.763	31.031	30.732
F. Ishii	M	52.162	68.021	250.4	39.95	26.569	78.707	43.785	34.922
Mean		51.764	67.650	249.1	35.28	30.611	74.576	45.001	29.575
Amateur golfer									
A. Furukawa	M	52.046	68.069	250.4	35.85	28.694	75.360	41.700	33.660
H. Shimada	M	49.192	57.859	212.8	23.32	12.816	68.421	42.095	26.326
T. Ando	M	45.348	62.650	230.4	24.31	21.182	62.722	31.847	30.875
Mean		48.862	62.859	231.2	27.83	20.897	86.834	38.547	30.287
M. Shimoyama	F	34.751	28.958	106.4	10.54	13.139	44.714	29.539	15.175

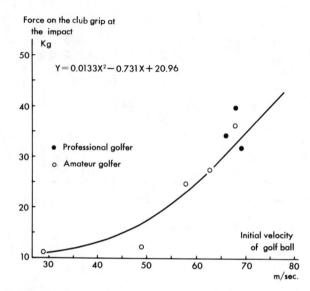

Force on the club grip at the impact

$Y = 0.0133X^2 - 0.731X + 20.96$

• Professional golfer

o Amateur golfer

Figure 2 — Relationship between the initial velocity of golf ball and the force at the grip of the club at impact.

In the present study, the flight of the ball could not be measured. However, the initial velocity of the ball was estimated to make it possible to estimate distance of the flight of the ball. The relationship of the initial velocity of the ball to some kinetic indexes are illustrated in Figures 2 and 3. The correlation between the initial velocity of the ball and the club head velocity just before the impact was highly significant ($r = 0.958$, $p < 0.001$). Figure 2 shows the relation between the initial velocity of the ball and the force at the grip of the club at impact. The relationship seems to be best expressed by a quadratic equation, suggesting that the force at the grip of the club tends to increase rapidly when the initial velocity of the ball gets into the range over 50 m/sec. The force as such exerted at the impact indicates the work done by the stroke, and Figure 3 shows the relationship between the initial velocity of the ball and work done at impact. As shown in Figure 3, the relation can be expressed by a quadratic equation, $y = 0.027 X^2 - 1.99 X + 45$. It is suggested that, when the initial velocity of the ball exceeds 50 m/sec, the work done increases rapidly. As the initial velocity of the ball was higher in the professional golfers, the work done was also larger than by the amateurs. The relationship between the initial velocity of the ball and the centripetal force on the arm at the impact was statistically significant ($r = 0.820$, $p < 0.02$).

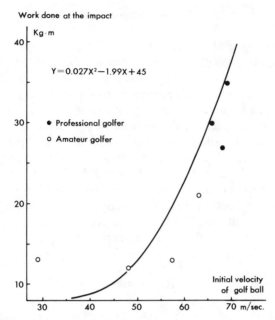

Figure 3 — Relationship between the initial velocity of golf ball and the work done in the moment of impact.

Changes in EMG During the Golf Swing

Electromyograms of the arm-, back-, and leg-muscles were observed on the records of a professional golfer, a low handicap player, and a beginner. It can be noticed that the discharge bursts of nine muscles tend to be concentrated at the moment of impact in the case of the professional golfer, and it was noteworthy, in general, that the time of motion from the start of takeaway to the impact was fairly constant at 0.90 sec in repetitions of the motions of the swing. Only a small number of bursts were seen to be concentrated at the impact in M. biceps brachii dextra, M. triceps brachii sinsistra, and the back muscles. In case of the amateur, on the other hand, the reproducibility of the time of swing motion from the start of takeaway to impact was low even in the low handicapped golfer. Moreover, in the case of the amateur, tonic discharges appeared continuously even during the preparatory phase of motion, viz. address, and often thereafter. That is to say, the discharge bursts were not always concentrated at the moment of impact, although the total amount of discharge in each muscle was larger in the amateur than in the professional golfers.

Conclusion

In the case of the golf swing, the important points seem to be the accuracy of the direction of each stroke and the flight of the ball. Regarding the former, an approach to it was tried by analyses of films of the patterns of motion and EMG records of nine muscles during the swinging motion. The latter was estimated by means of the initial velocity of the ball just after impact and the force at the grip of the club at impact. It can be stated in conclusion that in cases of highly skilled players, such as first class professional golfers, the voluntary motion of the golf swing is performed by rational and purposeful control of the muscle tonus, and the patterns of motion are precisely stabilized. The motion as a whole is fairly reflexive. Besides, professional golfers demonstrate a highly efficient mechanism and excellent kinetic features of motion to produce sufficient energy with a small amount of nerve impulses.

References

BASMAJIAN, J.V. 1978. Muscles Alive. Their Functions Revealed by Electromyography. Williams and Wilkins, Baltimore.

BULLOCK, M.I., and Harley, I.A. 1972. The measurement of three dimensional body movements by the use of photogrammetry. Ergonomics 15(3):309-322.

CARLSOO, S. 1967. A kinetic analysis of the golf swing. J. Sports Med. 7(2):76-82.

MASUDA, M., and Shibayama, H. 1971. A kinesiological study of golf swing. Bull. Phys. Fitness Res. Inst. 21:1-27.

MATSUI, H. 1972. The Foundations of Biomechanics. Kyorin Shoin Ltd., Tokyo. (in Japanese)

MILLER, D.I., and Nelson, R.C. 1973. Biomechanics of Sport. Lea & Febiger, Philadelphia.

SHIBAYAMA, H., and Ebashi, H. 1974. On the innervation of antagonist in voluntary movement from the viewpoint of electromyogram. Bull. Phys. Fitness Res. Inst. 27:1-9.

SUKOP, J. 1978. Application of biomechanics in the controlling system of athletes' sports movement activity. In: E. Asmussen and K. Jorgensen (eds.), Biomechanics VIB, pp. 5-14. University Park Press, Baltimore, MD.

Magnitude of Ground Reaction Forces
while Performing Volleyball Skills

Marlene J. Adrian and Cynthia K. Laughlin
Washington State University, Pullman, Washington, U.S.A.

With the advent of new techniques and new "mechanics" in the various skills of volleyball, there has arisen a need to investigate the kinetics of the skills. Several investigations (Cox, 1978; 1980) have attempted to determine the fastest locomotor patterns for moving into the block, but not the resulting impulse of the projection of the body. Although ground reaction forces of walking, jogging, running, and jumping (Payne et al., 1968; Ramey, 1972; Elsheikh, 1975) have been reported, only preliminary data have been reported in volleyball.

The purpose of this investigation was to determine the magnitude of ground reaction forces while performing selected volleyball movements.

Methods and Procedures

The subjects utilized were 15 female volleyball players at Washington State University who volunteered to participate in the study. The mean age, weight, and heights were 19 years, 623 N (140 lbs), and 170 cm (67 in), respectively.

The testing took place in an artificially-surfaced field house. A strain gauge-type force platform, .6096 m × .9144 m (2 ft × 3 ft) was imbedded in the floor of an indoor athletic arena. Two standards and a volleyball net were placed in front of the platform to simulate an actual game situation.

An electrogoniometer was strapped to the right knee of each subject while she performed two trials each of a stationary and moving block, a moving spike, and four variations of lunging movements utilizing the forearm pass (dig). Figure 1 depicts these movements.

All final actions of these movement patterns were performed on the force platform. Simultaneous recordings of the angular displacement at

Figure 1 — Volleyball movement patterns selected for analysis, A = block, B = spike, C = lunge (see next page).

C

Figure 1 (continued).

the right knee and the ground reaction forces (x,y,z) were obtained on a Honeywell 1508A oscillograph operating at 8 in/sec (.2 m/sec). As the subject performed the trials an assistant held the wires connecting the electrogoniometer to the oscillograph and another assistant tossed the ball to the appropriate site for the execution of the movement patterns by the subject. Observations of the movement patterns were made by the investigators to determine the general kinematic characteristics of the patterns.

Results and Discussion

Results indicated that the average degree of flexion at the knee was greatest during the front lunge (see Table 1). Except for the ball-of-the-foot situations, the lunge placed the performer in a mechanical advantage for recovering from the lunge. The thigh was near the horizontal and the lower leg near the vertical for the heel and flat foot landing situations.

In the ball-of-the-foot situation, the thigh remained near the horizontal, but the lower leg deviated from the vertical to a greater extent than in the other situation. Greatest variation in the angles at the knee and the force-time recordings were observed during the equilibrium phase of the lunge of the ball-of-the-foot condition. This suggests greater instability and possible risk of injury at the knee or ankle.

Table 1

Maximum Angle of Flexion at the Knee During Execution of Walking (W), Stationary Block (BS), Moving Block (BM), Spike (S), Front Lunge, Flat Foot Landing (FL_F), Front Lunge, Heel Landing (FL_H), Front Lunge, Ball Landing (FL_B), Side Lunge (SL)

Subject	Trial	W*	BS	BM	S	FL_F	FL_H	FL_B	SL
1	1	92°	66	70	86	30	40	12	60
	2		60	64	80	48	30	64	80
2	1	114°	98	114	116	76	91	84	98
	2		92	106	104	90	90	92	118
3	1	112°	104	100	98	136	138	138	124
	2		104	98	108	128		116	126
4	1	108°	106	86	98	84	55	78	70
	2		102	50	102	76	72	78	
5	1	118°	120	120	102	80	87	92	98
	2		112	114	84	84	84	94	112
6	1	94°	92	90	112	76	62	58	94
	2		94	92	100	75	77	92	92
7	1	106°	90	104	108	83	100		56
	2		80	84	110	74	74		58
8	1	104°	116	118	114	103	100	92	88
	2		108	114	116	89	108	86	78
9	1	92°	76	96	90	57	38	58	27
	2		82	90	98	50	28	59	42*
10	1	108°	108	94	98	84	78	76	92
	2		100	96	94	82	72	70	86
11	1	104°	98	120	104	74	76	70	86
	2		92	104	98	75	72	68	70

Table 1 (continued)

12	1	100°	118	126	130	74	88	90	74
	2		128	120	122	90	76	92	92
13	1	108°	94	108	104	70	60	78	80
	2		116	104	102	72	68	60	86
14	1	112°	118	126	128	100	106	98	108
	2		112	126	124	—	116	80	118
15	1	106°	90	108	94	74	62	62	74
	2		92	88	106	94	94	—	90
Mean		105.2	98.933	101	104.33	80.28	77.31	79.15	85.41

*Mean of 3 or more steps

The force time histories and goniograms for the knee of one subject for all testing situations are shown in Figure 2. Force-time patterns were unique for the moving block, stationary block, and spike. Greater forward-backward forces occurred during the extension of the latter two techniques. This resulted from the forward momentum prior to contact with the force platform. Most performances were higher in these forces during the retardation than the propulsive phase. Peak (short duration maximal) and maximal vertical forces were least for the stationary block and greatest for the spike (see Table 2). The greatest peak forces in the vertical direction occurred during the spiking situation. These forces averaged almost 5 × BW.

Peak vertical forces during the dig situations were less than ½ that shown for the stationary block (see Table 3). Subject variation was high, however, for the dig situation, as well as the blocking and spiking situations. Although commonalities existed among the force-time histories for all the dig situations, some major differences were noted. As evidenced by the slope of the goniograms the angular velocities during the extension phase of the stationary and moving block and the spike were

FRONT   LUNGE - BALL

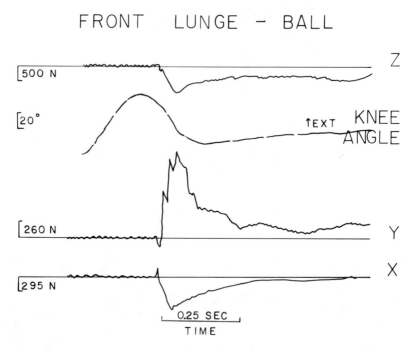

Figure 2—Force time histories and goniograms of volleyball patterns of one subject. Scales are given with zero baselines for vertical force (Z), anteroposterior force (X), and lateral forces (Y). Scale and direction of extension (↑ ext) are given for knee angle. The most upward trace in the stationary block represents the 180° angle for all patterns.

Figure 2 (continued)

STATIONARY   BLOCK

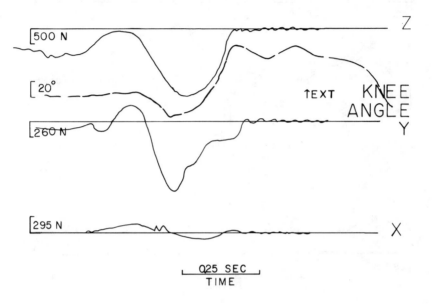

FRONT   LUNGE – FLAT

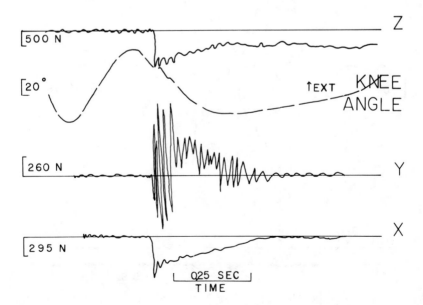

Figure 2 (continued)

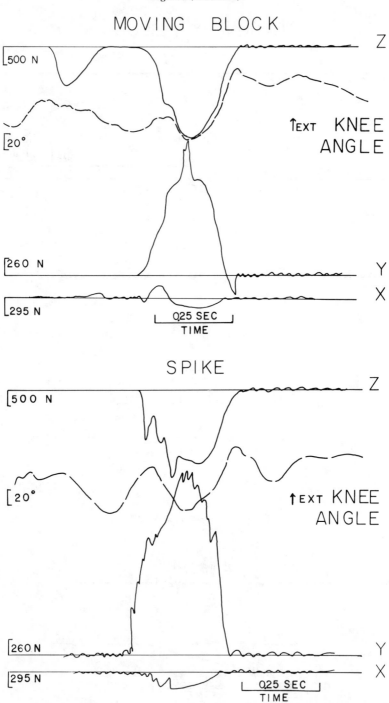

Figure 2 (continued)

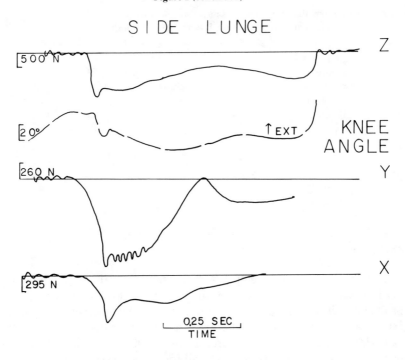

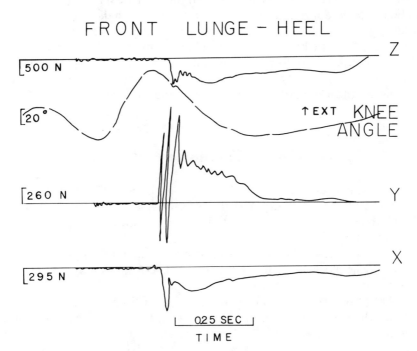

Table 2

**Characteristics of Stationary Block (BS), Moving (BM), and Spike (S)
as Executed by 15 Women Intercollegiate Volleyball Players**

	Number of degrees of extension at knee after take off	Peak vertical force	Height center of gravity was raised above standing height (m)	Average angular velocity at the knee during extension phase (rad/sec)
BS	9	1869N (3 × BW*)	.28	22
BM	7	2322.9N (3.7 × BW)	.32	24
S	10	2959.25N (4.8 × BW)	.38	25

*BW equals body weight

greater than the angular velocities of flexion during the digs. The latter motion was a force-regulation action, whereas the extension was a force-producing action. Table 2 shows the relationship of the velocities of leg extension to vertical forces. There was a direct relationship between leg extension velocity and vertical force, as well as displacement of the center of gravity. Usually peak (short interval) forces in the vertical (z) direction did not occur during the trials of the front lunge with the ball-of-the-foot landing. The impact forces were applied gradually. Variations among performers existed for the anterior-posterior forces (x) with about ½ of the subjects showing negligible (x-forward/backward) peak forces. Negligible right-left (y) forces appeared for all the front lunge conditions. At the instant of the dig (impact of the ball with the arms) a force to the right was noted.

Conclusion

This research approach provides a viable means for investigating the amount of risk of injury and the effectiveness of techniques used in volleyball. Results suggest a need to further investigate the foot-leg position at impact.

Similar force-time histories were noted for the stationary block as for the moving block. This suggests that skilled performers adapt to the goal of the task.

Although differences in peak forces did not exist among the dig situations, there were instances in which shearing forces were twice the body weight. The design of the heel contour, and sole-counter-interface for

Table 3

Ground Reaction Forces during Selected Volleyball Movement Patterns

Type of movement	Range and means	Vertical (z) force	Lateral (y) force	Anterior/posterior (x) force
Front lunge flat	Range	574.67 to 1253.83	40.49 to 378.25	291.25 to 770.96
	Means	876.81	98.28	518.49
Front lunge ball	Range	574.67 to 1279.95	68.09 to 287.47	274.12 to 805.23
	Means	862.47	108.77	513.48
Front lunge heel	Range	522.43 to 1306.08	18.69 to 400.95	239.86 to 822.36
	Means	807.15	132.70	487.09
Side lunge	Range	626.92 to 1593.41	105.91 to 446.34	291.25 to 770.56
	Means	915.99	194.50	477.19
Stationary block	Range	809.77 to 2768.88	L45.35 to 286.58 R37.83 to 181.56	B85.66 to 394.05 F17.13 to 188.46
	Means	1871.59	L 135.33 R 57.76	B 182.95 F 95.70
Moving block	Range	1593.41 to 3813.74	90.78 to 1240.66	B65.53 to 805.23 F17.13 to 963.20
	Means	2347.01	468.70	B 322.27 F 198.01
Spike	Range	1802.38 to 3970.47	378.25 to 1331.44	205.59 to 885.11
	Means	2658.36	792.89	470.35

control, and the sole and heel for attenuation appear to be important to the volleyball situation.

Acknowledgment

The authors thank Elton Huff for the fabrication of the accelerometer used in this study.

References

COX, R.H. 1978. Choice response time speeds of the slide and cross-over steps as used in volleyball. Research Quarterly 49:430-436.

COX, R.H. 1980. Response Times of Slide and Cross-Over Steps as Used by Volleyball Players. Research Quarterly for Exercise and Sports 51:563-567.

ELSHEIKH, M. 1975. A Three-Dimensional Model of the Ideal Technique in the Broad Jump. Ph.D. Thesis. Washington State University, Pullman, WA.

PAYNE, A., Stater, W., and Telford, T. 1968. The Use of Force Platform in the Study of Athletic Activities. Ergonomics 11(2):123-143.

RAMEY, M. 1972. Effective Use of Force Plates for Long Jump Studies. Research Quarterly 43:247-252.

Discovering Biomechanical Principles of Batting in Cricket

Ken Davis

Deakin University, Australia

The purpose of this paper is to evaluate the effectiveness of coaching points given in a coaching manual on batting and to discuss some biomechanical considerations that might produce more definite technique guidelines.

The results displayed in this presentation have not been measured under well controlled conditions. The purpose of the experience was to find simple techniques that could be used on the coaching field to demonstrate the difference in certain aspects of the skill and to provide a basis for discussion. Laboratory activities will be demonstrated for the grip and stance that will provide students with some answers to some of the 'grey' areas of these aspects of batting.

The Grip — Part I

Coaching Manual

Grip the bat with both hands together near the middle. The 'V's formed by the forefinger and thumb of each hand should be aligned with the spine of the bat and also point up their respective arms at the shoulder. Hold the bat firmly but not tightly.

Biomechanical Considerations.   The V-grip ensures more mass behind the bat and therefore better force production.

Activity 1.   Try timing hit for length of gym with two different grips. The ball is lobbed in the air for the batsman to hit the ball on the full. A

Figure 1 — Two kinds of grips: (right) v-grip according to coaching manual and (left) fingers on top of bat.

front drive is to be played and the ball is timed from contact with the bat till it reaches the end of the hall (or wall if outdoors). The two grips are illustrated below: (i) V-Grip as per coaching manual (on right), (ii) Fingers on top of bat.

Observations of Coaches

1. The 'V' grip enabled the batsman to provide and inject more power at impact.

2. The 'fingers behind the bat' grip tended to hit the ball in the air and therefore too much bottom hand was being used.

3. With the 'fingers behind the bat' grip the bottom hand tended to medially rotate and the bat came across the line of the ball.

4. With the 'fingers behind the bat' grip the arc of the bat is restricted and therefore there is a reduction in power.

5. The 'V' grip—the players felt they had more control.

6. In defense it was felt that the fingers behind the bat grip could be effective during defense shots as it kept the front elbow high.

The Grip — Part II

Coaching Manual

Hold the bat firmly but not tightly.

Biomechanical Considerations

Logically at impact with ball the hands should firmly control the bat so that some momentum is not lost in the impact phase. One would suspect that both hands should be gripping firmly at contact though this need not be the case in the preparatory phase of batting.

Table 1

Results Comparing Two Grips

Type of Grip	Mean Time (N = 20)
'V' Grip	2.17
Grip with fingers behind bat	2.47

Table 2

Results Comparing Tightness of Grip

Type of Grip	Mean Time (N = 15) (sec)
Both hands loose throughout	2.32
Top hand on bat tight, bottom hand loose	2.25
Both hands tight throughout	1.80
Both hands loose on backswing and tight on forward swing	1.56

Activity 2. Try timing hit for length of gym. (i) Hands gripping the bat loosely throughout. (ii) Top hand tight, bottom hand light grip. (iii) Both hands tight throughout. (iv) Hands loose on backswing and tight on forward swing.

Observations of Coaches. Although condition 4 was found to be superior many players felt that they had little control of their back lift and may have been in trouble playing that way in a match situation.

The Stance — Weight Distribution

Coaching Manual

Weight evenly distributed.

Biomechanical Considerations

The more weight there is on a foot the more difficult it is to move. Perhaps there could be some value in varying the distribution of weight for different types of bowlers or conditions. If, for example, a bowler is pitching short consistently and preventing the batsman from getting on to the front foot then it may be advantageous to unweight the back foot to facilitate backward movement. The reverse could occur when facing a spinner who tends to flight the ball and bring you on to the front foot

Figure 2—Unweighting the back foot (right).

(unweight the front foot in this case). An illustration of unweighting the back foot (right) is demonstrated in Figure 2.

Activity 3. Play a spinner with front foot unweighted and a bouncer with back foot unweighted. High speed film could be used to assess the speed of movement in these cases compared to the even distribution of weight.

Observation of Coaches

1. Appeared sound enough mechanically.
2. Once the ball flight path was known, the player tended to unweight automatically.
3. With the feet not being evenly weighted, many players felt unstable before playing shots.
4. The principle of weighting/unweighting would limit other shots that could be played.
5. Most felt an initial 'readiness' movement is good, but it is not desirable to overcommit oneself.
6. Players are committed too early.
7. Moving back required two movements.
8. Loss of comfort.
9. There doesn't seem to be any need against a spinner.

Width of Stance

Coaching Manual

The feet should be slightly apart.

Biomechanical Considerations

The stance should cater for both stability and ease of movement (mobility). As the base of support becomes larger stability improves. However, mobility becomes increasingly difficult, so a compromise between these variables needs to be sought. A taller player should place his feet wider than a player of shorter stature.

Activity 4. Try both wide stance and stance with feet together (vary speed of delivery). Recording of player reactions to each stance could be made on video, and high speed film could again be used to assess the speed of movement and balance. A demonstration of variation of width of stance is illustrated in Figure 3.

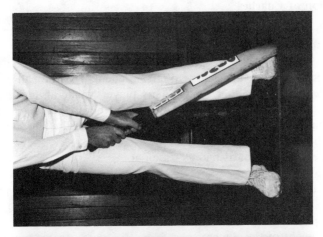

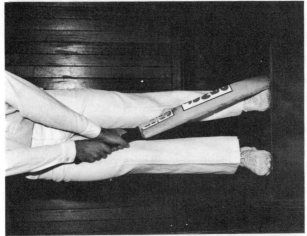

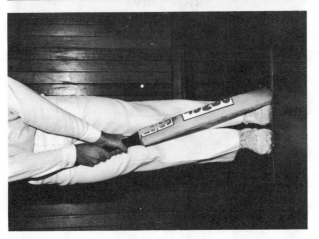

Figure 3 – Variations in the width of stance.

Observation of Coaches

1. With feet together in the stance players felt that they were late when getting into position.

2. With feet very wide apart — batsmen felt that they couldn't move forward or backward sufficiently to alter the length of the ball.

3. With the feet together stance many players made an unconscious movement first to gain a wider base of support before playing a stroke.

4. Optimum width of stance — the outer borders of the feet should be in line with the hips.

5. Close together not as bad as wide part.

A similar approach was adopted for the following aspects of cricket batting:

1. The placement of the bat in the stance.
2. The height of the stance.
3. Closed or open stance.
4. The first movement of the feet.
5. The angle of the front leg in drive shots.
6. Weight transference in the back foot drive.
7. Methods of 'backing-up' for running between wickets.

Conclusion

The session provided valuable insights into techniques of cricket batting and had many of the senior coaches questioning some of the established principles of coaching cricket skills.

Reference

TYSON, F. 1976. Complete Cricket Coaching. Thomas Nelson Australia Limited.

Knee Shear Forces during a Squat Exercise using a Barbell and a Weight Machine

James G. Andrews and James G. Hay
University of Iowa, Iowa City, Iowa, U.S.A.

Christopher L. Vaughan
University of Cape Town Medical School,
Cape Town, South Africa

The joint shear force is that component of the resultant force transmitted across a joint which tends to transversely displace one joint segment relative to the other. Resistance to shear may be provided by the bones and/or by the soft-tissue structures surrounding the joint. At the knee, where shear is resisted primarily by the ligaments, the value of the joint shear force during any activity serves as an important measure of the exposure of that joint to injury.

The purpose of this study was to compare knee shear force values during squat exercises performed at the same rate of lifting and with comparable loads, using a barbell and a weight machine. Adopting the preferred shear force definition proposed by Andrews et al. (1980), the knee shear force was defined as the component of the knee resultant force which is perpendicular to the longitudinal axis of the shank segment.

Methods

Three experienced male weightlifters performed squat exercises with 40%, 60%, and 80% of their four-repetition, free-speed maximum (4RM), at three lifting rates, using a barbell and a Universal weight machine. The lifting phases occurred during 1, 2, or 3 sec intervals, whereas all lowering phases occurred during 2 sec intervals. A metronome and voice commands were used to help each subject maintain the proper tempo.

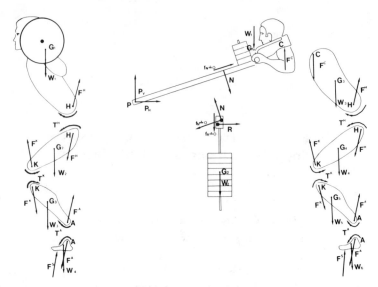

Figure 1—Systems models. (a) = planar, 4-segment, barbell squat exercise model, (b) = planar, 6-segment, machine squat exercise model.

Planar, bi-laterally symmetric system models (see Figure 1), and the corresponding Newtonian equations of motion, were used to represent the subject and apparatus during the exercise. Frictional effects in the exercise apparatus were assumed to be negligible, and the pad contact force acting on the subject's shoulder was assumed to act vertically. Body segment parameters were estimated using the mean values of Clauser et al. (1969) and Whitsett (1963), and the inertial properties of the exercise apparatus were determined from simple laboratory experiments. Kinematic data were recorded during the third repetition of a four repetition set using a Locam camera operating at 50 frames/sec. These data were reduced using a Vanguard Motion Analyzer and cubic spline functions. The inverse dynamics problem was solved to determine the variable knee resultant force magnitude F^k ($F^k = |F^k|$), and the variable knee shear force based on shank orientation F_s^k ($F_s^k = |F^k x \, u_s|$ where u_s is a unit vector parallel to the longitudinal axis of the shank, and x denotes the cross-product).

Results and Discussion

Results for Subject #2 performing at the slow (a) and fast (b) exercise rates with 80% of the 4RM, and using the barbell and weight machine, are shown in Figures 2 and 3, respectively. These results are typical of those obtained for all three subjects at all three load levels. They indicate that, for both the barbell and machine squat exercises, slow lifting rates

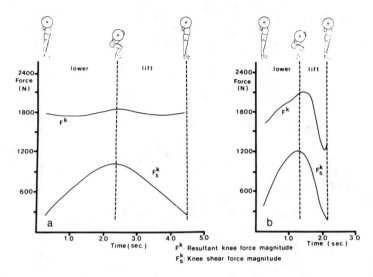

Figure 2 — Barbell squat exercise of subject #2 with 80% of the 4RM. (a) = slow exercise rate, (b) = fast exercise rate.

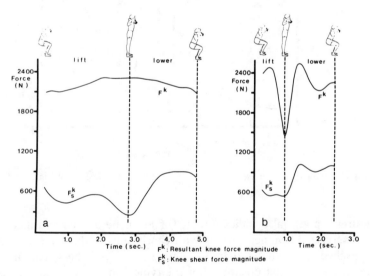

Figure 3 — Machine squat exercise of subject #2 with 80% of the 4RM. (a) = slow exercise rate, (b) = fast exercise rate.

led to essentially constant F^k values, whereas fast lifting rates led to appreciable variations in F^k due to inertial (acceleration) effects. Results obtained at the intermediate lifting rate, again for all subjects at all load levels, differed only slightly from those obtained at the slow exercise rate.

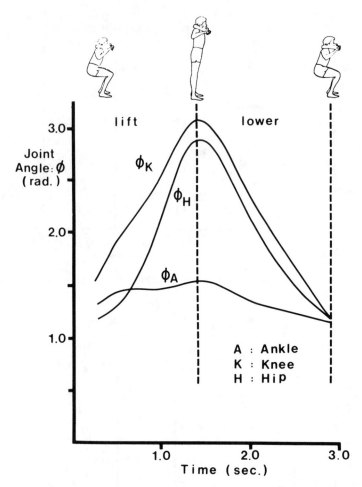

Figure 4 — Machine squat exercise of subject #1 with 80% of the 4RM. Joint angles vs. time.

Figures 2 and 3 also indicate that, for both the barbell and machine squat exercise, F_s^k values varied significantly during the exercise cycle, and reached peak values of approximately 50% of the maximum F^k value at the full squat position. For the typical machine squat exercise, however, force levels were found to be approximately 30% to 40% higher than for the comparable barbell squat exercise. Hence, greater strength development may accompany the use of a weight machine, although there may be greater risk of injury due to shear at certain joints.

Results obtained for all subjects using the barbell and the weight machine at a given rate indicated that both F^k and F_s^k values increased with load, as expected. For all exercise conditions, the maximum values of F^k and F_s^k occurred when the greatest load was lifted at the fastest exercise rate. The greatest F_S^k value obtained was 1350 N and occurred during the machine squat exercise. This value compared favorably with Ariel's (1974) value of 1590 N during a deep knee bend with a heavy load.

Finally, and somewhat more subtly, Figures 2 and 3 indicate a simpler and more symmetric variation in the results obtained during the barbell squat exercise than during the machine squat exercise. This interesting and somewhat surprising discrepancy in system kinetics led to a closer examination of system kinematics. In particular, the joint angles (ϕ) at the hip (H), knee (K), and ankle (A) were examined for both the barbell and machine squat exercise, to see how they varied over the exercise cycle. The results for the barbell squat exercise showed a simple monotonic variation during both the lifting and lowering phases for all three joint angles, for all subjects, at all loads, and for all exercise rates. For the machine squat exercise, the hip and knee angles also showed this same simple monotonic variation during both phases of the exercise cycle for all subjects, at all loads, and for all exercise rates. The ankle angle, however, for two of the three subjects, varied in an essentially non-monotonic manner during eight of the 18 lifting phases of the exercise cycles studied. This non-monotonic variation in ϕ_A during lifting was not consistently related to either the load level or to the exercise rate, although five of the eight cases occurred with 80% of the 4RM.

A typical non-monotonic variation in ϕ_A during the lifting phase of a machine squat exercise at the fast rate with 80% of the 4RM is shown in Figure 4 for Subject #1. This subtle and unexpected variation may have been due to a reduced balance or stability requirement, which permitted variable body placement relative to the machine, and caused subsequent variations in the subject's performance.

References

ANDREWS, J.G., Hay, J.G., and Vaughan, C.L. 1980. The concept of joint shear. Proc. of the 2nd Big Ten CIC Phy. Ed. Body of Knowledge Biomechs. Symp. Bloomington, Indiana.

ARIEL, G.B. 1974. Biomechanical analysis of the knee joint during deep knee bends with a heavy load. In: R.C. Nelson and C.A. Morehouse (eds.), Biomechanics IV, Int'l. Series on Spts. Sci., Vol. 1, pp. 44-52. University Park Press, Baltimore, MD.

CLAUSER, C., McConville, J., and Young, J. 1969. Weight, volume and center of mass of segments of the human body. AMRL-TR-69-70, Wright-Patterson AFB, Dayton, Ohio.

WHITSETT, C.E. 1963. Some dynamic response characteristics of weightless man. AMRL-TR-63-18, Wright-Patterson AFB, Dayton, Ohio.

Effectiveness and Efficiency during Bicycle Riding

M.A. Lafortune and P.R. Cavanagh
The Pennsylvania State University
University Park, Pennsylvania, U.S.A.

Most studies on bicycle riding have dealt with either the physiology or the biomechanics of the activity. The physiological experiments have typically investigated the effects of pedalling frequency and load setting on heart rate and oxygen consumption (Åstrand and Rodahl, 1970), as well as the efficiency of bicycle riding (Dickinson, 1929; Gaesser and Brooks, 1975). On the other hand, biomechanical studies have been concerned with the pattern of force application, bilateral asymmetry and the effectiveness of pedaling style (Hoes et al., 1968; Daly and Cavanagh, 1976; Gregor, 1976; Davis and Hull, 1980; Lafortune and Cavanagh, 1980). This study was designed to examine both the biomechanical and physiological effects of changing the shoe-pedal interface.

Methods and Procedures

Twenty college students ranging in weight between 510 and 870 N. were used as subjects in the present study. They rode a 10-speed bicycle mounted on a stand for two periods of five min at a power output of approximately 155 watts and a pedalling frequency of 60 rev/min. The stand simulated both resistance and inertia encountered by a cyclist on the road. Two different shoe-pedal interfaces were randomly assigned to the riders during each riding period; the first was rubber surface pedals and leather soled shoes (RLS) while the second was conventional metal surface pedals with cleated shoes and toe clips (MCS).

During the last min of each ride, the expired gases were analyzed by a Metabolic Rate Monitor to determine the O_2 consumption ($\dot{V}O_2$) of the rider. The 11 channels of analog data required to compute the biomechanical parameters were sampled at a frequency of 195 Hz per channel by a PDP 11/34 laboratory minicomputer. The measuring devices

Figure 1 — Measuring devices on the bicycle. A = strain gauged pedal, B and C = Potentiometers, and D = top dead center detector.

mounted on the instrumented bicycle recorded normal and tangential force components, sine and cosine of pedal angle with respect to the crank for both left and right sides, sine and cosine of the crank angle to the bicycle frame (see Figure 1), and the top dead center (TDC) of the right crank.

The tangential (F_T) and normal (F_N) force components measured directly from the pedal enabled the magnitude (F_R) and direction (θ) of

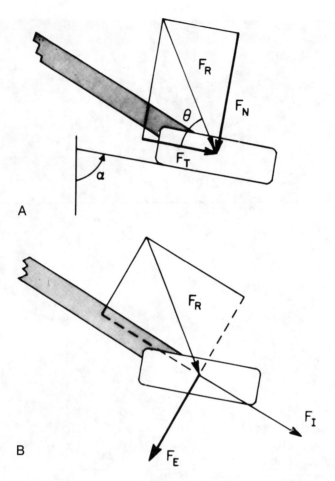

Figure 2—A: Force components measured (F_N and F_T), resultant force (F_R), angle of resultant force to the pedal (θ) and foot angle (α). B: Force component perpendicular to the crank (F_E) and component along the crank (F_I).

the resultant forces to be calculated. The angles between the pedals and the vertical (α), defined as the foot angles, were obtained by combining the angles of the pedals to the cranks with the angles of the cranks to the vertical (see Figure 2A). Subsequently, F_R was resolved into two components (see Figure 2B); F_E, the component at right angles to the crank, and F_I, the component along the crank. F_E therefore represents the force component used to produce torque while F_I is the component ineffective in producing torque.

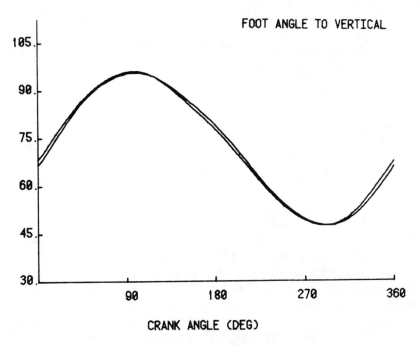

Figure 3—Foot angles vs. crank angle for both MCS and RLS conditions.

The Index of Effectiveness (IE) represents the percentage of the applied linear impulse that is used to generate angular impulse in the interval $(t_i - t_o)$:

$$\text{I. E.} = \frac{\int_{t_o}^{t_i} F_E \ dt}{\int_{t_o}^{t_i} F_R \ dt} \times 100 \tag{1}$$

The Index of Effectiveness can be further subdivided into IE_N and IE_T which are the percentages of the linear impulses of F_N and F_T used to generate angular impulse.

Finally, the Net Efficiency (NE) of the subjects was calculated from the biomechanical and physiological results as follows:

$$NE = \frac{W}{E_c - E_R} \tag{2}$$

where W is the external work accomplished, E_c the energy expended during cycling and E_R the energy expended at rest.

Results

Mean pedal angles (α) for both MCS and RLS conditions are shown in

Table 1

Force Components Parameters

Parameters	Units	Conditions		p < 0.05
		RLS	MCS	
Peak F_N	N	301.8	291.1	sig.
Peak F_T	N	51.2	62.5	sig.
$\int F_N dt$	Ns	136.7 sig.	140.3	sig.
$\int F_T dt$	Ns	21.9	27.9	sig.

Figure 3. The similarity between the two conditions indicates, somewhat surprisingly, that the subjects did not exploit the toe clips and cleats to change the orientation of the pedal. It should, however, be recalled that the subjects were recreational cyclists who had not had extensive practice with either condition.

When the force components (F_N and F_T) were considered, significant differences were found between interface conditions in peak forces and linear impulses as shown in Table 1. Peak F_N and normal impulses were greater in the RLS condition while peak F_T and tangential impulses were greater in the MCS condition.

These findings suggest that although pedal orientation (α) was not changed, forces were applied differently. The riders apparently took advantage of the high slip resistance offered by the MCS condition to apply more tangential force to the pedals.

In Figure 4, mean data from all subjects in the MCS condition are shown (mean of 2000 cycles). The resultant force F_R is drawn in its correct orientation with the pedal at 18° increments. Peak F_R (350N) occurs at a crank angle of 108°. Note that the orientation of the force vector with the crank is far from ideal at this time. It is also apparent that the subject is still pushing down on the pedal during the recovery phase (positive F_R in 180-360°). This indicates that under the conditions studied, part of the propulsive impulse is used to lift the recovery leg.

In order to examine the effective force, the forces applied by both legs were combined. The mean resultant force from both legs and the mean effective force from both legs for all subjects in the MCS condition are shown in Figure 5. The shaded area represents the 'unused force' when torque production is considered. Effective force is largest (225N) when the cranks are almost horizontal (94° and 274°) while resultant force (340N) peaks some 15° later. One can see that the pattern of F_R is closer to the feeling of constant effort perceived by the riders than that of F_E. Also the large differences in the F_E/F_R ratio (.16 at 0° and 180° and .73

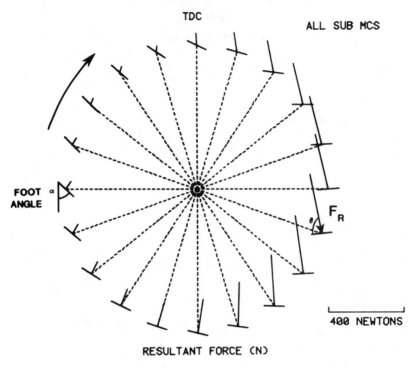

Figure 4—Resultant force for all subjects and all cycles for the MCS condition.

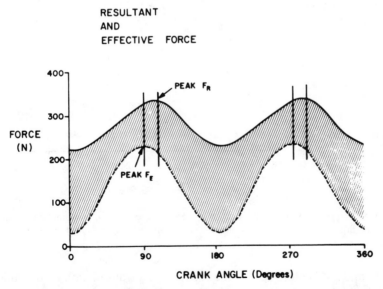

Figure 5—Resultant and effective force vs. crank angle for all subjects in the MCS condition. The shaded area represents the unused force.

Table 2

Indexes of Effectiveness

Parameters	Conditions		p < 0.05
	RLS	MCS	
IE	47.6%	48.1%	*
IE_N	48.4%	46.4%	sig.
IE_T	12.7%	23.4%	sig.

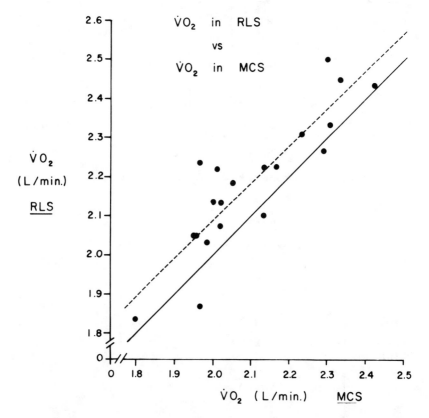

Figure 6 — Oxygen consumption for the RLS condition vs. oxygen consumption for the MCS condition. The solid line indicates an equal ratio while the dashed line shows the linear regression in the present experiment.

at 85° and 265°) indicate that the conventional transmission used in this experiment leaves much to be desired in the effective coupling of man and machine.

The Index of Effectiveness (IE) which indicates the percentage of the linear impulse used to produce angular impulse was not significantly different between conditions (see Table 2). When the components of force were examined, however, the normal component was significantly more effective in the RLS conditions, while the tangential component was significantly more effective in the MCS condition. Further analysis showed that on average 98.5% of the propulsion could be attributed to the normal component in the RLS condition, while 92.2% came from the same source in MCS.

When the oxygen consumption ($\dot{V}O_2$) was examined, it was found that a significantly lower amount of oxygen was used in the MCS condition than in the RLS condition (2.03 l/min vs. 2.08 l/min.). This is shown in Figure 6 where the regression line of $\dot{V}O_2$ in RLS condition vs. $\dot{V}O_2$ in MCS condition lies above the equal ratio line. All but three subjects used more oxygen to ride in the RLS condition. However, only a low correlation ($r = 0.43$) was found between net physiological efficiency and Index of Effectiveness (the correlation included all subjects and trials) indicating that even though some subjects were applying less force to work at the same power, they were not consequently more efficient bicycle riders.

Conclusion

The mean oxygen consumption was significantly higher with the RLS interface than with the MCS. In the absence of differences in IE, the lower net efficiency of the RLS condition may be explained by the fact that an extra amount of muscular activity was needed to prevent the foot from slipping off the pedal. It is also possible that the use of more tangential force tends to decrease the O_2 consumption by providing a better load sharing between the various leg muscles.

A major finding of this study was that only a weak correlation existed between the Index of Effectiveness and Net Efficiency. In this group of riders, therefore, a high value for IE did not necessarily mean high NE. Presumably, upper extremity and trunk muscle activity, muscle morphology, postural and other factors not measured in this experiment accounted for a larger proportion of the variance in net efficiency than the effectiveness of force application to the pedals.

References

ÄSTRAND, P.O., and Rodahl, K. 1970. Textbook of Work Physiology. McGraw-Hill Book Company, New York.

DALY, D.J., and Cavanagh, P.R. 1976. Asymmetry in bicycle ergometer pedalling. Medicine and Science in Sports 8(3):204-208.

DAVIS, R.R., and Hull, M.L. 1980. Measurement of foot-pedal loads during bicycling. Progress Report, Dept. of Mech. Eng., University of California, Davis.

DICKINSON, S. 1929. The efficiency of bicycle pedalling as affected by speed and load. Journal of Physiology 67:242-55.

GAESSER, G.A., and Brooks, B.A. 1975. Muscular efficiency during steady-rate exercise: effects of speed and work rate. Journal of Applied Physiology 38(6):1132-39.

GREGOR, R.J. 1976. A biomechanical analysis of lower limb action during cycling at four different loads. Unpublished doctoral dissertation, The Pennsylvania State University.

HOES, M.J.A.J.M., Binkhorst, R.A., and Smeedes-Kuyl, A.E.M.C. 1968. Measurement of forces exerted on pedal and crank during work on a bicycle ergometer at different loads. Internationale Zietschrift fur Angeandte Physiologie, Einschliesslich Arbeitsphysiologie 26:33-42.

LAFORTUNE, M.A., and Cavanagh, P.R. 1980. Force effectiveness during cycling. Medicine and Science in Sports 12(2):95.

III.
TRAINING

Load, Speed and Equipment Effects in Strength-Training Exercises

James G. Hay and James G. Andrews
University of Iowa, U.S.A.

Christopher L. Vaughan
University of Cape Town, South Africa

Kiyomi Ueya
Yamanashi University, Japan

Strength and strength development have been the subject of an enormous number of studies. With the notable exception of those devoted to the force-velocity and length-tension relationships in muscle, however, few of these studies have been concerned with biomechanical aspects of the subject. The contribution which biomechanics might make to our understanding of strength and strength training is thus largely unexplored.

It is widely-accepted (perhaps even universally-accepted) that the best way to increase the strength of a muscle is to subject it to near-maximum loads and to increase these loads progressively as the strength of the muscle increases. The assumption inherent in this procedure is that changes in the external load bring about corresponding changes in the demand placed on the muscles involved. This seemingly logical assumption has apparently been accepted in strength-training circles as an article of faith. As far as we have been able to ascertain, it has not been the subject of scientific investigation.

The recent development of isokinetic devices (which permit the performer to exert force against a bar moving at a fixed speed through a predetermined range) has revived interest in the role that the speed of exercise plays in strength development. Katch et al. (1975), Van Oteghan (1975), Davies (1977), and Gettman and Ayres (1978), for example, have all conducted studies in which isokinetic devices were used to evaluate the influence that the speed of exercise has on strength development.

Numerous 'isotonic' strength-training machines have been developed

Figure 1—Universal leg squat machine.

in recent years as alternative to the traditional barbells and dumbbells. Although these machines undoubtedly provide a more compact, convenient, and safer form of external loading than do barbells and dumbbells—and this, it might be argued, is all the justification they need—their existence has raised the inevitable question of whether they are also superior with respect to strength development.

The purpose of this study was to address each of the questions raised in the preceding paragraphs and, specifically, to determine the effects that variations in the external load, the speed of lifting, and the equipment used, have on the joint torques exerted at the hip, knee, and ankle joints during the performance of a squat exercise.

Procedures

Experimental Procedures

Three adult males with extensive experience in weight-training were used as subjects in the study. A barbell and a Universal Leg-Squat machine (see Figure 1) were used to provide the external loads against which the subjects worked.

The subjects were asked to perform a series of lifts using three different loads (40%, 60%, and 80% of their four-repetition free-speed maximum, 4RM) and three different speeds of lifting (slow, medium, and fast). The duration of the lifting phase of the exercise was set at 3 sec, 2 sec, and 1 sec, for the slow, medium, and fast speeds of lifting, respectively. The duration of the lowering phase was set at 2 sec in all cases.

The subjects performed lifts under a total of 18 different conditions of load, speed and equipment — 3 loads × 3 speeds × 2 pieces of equipment (barbell and machine) — over two days. Four repetitions were performed under each condition and the third of these was filmed with a high-speed motion-picture camera. The exposed film was then analyzed to determine the kinematic characteristics of the motion of each segment of the system.

Analytical Procedures

The subject-plus-equipment system was considered in each case to be planar and bilaterally symmetric and to consist of a series of rigid segments linked by frictionless pin joints. Free body diagrams were drawn and used as a basis for writing the equations of motion for each segment. These equations were then solved — using the known or estimated inertial characteristics of the subject-plus-equipment system and the observed kinematics of its motion — to yield the resultant torques at the hip, knee, and ankle joints of the subject.

Results

Joint Torques

The equations of motion for the upper body-plus-barbell segment shown in Figure 2 are:

$$\Sigma F_x = F_x^H = m\ddot{x} \tag{1}$$

$$\Sigma F_y = F_y^H - mg = m\ddot{y} \tag{2}$$

$$\Sigma M_z^G = -T_z^H + (F_x^H y_1) + (F_y^H x_1) = I_z^G \ddot{\theta} \tag{3}$$

where F_x^H and F_y^H are, respectively, the x and y components of the resultant joint force at the hips; x_1 and y_1 are the moment arms of these forces relative to a transverse axis through the center of gravity G of the segment; I_z^G is the moment of inertia of the segment relative to the same axis; and m, $\ddot{x}$, $\ddot{y}$, and $\ddot{\theta}$ are, respectively, the mass, the x and y components of the acceleration of G and the angular acceleration of the segment. These

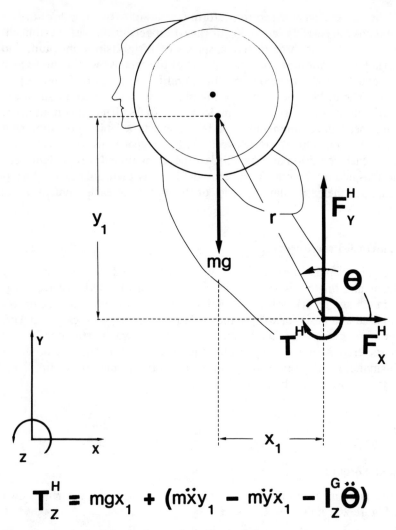

$$T_z^H = mgx_1 + (m\ddot{x}y_1 - m\ddot{y}x_1 - I_z^G\ddot{\theta})$$

Figure 2—Free body diagram of upper body-plus-barbell segment.

equations can be rearranged to yield the following expression for the torque at the hips:

$$T_z^H = mgx_1 + (m\ddot{x}y_1 + m\ddot{y}x_1 - I_z^G\ddot{\theta}) \qquad (4)$$

(The torque at the hips in the machine squat, and the torques at the knees and ankles in both barbell and machine squats, can be expressed in similar fashion.)

The effects that the accelerations of the segment have on the joint torque are reflected in the $(m\ddot{x}y_1 + m\ddot{y}x_1 - I\ddot{\theta})$, or inertial, terms of this equation. The contribution that these inertial terms made to the total hip torque was generally less than 10%, except at the ends of the lifting and lowering phases. At these points in the exercise, the contributions of the inertial terms were, as expected, substantially higher. Similar results were obtained for the inertial contributions to the knee and ankle torques.

The results obtained for the magnitudes of the contributions made by the inertial terms were generally small enough to permit the subject-plus-barbell (or machine) system to be considered quasi-static. This was especially true, of course, for those parts of the exercise—the middle parts of the lifting and lowering phases—where the accelerations were small.

Given that F_x^H and F_y^H are essentially constant under quasi-static conditions (the former approximately equal to zero, and the latter approximately equal to the weight of the segment above), the torque at the hips is almost completely dictated by the length of the moment arm x_1. Further, since x_1 is equal to $r(\cos\theta)$, where r is a constant and θ is an indicator of the inclination of the trunk, for any given load the torque at the hips is effectively determined by the inclination of the trunk. If the trunk is near vertical, the torque at the hips is small; if the trunk is well forward of the vertical, the torque at the hips is large.

Under these same quasi-static conditions, the torque at the knees is a function primarily of the torque at the hips and the angle of inclination of the thigh, with the torque at the hips usually the dominant influence. Although the torque at the ankles was clearly influenced by the torque at the knees, the sources of variation in the ankle torque were by no means as obvious as in the previous two cases. This is due in part, no doubt, to the smaller range of motion and smaller torques involved.

Effects of Load

The effects that variations in the external load have on the joint torques required to perform a squat exercise were first determined theoretically. For this purpose, the coordinate data obtained from film and a series of assumed values for the external load were run with a computer program to determine the corresponding values for the joint torques. Thus, for example, use of the coordinate data for the lifting phase of a machine squat performed with a 40% 4RM load and assumed loads of 60%, 80%, and 100% 4RM, yielded the computed values for the joint torques shown in Figure 3. These typical results indicated that the torques exerted at any given joint angle are essentially a linear function of the external load.

The effects that variations in the external load have in practice—as distinct from the effects observed in a theoretical analysis—were determined by simply having the subjects perform the lifts with different loads and analyzing their performances. Although increases in the external load were usually accompanied by increases in the torques recorded for the hip, knee, and ankle joints, the results obtained were not always con-

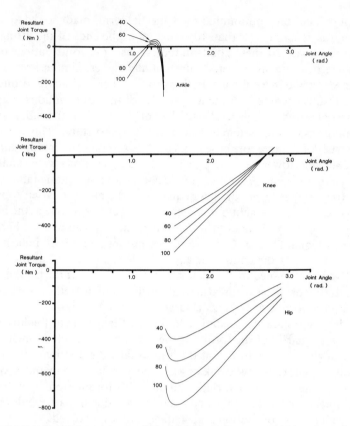

Figure 3 – Resultant-joint-torque vs. included-joint-angle curves showing theoretical effects of increasing the external load. Curves labelled 40 are for experimental data obtained when one subject performed a squat with the weight machine with a load equal to 40% of his 4-repetition maximum (4RM). Curves labelled 60, 80 and 100 are for theoretical data obtained by assuming the lift was actually performed with loads of 60%, 80%, and 100% of the subject's 4RM.

sistent with the theoretical analysis. The most noticeable departure from theory was in the torques recorded for the knee joint. Consider, for example, the knee joint torques shown as a function of joint angle in Figure 4. This figure shows the knee joint torques actually recorded during machine squats with 40%, 60%, and 80% 4RM loads and the knee joint torques that would have been recorded if the trials with the heavier loads had been performed in exactly the same manner as the trial with the lightest load. In this case, the muscles crossing the knee joint appear to have taken much less than their share of the increases in load from 40% to 60% and from 60% to 80% 4RM.

The muscles crossing the knee joint and those crossing the hip and ankle joints appeared to work in a compensating fashion. Thus, in this

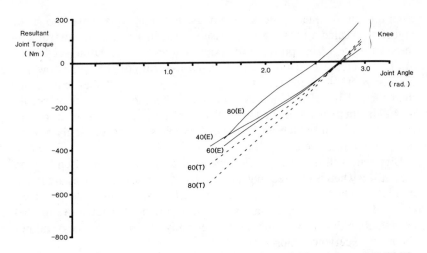

Figure 4—Comparison of knee joint torques for experimental (E) and theoretical (T) lifts by one subject performing a squat with the weight machine. Curves labelled 40, 60, and 80 are for lifts with 40%, 60% and 80% of the subject's 4RM.

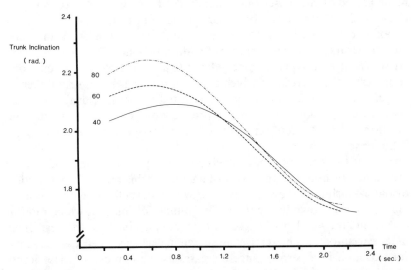

Figure 5—Variations in trunk inclination during the lifting phase of squats with the weight machine and loads of 40%, 60%, and 80% of the subject's 4RM.

case, and in several others like it, a less-than-expected increase in the torque at the knee joint was compensated for by a more-than-expected increase in the torques at the hips and ankles.

The kinematic characteristics of the motions that produce this shifting of load among the joints are of some importance from a practical standpoint. An examination of the kinetic and kinematic quantities used to compute the torque at the knees in the machine squat case shown in Figure 4 revealed a progressive increase in the inclination of the trunk at the start of the lifting phase as the external load was increased (see Figure 5). This increase in trunk inclination—a lowering of the trunk towards the horizontal—meant that the hip extensors assumed markedly more than their share of the increased load and the knee extensors markedly less. (The inclination of the trunk also appeared to play a central role in the other cases in which an increased load was shared unevenly among hip, knee and ankle extensors.)

Differences in the lifting technique used by a subject can often be seen very easily when presented in graphical form. These differences are much more difficult to detect in the real-life practical situation. Indeed, differences of a magnitude necessary to produce the results shown in Figure 4 are so subtle that they would almost certainly escape the notice of all but the most experienced observer.

There are some obvious ways in which one might reduce variability in the kinematics of a joint motion and thus improve the likelihood that an increase in external load will result in a corresponding increase in the demand placed on the muscles crossing that joint. One could select an exercise in which the only motion occurs at the joint of interest or, if this is impossible or undesirable for some reason, select that exercise in which motion occurs at the least number of additional joints. If two or more exercises involve an equal number of joints at which motions occur, one could select the exercise in which the performer has the least number of degrees of freedom. Finally, having selected the most appropriate exercise, one could take every available precaution to ensure that the starting position for each repetition of the exercise is the same in every respect.

Suppose, for example, that one wished to develop strength in the extensors of the knee by progressively increasing the external load. There are numerous exercises that might be used for this purpose. A selection of these—including the squat exercise with a barbell and with a weight-machine—are shown in Figure 6. The number of joints at which motion occurs and the number of degrees of freedom available to the performer in the execution of the exercise are shown in each case. Because the scope for variation in a performance increases as the number of these joints and the number of degrees of freedom increases, the seated knee extension (one joint, one degree of freedom) would appear to be the most suitable exercise and the squat with the barbell (three joints, three degrees of freedom) the least suitable exercise, for increasing the strength of the knee extensors. Further, and irrespective of which exercise is selected for use, the variation in performance from trial to trial will be reduced if the hips and feet are positioned consistently relative to the apparatus.

1 JOINT, 1 DEGREE OF FREEDOM

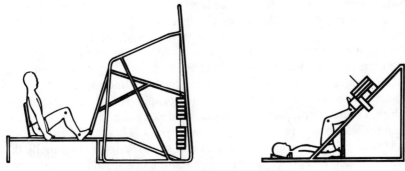

3 JOINTS, 1 DEGREE OF FREEDOM

3 JOINTS, 2 DEGREES OF FREEDOM

3 JOINTS, 3 DEGREES OF FREEDOM

Figure 6—The number of joints at which motion occurs and the number of degrees of freedom involved, in selected knee extension exercises. (Note: For the purpose of determining the number of degrees of freedom it is assumed: 1) that the performer does not introduce movements at joints other than those which the exercise was designed to involve, and 2) that, in those exercises where they press against the ground or some other surface, the soles of the feet remain firmly in contact with the surface throughout the exercise.)

Effects of Speed

Variations in the speed of lifting had different effects on the torques recorded at the hip, knee, and ankle joints. The torque at the hips was especially influenced by variations in the speed of lifting at the start of the lifting phase where it increased as the speed increased. The torque at the knees was generally little effected by variations in the speed of lifting. Finally, the torque at the ankles varied in two distinctly different ways as the speed of lifting varied. In the squat exercise performed with the barbell, the torque at the ankles — a plantar-flexor torque throughout — increased at the start of the lift as the speed of lifting increased but was otherwise little affected by variations in the speed. In the exercise with the machine, the dorsi-flexor torque recorded at the start of the lift and the plantar-flexor torque recorded at the end both varied little and in a rather inconsistent manner as the speed of lifting increased.

According to the force-velocity relationship for muscle, the force which a muscle, or a group of muscles, can exert at a fast speed of shortening is less than it can exert at a slower speed. Thus, if the torques required to perform the exercise at fast and slow speeds are essentially the same, the demand placed on the muscles at the fast speed is greater — that is, a greater percentage of their maximum — than it is at the slow speed. If the torques required to perform an exercise at a fast speed are greater than those required at a slower speed, the demand placed on the muscles at the fast speed is also greater than at the slow speed.

In the present study, the torques recorded at the fast speed were essentially the same as, or greater than, those recorded at the slower speeds. The fast speed of lifting was, thus, likely to have placed a greater demand on the muscles involved than did the slower speeds. With the possible exception of the torques at the hip joint at the start of the lift, however — which torques occasionally differed between speeds by 20% to 25% — the differences in the demand placed on the muscles would almost certainly have been relatively small. It would seem reasonable to assume, therefore, that any differences in the rate of strength development which might result from the use of a fast speed of lifting, rather than a slow speed, would also be small.

Effects of Equipment

The results obtained when the barbell and the weight-machine were used under comparable conditions differed in two major respects. The magnitudes of the extensor torques recorded at all three joints were substantially greater for the exercises performed with the machine; and, the general form of the joint torque-joint angle curves for the hip and ankle joints were decidedly different.

Because the torques recorded at the joints were primarily determined by the external load and the geometry of the motion, it is only logical to look to these factors for an explanation of the observed differences. The 4RM recorded for each subject was substantially greater for the lifts with the machine than for those with the barbell. Thus, when comparisons were made of the torques recorded at loads of, say, 80% of the 4RM, the external load in the machine case was considerably greater than in the barbell case. This difference favoring the machine was due in part to the fixed initial height of the shoulder pads of the machine—a height which especially favored the shorter subjects who did not assume as deep a squat position at the start of the lift as in the barbell case. It was also due in part, no doubt, to the subjects having greater confidence in their ability to maintain their balance and control the load when using the machine.

These two conclusions have important practical implications. The fixed initial height of the shoulder pads dictates to a large extent the position that the performer adopts at the start of the lift and, with short people, severely limits the range of motion at the joints during the lift. This is an important consideration with respect to the use of such machines by those whose statures are markedly less than that of the United States males for whom the machine was originally designed. The ideal solution to this problem would be to redesign the machine so that the starting position of the shoulder pads could be adjusted with the same ease as the load is adjusted in the present design.

The additional confidence which use of the machine appears to afford, and the increased loads that can be lifted as a consequence, implies that such use permits a greater demand to be placed on the extensor muscles of the hip, knee, and ankle joints than does the use of a barbell—given, of course, that the speed of lifting, ranges of joint motion, etc., are comparable. This finding would appear to be of considerable practical significance because it is generally accepted that working a muscle at close to its maximum capacity is a prerequisite to increasing its strength.

Conclusions

Several conclusions appear to be warranted on the basis of this study:

1. The torques exerted at the hip, knee, and ankle joints during the course of a squat exercise with a given load are primarily determined by the inclination of the trunk of the performer.

2. The joint torques exerted during a squat exercise are extremely sensitive to variations in lifting technique. For this reason, increasing the external load is not an effective way of increasing the demand placed on the extensors of these joints unless the kinematics of the motion are held constant from trial to trial.

3. Differences in the speed of lifting within the range used in this study

appear to make little difference in the demand placed on the muscles.

4. Heavier loads can be lifted with the machine than with the barbell. This appears to be due in part to the restricted range of motion permitted by the design of the machine. It might also be due in part to performers having greater confidence in their ability to maintain balance and to control the load when using the machine.

References

DAVIES, A.H. 1977. Chronic effects of isokinetic and allokinetic training on muscle force, endurance, and muscular hypertrophy. Ph.D. Dissertation, Temple University.

GETTMAN, L.R. and Ayres, J. 1978. Aerobic changes through 10 weeks of slow and fast-speed isokinetic training. Med. Sci. Sports 10:47.

KATCH, F.I., Pechar, G.S., Pardew, D., & Smith, L.E. 1975. Neuromotor specificity of isokinetic bench press training in women. Abstract, Med. Sci. Sports 7:77.

Total Telemetric Surface EMG of the Front Crawl

J.P. Clarys, C. Massez, M. Van Den Broeck,
G. Piette, and R. Robeaux
Vrije Universiteit, Brussel, Belgium

In 1935, Karpovich stated that 44 different muscles are active in swimming the front crawl and recently Weineck (1981) in his 'sportanatomy' book listed 30 muscles active in the front crawl. Between Karpovich and Weineck many other authors have made attempts to describe the muscle participation in this swimming stroke and most of these descriptions are based on empirical and functional anatomy.

Electromyographically, the front crawl has been studied by Ikai et al. (1964—15 muscles), Lewillie (1968; 1973—2 muscles), Vaday and Nemessuri (1971—20 muscles), Clarys et al. (1973—4 muscles), Maes et al. (1975—6 muscles), Belokovsky (1967—10 muscles), and Piette and Clarys (1975—5 muscles).

All these publications are very informative but do not allow for an overall view of the electrical activity during muscle action while swimming the front crawl. Due to different technical approaches and lack of pattern standardization, comparison is very difficult. Hence, the practical uses of these data are relatively small.

Using the methodological investigations of Lewillie (1967; 1968; 1971) and based on previously published preliminary research (Clarys et al., 1973; Maes et al., 1975; Piette and Clarys, 1975) an attempt was made to produce a total experimental image of all superficial body muscles presumed to be electrically active during the front crawl movements (excluding the smaller hand, feet, and head muscles). In order to make the data of practical use and provide the possibility for comparison of these data, the results are presented as normalized pattern diagrams based on the non-dimensional expression of integrated EMG patterns.

Procedures

Working with a telemetric transmission method, the electrical potentials were broadcasted from a transmitter, fixed on the subject, to an amplifier at the edge of the swimming pool and recorded via a two-channel recorder on a Elema-Schönander six-channel minograph 81. This test set up and its characteristics have been described by Lewillie (1967) and Clarys et al. (1973) and technically improved by Robeaux (unpublished material).

The surface electrodes were fixed with extreme care using two kinds of adhesive bandage (tensaplast and Leukoplast) before being covered with a waterproof (Stomasol 3M) adhesive disc and (Nobecutane) plastic varnish. Before fixation of the electrodes the 'motor point' of each muscle was defined. Unfortunately, the localization of these high potential points was not technically adequate for efficient identification in all muscles. Therefore, the electrodes were cutaneously fixed in the topographical mid-point of the muscle surface area, independent of the motor point localization and according to the recommendations of Basmajian (1967) and Goodgold (1974).

A total of 25 muscles were investigated covering the overall surface area of the human body. The selection criteria being described previously (Piette and Clarys, 1975), these muscles were: m. biceps brachii, m. triceps brachii, caput longum and caput laterale, m. deltoideus pars anterior, pars lateralis and pars posterior, m. extensor digitorum, m. trapezius sinistra and dextra, m. sterno-cleido mastoideus, m. obliquus externus sinistra and dextra, m. gastrocnemius caput laterale, m. rectus femoris, m. semi tendinosus, m. flexor carpi ulnaris, m. pectoralis major pars clavicularis and pars sternocostalis, m. latissimus dorsi, m. rectus abdominis pars superior and pars inferior, m. gluteus maximus pars superior and pars inferior, m. tibialis anterior.

There were 60 subjects who were studied (30 competitive swimmers and 30 swimmers with good technical skills). Swimming speed was standardized for all subjects by means of a series of successive lights, fixed 1 m below the water surface.

Before the actual data collection, the amplification apparatus was adapted to the signal intensity of each subject, while the actual EMG registration consisted of the integrated registration of: 1) the dynamic swimming contraction at a known speed of 1.7 m/sec^{-1} (D.C.), 2) the (relative) isometric maximal contraction (I.C.), and 3) a constant calibration value (C.V.).

The surface areas of the integrated patterns were also measured (Piette and Clarys, 1975). The expression of microvoltage results in terms of surface area simplifies further calculation and allows for a normalized dynamic contraction index:

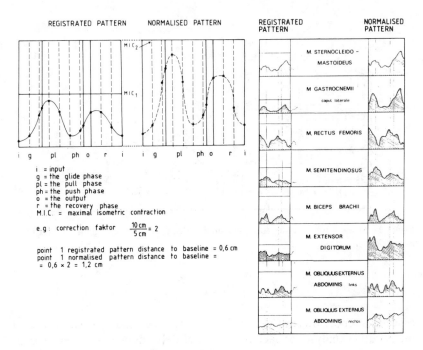

Figure 1 — Pattern normalization procedures.

$$[\text{N.D.C.} = \text{D.C. (cm}^2/\text{sec})/\text{C.V. (cm}^2/\text{sec})] \qquad (1)$$

and a normalized (relative) isometric maximum index

$$[\text{N.I.C.} = \text{I.C. (cm}^2/\text{sec})/\text{C.V. (cm}^2/\text{sec})] \qquad (2)$$

through which muscle activity can be presented as percentage of the isometric maximum

$$(\frac{\text{N.D.C.}}{\text{N.I.C.}} \times 100) \qquad (3)$$

This non-dimensional expression of integrated EMG patterns allows for a comparison of muscle activity between subjects of totally different swimming abilities, enabling one to establish an overall image of muscle contraction for the front crawl movement (see Figure 1).

For each part of the investigation the m. bicep brachii was used as a reference muscle in order to clarify the chronological order of the contractions within the different phases of the crawl movement. One arm cycle was divided according to the distribution of the movement pattern (Vaday and Nemessuri, 1971).

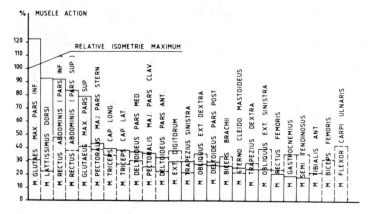

Figure 2 — Mean muscle activity equal swimming velocity as a percentage of the maximum isometric contraction.

Results

Since this report was not written in order to give detailed qualitative and quantitative analyses of each muscle separately, an overall review of contraction characteristics is shown in Table 1. All muscles had a greater or smaller constant activity during the swimming cycle regardless of swimmer technique. Twenty out of 25 muscles were labeled with two contraction peaks, generally during the input gliding or pull phase and the push phase, indicating a relative relaxation during the recovery phase. The variability of the m. deltoideus pars posterior relates to 'high' and 'wide' recoveries. The irregularity of the electromyogram of the leg muscles can be explained by the individual variations in percentages of mean muscle activity of the leg kick; a comparison of competitive and non-competitive swimmers is illustrated in Figure 2, indicating the extreme importance of the trunk and pelvis muscle activity in swimming the competitive front crawl.

It is common knowledge that the main propulsive force in swimming, as seen in the front crawl, is produced by the arm and shoulder action (De Goede et al., 1971). It appears, however, that the trunk muscles, including the m. gluteus maximus, clearly have a more important activity than the arm and shoulder musculature, confirming partial statements of Piette and Clarys (1975). Although comparable EMG patterns were found between both investigating groups, a difference was noticeable in the trunk-pelvis and leg muscle work intensity. All these observations stress the importance of correct use and specific training of muscles to improve performance in front crawl swimming.

As it was the main purpose of this study to produce reference data for practical use and comparison, the average results of each muscle for both

Table 1

Average Front Crawl Muscle Action as
Percentages of the Relative Maximal Isometric Contraction

Muscle	Most frequent N° of contraction + pattern	% muscle action/cycle competition swimmers	% muscle action/cycle non-competition swimmers
m. extensor digitorum	2	30.23	38.35
m. flexor carpi ulnaris	2	*	43.10
m. tricep caput laterale	3	38.78	28.92
m. tricep caput longum	2	39.47	35.64
m. bicep brachii	2	27.00	34.61
m. deltoideus pars ant.	2	33.12	23.43
m. deltoideus pars med.	2	37.33	31.97
m. deltoideus pars post.	3*	27.58	36.33
m. sterno cleido mast.	2	25.54	21.81
m. trapezius sinistra (respiration right)	2	29.91	22.57
m. trapezius dextra (respiration right)	1	24.51	29.54
m. latissimus dorsi	2	92.34	23.66
m. pectoralis major (clav.)	2	36.75	26.10
m. pectoralis major (sterno)	2	43.27	39.66
m. rectus abdominis (pars superior)	2	83.13	37.86
m. rectus abdominis (pars inferior)	2	91.96	48.33
m. obliquus ext. sinistra	2	24.49	39.92
m. obliquus ext. dextra	2	28.64	35.84
m. gluteus maximus (pars sup.)	2	79.52	41.40
m. gluteus maximus (pars. inf)	2	122.41	31.18
m. rectus femoris	2*	21.61	24.57
m. semitendinosus	2*	18.53	36.57
m. tibialis anterior	--*	--*	22.70
m. gastrocnemius cap. lat.	2*	19.13	36.13
m. bicep femoris	--*	--*	32.50

*These muscles show a highly variable contraction pattern.

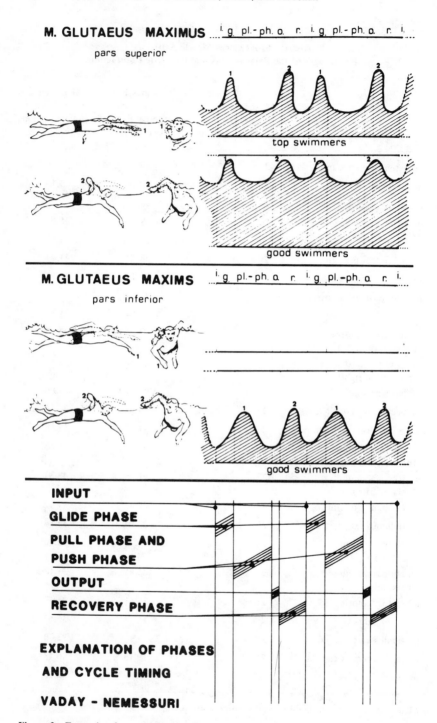

M. GLUTAEUS MAXIMUS
pars superior

top swimmers

good swimmers

M. GLUTAEUS MAXIMS
pars inferior

good swimmers

INPUT
GLIDE PHASE
PULL PHASE AND
PUSH PHASE
OUTPUT
RECOVERY PHASE

EXPLANATION OF PHASES
AND CYCLE TIMING

VADAY - NEMESSURI

Figure 3—Example of a normalized reference pattern diagram.

the competitive and non-competition level are schematically illustrated as 'normalized reference pattern diagram' in Figure 3. The data show a test repeated over two cycles, including contraction timing.

Referring to the different levels of muscle participation during total body movement, these normalized pattern diagrams allow for a better selection of specific 'dry land power training' exercises and it will provide the necessary information to the coach and swimmer to enable them to establish a specific training program. For this purpose, Olbrecht and Clarys (1980) have investigated a series of specific 'dry land power training' exercises using different power training devices (such as callcraft, roller board, isokinetics, spring systems, etc.).

Using the 'normalized reference pattern diagrams' as bases for analyses, a combination of 'dry and wet' EMG patterns was established.

Included in these analyses were: 1) an amplitude-differential curve, 2) a slope study and direction coefficient data and the influence of amplitude and time shifting on both dry and wet muscle activity.

The results of this validation study of dry power training will be presented at the 4th International Symposium on Biomechanics on Swimming (June '82, Amsterdam, the Netherlands).

References

BASMAJIAN, J.V. 1967. Muscles Alive. The Wilkins and Williams Co., Baltimore.

BELOKOVSKY, V.V. 1967. An analysis of pulling motions in the crawl arm stroke. In: L. Lewillie and J.P. Clarys (eds.), First International Symposium on Biomechanics in Swimming, pp. 217-221. Université Libre de Bruxelles, Brussels.

CLARYS, J.P., Jiskoot, J., and Lewillie, L. 1973. A kinematographical, electromyographical and resistance study of water polo and competition front crawl. In: S. Cerquiglini, A. Venerando, and J. Wartenweiler (eds.), Biomechanics III, pp. 446-452. S. Karger Verlag, Basel.

DE GOEDE, H., Jiskoot, J., and Van der Sluis, A. 1971. Over stuwkracht bij zwemmers (On the propulsion of swimmers). De Zwemkroniek 48(2):77-90.

GOODGOLD, J. 1974. Anatomical Correlates of Clinical Electromyography. The Williams & Wilkins Co., Baltimore.

IKAI, M., Ishii, K., and Miyashita, M. 1964. An electromyographical study of swimming. Res. J. Phys. Educ. 7:47-54.

KARPOVICH, P.V. 1935. Analysis of propelling force in the crawlstroke. Res. Quart. 6:49-58.

LEWILLIE, L. 1967. Analyse télémetrique de l'electromyogramme du nageur (Telemetric analysis of the electromyogram of the swimmer). Trav. Soc. Med. Belge l'Educ. Phys. Sport, fasc. 20:174-177.

LEWILLIE, L. 1968. Telemetrical analysis of the electromyograph. In: J. Wartenweiler, E. Jokl, and M. Hebbelinck (eds.), Biomechanics I, pp. 147-148. S. Karger Verlag, Basel.

LEWILLIE, L. 1971. Quantitative comparison of the electromyograph of the swimmer. In: L. Lewillie and J.P. Clarys (eds.), First International Symposium on Biomechanics in Swimming, pp. 155-159. Université Libre de Bruxelles, Brussels.

LEWILLIE, L. 1973. Muscular activity in swimming. In: S. Cerquiglini, A. Venerando, and J. Wartenweiler (eds.), Biomechanics III, pp. 440-445. S. Karger Verlag, Basel.

MAES, L., Clarys, J.P., and Brouwer, P.J. 1975. Electromyography for the evaluation of handicapped swimmers. In: J.P. Clarys and L. Lewillie (eds.), Swimming II, pp. 268-275. International Series on Sports Sciences, Vol. 3. University Park Press, Baltimore.

OLBRECHT, J., and Clarys, J.P. 1980. Specifieke krachttrainingsoefeningen voor de borstcrawl: concept of realiteit? (Specific power training exercises for the front crawl: concept or reality?). Unpublished Masters thesis, Vrije Universiteit Brussel.

PIETTE, G., and Clarys, J.P. 1975. Telemetric EMG of the front crawl movement. In: J. Terauds and W. Beddingfield (eds.), Swimming III, pp. 153-159. International Series on Sport Sciences, Vol. 8. University Park Press, Baltimore.

ROBEAUX, R. Unpublished material.

VADAY, M., and Nemessuri, N. 1971. Motor pattern of free style swimming. In: L. Lewillie and J.P. Clarys (eds.), First International Symposium on Biomechanics in Swimming, pp. 1961-1971. Université Libre de Bruxelles, Brussels.

WEINECK, J. 1981. Sportanatomie Beitrage zur Sportmedizin, Band 9. Perimed Fachbuch — Verlaggesellschaft, mbh, D-8520 Erlangen.

The Effect of Arm Elevation on
Changes in Muscle Work Capacity with Training

T. Fukunaga, K. Hyodo,
T. Ryushi, A. Matsuo, H. Yata, and M. Kondo
University of Tokyo, Tokyo, Japan

When the position of an arm or leg is altered in a vertical plane, the blood circulation in the extremity is affected by the change in hydrostatic pressure. For example, muscle blood flow in the arm decreases with arm elevation because of decrement of hydrostatic or transmural pressure in the vessels (Åstrand et al., 1968; Hartling et al., 1976). The changes in blood circulation with the muscle may have an influence upon work performance and also upon effects of training.

The purpose of this study was to compare the differences of training effects with the arm positioned at heart level and with the arm elevated vertically.

Method

Healthy men (n = 9), aged 20 to 22 yrs, volunteered to act as subjects. All subjects were required to perform dynamic hand grip contractions by raising a weight equivalent to ⅓ of maximum strength up to 2 cm high at a rate of 30/min with a metronome until exhaustion.

Training exercises were performed at two different arm positions. Four subjects were trained with the arm vertically elevated (U-group), and another five subjects with the arm at heart level (L-group). Training was performed 3 times/week for 3 months. Before and after the training, maximum work performance, blood flow, oxygen uptake, and blood lactate were measured at both arm positions. The forearm blood flow was determined by venous occlusion plethysmography using Whitney's strain gauge. Venous blood samples of 5 ml were taken with teflon catheters by inserting them percutaneously into the median cubital vein which drains the deep tissues. The recordings of blood flow and collections of venous blood samples were done in a rest condition and at the 1st, 2nd, 5th, and 10th mins of the recovery period. O_2 and CO_2 contents of the blood samples were determined by the manometric method designed by Van Slyke and Neil (1924) and pH, pO_2, and pCO_2 determined by IL meter. The calculation of forearm oxygen uptake was performed as follows: forearm oxygen uptake (ml/100 ml • min) = blood flow (ml/100 ml • min) × a-v O_2 difference (vol %). An arterial blood sample of 5 ml was taken from the brachial artery during the resting period. Blood lactate concentration was determined by the enzymatic method (Lactate-UV-Test).

Results and Discussion

The maximum number of contractions in the hand grip exercise with the arm elevated vertically was 22% lower than that with the arm at the heart level. Exercise blood flow decreased by 43% in the elevated arm. On the other hand, the oxygen uptake of the forearm was not affected by arm elevation because of the increase in the a-v O_2 difference which was estimated to be 2.5 ml/100 ml • min. Lactate concentration of venous blood in arm elevated position was 35.6% higher than that for the heart level condition. Hartling et al. (1976) reported that resting leg muscle blood flow decreased by about 50% in a body position with the legs elevated. With arm elevation, a decrease of forearm resting blood flow accompanied with an increment in the a-v O_2 difference resulting in constant oxygen uptake was observed by Holling (1957). For the exercising forearm muscle, Wahren (1966) reported that in the elevated arm, the blood flow and deep venous oxygen saturation were lower and lactate production was larger than with the arm horizontal. These are in agreement with the observations of the present study. Therefore, it may be considered that the lower work performance at the elevated arm position in spite of the same oxygen uptake is attributed to the increased lactate production and decreased dilution because of lowered blood flow.

The rates of increase of blood flow, a-v O_2 difference, oxygen uptake, and number of contractions due to muscle training were compared in both training groups, as shown in Figure 1. Both oxygen uptake and number of contractions increased more in the U-group than in the L-group. The blood flow indicated the same increments in both groups for the exercise test. It was observed that of the increment of blood flow affected by training in the arm elevation the U-group was remarkably larger than the L-group. Such increase in exercise muscle blood flow caused by training as seen in this study corresponds to a previous report by Taguchi (1969). That is, the increase of blood flow is due to capillarization of the muscle, which raises the improvement of endurance

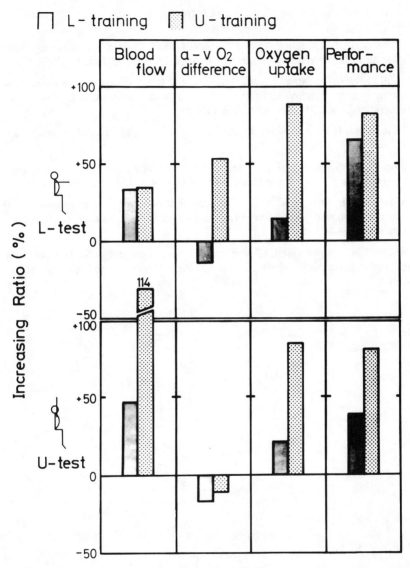

Figure 1—Rates of increase of blood flow, a-v O₂ difference, oxygen uptake, and performance by two different types of muscle training.

capacity of muscle. In the present study, it was of interest to note that more improved capillarization in the forearm muscle was induced by the training with the arm elevated vertically.

References

ÅSTRAND, I., Guharay, A., and Wahren, J. 1968. Circulatory responses to arm exercise with different arm positions. J. Appl. Physiol. 25(5):528-532.

HARTLING, O., Noer, I., and Trap-Jensen, J. 1976. Leg muscle blood flow during reactive hyperemia. Effect of different body positions, and of subatmospheric pressure. Pflügers Arch. 366:131-135.

HOLLING, H.E., and Verel, D. 1957. Circulation in the elevated forearm. Clin. Science 13:197-213.

TAGUCHI, S. 1969. Training effect of muscular endurance with respect to muscle oxygen intake in human forearm (in Japan). Res. J. Phys. Educ. 14:19-27.

VAN SLYKE, D.D., and Neill, J.M. 1924. The determination of gases in blood and other solution by vacuum extraction and manometric measurement. J. Biol. Chem. 61:523-573.

WAHREN, J. 1966. Quantitative aspects of blood flow and oxygen uptake in the human forearm during rhythmic exercise. Acta Physiol. Scand. 67:1-93, Suppl. 269.

Comparison of Sprint Running
in the Trained and Untrained Runners
with Respect to Chemical and Mechanical Energy

Mitsuru Saito

Toyota Technological Institute, Nagoya, Japan

Tetsuo Ohkuwa

Nagoya Institute of Technology, Nagoya, Japan

Yasuo Ikegami and Miharu Miyamura

Nagoya University, Nagoya, Japan

It has already been reported that individual differences in sprint running are related to differences in muscle fiber distribution and energy supply (Gollnick and Hermansen, 1973; Costill et al., 1976; Komi et al., 1977; Thorstensson, 1977). Performance in sprinting, however, depends not only on the rate of energy supply, but also on the conversion process of mechanical energy which produces the propelling force efficiently. The present study, therefore, was undertaken to elucidate the differences in performance in 400 m sprinting in trained and untrained runners with respect to chemical energy and mechanical work.

Methods

Five trained runners and three untrained male students participated in this study as subjects. Their mean age, height, weight, and maximum oxygen uptake per kg of body weight were 19.8 yr, 170.6 cm, 56.4 kg, and 63.9 ml/kg • min for the trained group, and 21.0 yr, 170.3 cm, 59.5 kg, and 50.8 ml/kg • min for the untrained group, respectively.

After each subject rested in the supine position for 40 min, warm-up and 400 m sprinting were performed on the en-tout-cas 400 m track. The subject again rested in a supine position for 70 min after the 400 m sprinting. In order to measure oxygen uptake and oxygen debt, expired gas was collected in Douglas bags at rest, during exercise, and during recovery. Gas analysis was carried out with an O_2 analyzer and an infrared CO_2 analyzer.

Blood samples of about 5 ml were drawn from the antecubital vein at rest, during warm-up, and during recovery. Blood lactate (LA) and creatine phosphokinase activity (CPK) were determined by means of an enzymatic method (Hohorst, 1962; Rosalki, 1967).

The running form for each subject at 170 m (hereafter called Condition I) and 350 m (hereafter called Condition II) from the starting point was filmed at 100 frames/sec by a 16 mm cine camera (Photo-sonics, 16 mm-1P, USA). Film analysis was performed on a film motion analyzer (NAC, MC-OB, Japan). The center of gravity during running was calculated using Matsui's method (Matsui, 1958). Furthermore, the steps during 400 m sprinting were obtained from 8 mm film which was taken at a camera speed of 24 frames/sec by an 8 mm cine camera (ELMO, 8S-60, Japan). Since speed and height of center of gravity vary for one step, mechanical work for one step under Conditions I and II were calculated from the kinetic energy of the center of gravity ($\frac{1}{2}mv^2$) and the change in potential energy (mgh) for one step as follows:

$$\text{mechanical work/step} = (\tfrac{1}{2}mv^2 + mgh)_{max} - (\tfrac{1}{2}mv^2 + mgh)_{min} \quad (1)$$

where m = body mass, v = velocity of center of gravity, g = gravity acceleration (9.8 m/sec^2), and h = height of center of gravity.

Results

Figure 1 shows the results of the mean velocity, oxygen uptake during running, and oxygen debt both in the trained and untrained groups. Average values of the mean velocity in 400 m sprinting were 7.14 m/sec for the trained group and 5.47 m/sec for the untrained group. This difference was significant statistically ($p < 0.001$). But there was no difference between the two groups in oxygen uptake during sprinting or in oxygen debt.

Peak values of blood lactate (LA) and creatine phosphokinase activity (CPK) during recovery are indicated in Figure 2. Peak blood lactate after 400 m sprinting was higher in the trained group than in the untrained group, but the difference was not statistically significant. Peak value of CPK was significantly ($p < 0.01$) higher in the trained as indicated in Figure 2.

Table 1 demonstrates the step length, step frequency, mean velocity of the center of gravity, and the mechanical work per step under Conditions I and II. The step length, step frequency, and velocity of the center of

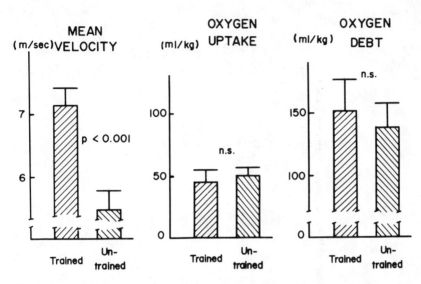

Figure 1 – Average values of mean velocity, oxygen uptake per kg of body weight, and oxygen debt per kg of body weight in 400 m sprinting in trained and untrained groups.

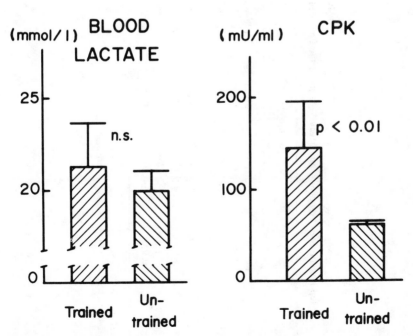

Figure 2 – Average value of peak blood lactate and creatine phosphokinse activity after 400 m sprinting in trained and untrained groups.

Table 1

Step Length, Step Frequency, and Mean Velocity of the Center of Gravity and Mechanical Work per Step at Condition I and II

Group	Condition I				Condition II			
	Step length (m)	Step frequency (steps/sec)	Mean velocity (m/sec)	Mechanical work (joule)	Step length (m)	Step frequency (steps/sec)	Mean velocity (m/sec)	Mechanical work (joule)
Trained n = 5	2.06 ±0.09	3.85 ±0.12	8.01 ±0.38	264 ± 23	1.84 ±0.12	3.51 ±0.10	6.57 ±0.30	249 ± 35
Untrained n = 3	1.82* ±0.02	3.49^{ns} ±0.29	6.18** ±0.18	327^{ns} ± 62	1.50* ±0.14	3.08** ±0.16	4.72** ±0.56	243^{ns} ± 76

Values are mean ± SD

* $p < 0.05$; ** $p < 0.01$; ns = not significant

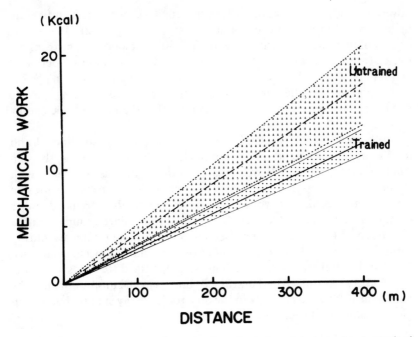

Figure 3—Relationship between distance and mechanical work in trained and untrained groups of runners.

gravity under Condition I were greater in the trained group. In Condition II, the mean value for each variable decreased as compared with Condition I.

Discussion

The energy needed for muscular work is derived from both aerobic and anaerobic energy processes. In this experiment, there were no significant differences between trained and untrained groups in oxygen uptake during 400 m sprinting or oxygen debt (see Figure 1). These results seem to indicate that energy supply from aerobic sources was nearly the same during 400 m sprint running for both groups.

Energy during heavy exercise is also derived from the splitting of energy rich substrates, i.e., glycogen, adenosine triphosphate (ATP) and creatinephosphate (CP). Although the amounts of ATP, CP, and lactate in the muscle were not determined, it was revealed that there was no significant difference between the two groups in peak blood lactate as shown in Figure 2. Therefore, the differences in performance in 400 m sprinting cannot be explained by the differences in energy obtained from glycolysis associated with the formation of lactic acid. It is of interest, however, that peak values of CPK in the trained group were significantly higher than those in the untrained group (see Figure 2). If creatine phosphokinase activity which is effused into the blood stream from active cells reflects the energy status of the cell as pointed out by Berg and Haralambie (1978), it is possible to assume that the differences in performances between trained and untrained runners in the sprinting may be due to the differences in CP and/or ATP rather than the aerobic energy process.

On the other hand, mechanical work per step under Condition I and total steps for the 400 m were less in the trained group, whereas, the step length under Condition II was greater in the trained as compared with the untrained group. If it is assumed that the subjects run 400 m with step lengths and mechanical work per step similar to that obtained under Condition I, mechanical (external) work in the untrained group increased with increased running distance (see Figure 3) because total work during sprinting is the product of the number of steps and the mechanical work per step. From these results, it is suggested that the differences in performances between trained and untrained runners in 400 m sprinting may be due to the difference not only in the anaerobic energy supply, but also in the mechanical efficiency of sprinting.

References

BERG, A., and Haralambie, G. 1978. Changes in serum creatine kinase and hexose phosphate isomerase activity with exercise duration. Eur. J. Appl. Physiol. 39:191-201.

COSTILL, D.L., Daniels, J., Evans, W., Fink, W., Krahenbuhl, G., and Saltin, B. 1976. Skeletal muscle enzyme and fiber composition in male and female track athletes. J. Appl. Physiol. 40:149-154.

GOLLNICK, P.D., and Hermansen, L. 1973. Biochemical adaptation to exercise: Anaerobic metabolism. In: J.H. Wilmore (ed.), Exercise and Sport Sciences Reviews 1, pp. 1-43. Academic Press, New York.

HOHORST, H.J. 1962. Methods of Enzymatic Analysis. Weinheim, West Germany. pp. 266-270.

KOMI, P.V., Rusko, H., Vos, J., and Vihko, V. 1977. Anaerobic performance capacity in athletes. Acta Physiol. Scand. 100:107-114.

MATSUI, H. 1958. Exercise and center of gravity. Taiku no Kagakusha, Tokyo, pp. 21-42. (in Japanese)

ROSALKI, S.B. 1967. An improved procedure for serum creatine phosphokinase determination. J. Lab. Clin. Med. 69:696-705.

THORSTENSSON, A. 1977. Muscle strength and fiber composition in athletes and sedentary men. Med. Sci. Sports. 9:26-30.

IV.
MODELLING
AND
SIMULATION

A.
Modelling

A Model of Bending Man for the Description of His Motor Performance of Body Sway

Ryoichi Hayashi, Akihide Miyake,
Satoru Watanabe, and Kazumi Umemoto
Gifu University School of Medicine, Japan

A gravicorder and some similar devices are used to measure the horizontal displacement of the center of gravity. The position measured by the force-platform and that of the line of gravity estimated by photographic records are quite different during dynamic activities (Murray et al., 1967) and this discrepancy has been studied by analyses of a model of the human body (Gurfinkel, 1973; Geursen et al., 1976). Such studies have suggested that it is the activity of muscles which may produce the difference, but the direct examination of muscular activity was not undertaken.

In this paper these problems were studied by relating the leg muscle activities with the magnitude and direction of the vertical supportive force 1) during quick body sway responding to auditory signals, and 2) when electrical stimuli were applied to the tibial or common peroneal nerve. The experimental results are discussed by comparing them with an analysis of the single reversed pendulum.

Theoretical Consideration and Experiments

Standing human posture has been theoretically treated as a single or double reversed pendulum (Gurfinkel, 1973; Geursen et al., 1976). The bending posture may be considered as a more appropriate posture than the erect standing when it is treated as the single reversed pendulum. Therefore, our subject bent forward with the knees "locked" and attempted to maintain a constant angle between the trunk and hip joint. The model consists of a weightless rod (length $= \ell$), with a point mass (m) at the top

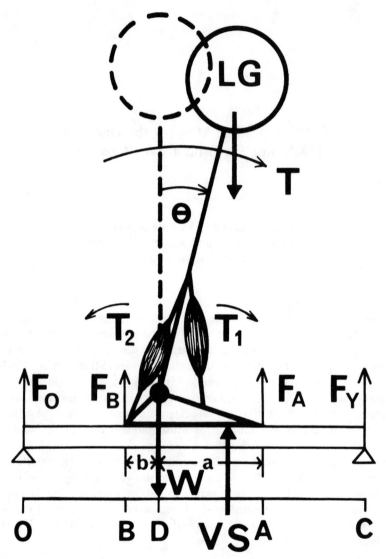

Figure 1 — One-link reversed pendulum. F = a reaction force at each point; T = a moment; θ = angle; LG = vertical projection of the center of gravity of mass; VS = vertical supportive force on platform; W = total weight at the axis.

and with the rotation axis perpendicular to the rod at the other end (Figure 1).

Experiments

Eight normal men (20 to 25 years old) were studied. The subject stood on the gravicorder with bare feet placed parallel to each other, and positioned 5 to 6 cm apart. Fmax shall be used hereafter to express the position of the body's center of gravity when the bending subject swayed maximally forwards. NP represents the neutral position at which the soleus muscle showed tonic activity but the tibialis anterior muscle was electrically silent.

In the first condition, the subject was asked to sway his body as quickly as possible from NP to Fmax or from Fmax to NP, following an auditory signal. In the second condition, the subject was asked to maintain the bending posture at NP against the muscle contraction evoked by electrical stimuli applied unilaterally to the tibial or common peroneal nerve. Stimulus duration was 1 msec. Stimulus strength was adjusted to evoke M-response without any larger H-response. The EMGs of the soleus and tibialis anterior muscles were taken bilaterally with surface electrodes, and at the same time the angle at the ankle joint was measured to estimate the excursion of the body's center of gravity.

Model

The single reversed pendulum is joined to the rigid stand which touches the plate at only two points, A and B. The reaction forces at each point are F_a and F_b. T is the total clockwise moment around the axis. T_1 and T_2 are the magnitudes of the clockwise and counter-clockwise moments, respectively. F_o and F_y are the reaction forces at fulcra ($\overline{OC} = L$) for the platform. Θ is the angular deviation from the vertical line and is positive in a clockwise direction. In this paper, we consider only the vertical component of the supportive forces reacting on the platform.

Suppose Y is the distance from the origin (0 point in Figure 1) to the position of the vertical supportive force (VS) on the platform:

$$Y = \frac{A \cdot F_a + B \cdot F_b}{W} ; \text{ where } W = m\,(g - \ell\ddot{\theta}\sin\theta - \ell\dot{\theta}^2\cos\theta) \qquad (1)$$
$$|W| = |F_a + F_b| = |F_o + F_y|$$

Let the equation of motion of the pendulum be

$$m\ell^2\ddot{\theta} - mg\ell\sin\theta = T, \text{ where } T = T_1 - T_2 \qquad (2)$$

Under the condition that the stand is immobile, and according to the balance condition around the points A and B, two equations hold:

$$-aW + (a + b)F_b = T \qquad (3)$$

$$-bW + (a + b)F_a = -T \qquad (4)$$

Substitute (3) and (4) for (1):

$$Y = D - \frac{T}{W}, \text{ where } D = \frac{Ab + Ba}{a + b} \tag{5}$$

Substitute (2) for (5):

$$Y = D + \frac{mg}{W}\left(\ell\sin\theta - \frac{\ell^2\ddot{\theta}}{g}\right) \tag{6}$$

| position of the axis | position change dependent on the shift of the mass | position change dependent on the acceleration |

From the balance condition around the origin and from the above equations:

$$L \cdot F_y = W \cdot Y; \text{ where } |W| = |VS| \tag{7}$$

$$F_y = \frac{D \cdot W}{L} + \frac{mg}{L}\left(\ell\sin\theta - \frac{\ell^2\ddot{\theta}}{g}\right) \tag{8}$$

In our experiments, the maximum ratio of fluctuation of W on body weight (mg) is less than 0.05.

Results and Discussion

Figure 2 illustrates one of the records obtained when the subject swayed his body following an auditory signal. An undershoot and an overshoot were observed in the records of Y and F_y at the beginning and at the end of the sway. By contrast, these fluctuations were not observed in the excursion of the angle around the ankle joint. In the initial phase of body sway from NP to Fmax, activity in the soleus muscle decreased while that in the tibialis anterior muscle was initiated. At the end of sway, the activity of the soleus muscle exceeded that of static holding at the Fmax position and then decreased to the static level; during this time the tibialis anterior muscle was almost silent. It may be considered that at the beginning of sway, these muscle activities correspond to the increase in T caused by an increase in T_1 and decrease in T_2 and from the equation (5) Y decreased. At the end of sway, muscle activities corresponded to the decrease in T through the decrease in T_1 and increase in T_2, and Y increased. The reverse was observed during the body sway from Fmax to NP. From these results, it is suggested that the muscular activities produce the body's forward or backward acceleration and cause the under- and over-shoot in the record of Y.

The angular velocity measurements showed characteristic triangular-shaped waves. These waves would seem to be identical to those observed in the rapid tracking experiments of the extremities (Smith, 1962; Wilde & Westcott, 1963). The characteristic triangular-shaped wave in the ve-

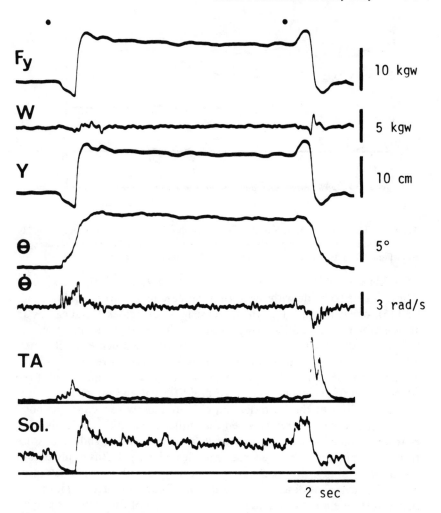

Figure 2 — Normal male responding to a sound signal (indicated by the dots) and swaying his body in an anterior-posterior direction. Fy = reaction force measured at the anterior fulcrum of gravicorder; W = total reaction force measured by gravicorder; Y = the position of the vertical supportive force; θ = angle of ankle joint; $\dot{\theta}$ = velocity of body movement. In each trace, the upper deflection indicates the following: Fy and W, increase of force; Y, anterior direction; θ, clockwise direction. The lower two traces show the integrated activities of the tibialis anterior (TA) and soleus (Sol.) muscles.

locity measurement shows that a stepwise change occurred in acceleration from positive to negative or vice versa. This suggests that the body sway is performed by the same kind of control process as that of a bang-bang servo during quick movement.

Figure 3(A) shows VS responses when the soleus muscle was contracted by an electrical stimulus applied to the tibial nerve. The muscle contraction caused a transient deflection of Y anteriorly, while the angle

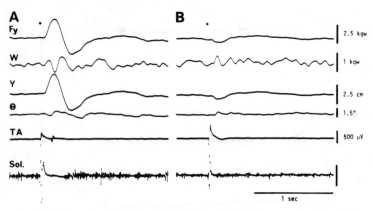

Figure 3—The effect of muscle contraction on supportive forces. Stimuli applied to the tibial (A) and common peroneal (B) nerve are marked by dots. Abbreviations are the same as in Figure 2, but muscle activities represent EMG data.

of ankle joint (Θ) decreased and the body shifted slightly backwards. These events coincide with the abrupt increase in T_2. From equations (2) and (5), one can notice that the increase in T_2 decreases T and causes an increase in Y. Figure 3(B) shows a response in the opposite direction was evoked by the electrical stimulation on the peroneal nerve which innervates the tibialis anterior muscle. This response corresponds to the abrupt increase in T_1. These observations show that the direction of the acceleration of body and that of VS are different and opposite.

The activities of the lower leg muscles are closely related to the position of the center of gravity when the subjects maintain statically each position from posterior to anterior in a leaning or a bending posture (Miyake et al., 1977). The recorded muscle activity patterns imply that in the present model, the acceleration is negligible and Y is a function of angle (Θ) such as is described in equations (2) and (6). This relationship between the leg muscle activities and the position of the center of gravity may be considered as a basic mode of postural control. In the quick body sway, the activities of the soleus and tibialis anterior muscles were switched on and off out of phase with each other. Therefore, it is concluded that the leg muscles are controlled and programmed in such a way as to allow the agonist-antagonist cooperation according to the mode of performance, thereby producing more effective movements.

This study has indicated that the line of gravity and the position of VS are distinctly different entities and these differences are produced by muscle activities. It is suggested that through the combined use of gravicorder records and measurements of muscle activities, meaningful experiments are conducted for the study of postural control and motor performance during quasi-static standing or dynamic actions.

Summary

The vertical supportive force (VS) was measured with a gravicorder during body sway in eight normal males. The activities of lower leg muscles and the angle of the ankle joint were recorded simultaneously. To enable the development of a simplified model of a standing human, represented by a single reversed pendulum, the subject was asked to bend forward without flexing the knees and to minimize movement between the upper part of the body and the legs. In response to auditory signals, the subject rapidly swayed his body around the ankles in an anterior-posterior direction. The results obtained have been compared with an analysis of the model.

Excursions were distinctly different between VS and the body's center of gravity; the former fluctuated while the latter moved smoothly. These fluctuations were attributed to the direction of body acceleration produced by the leg muscles. This conclusion was confirmed by examinations of 1) the VS responses to the muscle contraction evoked by electrical stimuli, and 2) results of model analysis. In quick sway, the shape of the 1st differential of the ankle joint's angle was triangular, and from this pattern, estimations of the direction of the acceleration indicated stepwise changes from positive to negative or vice versa. These results suggested that a quick body sway is performed as a maximum-effort minimum-time optimum bang-bang control.

References

GEURSEN, J.B., Altena, D., Massen, C.H., & Verduin, M. 1976. A model of the standing man for the description of his dynamic behaviour. Agressologie 17(B):63-69.

GURFINKEL, V.E. 1973. Physical foundations of stabilography. Agressologie 14(C):9-14.

MIYAKE, A., Watanabe, S., Jijiwa, H., Yamaji, K., & Komachi, K. 1977. EMG and fluctuation of center of gravity with posture in man. Electroenceph. Clin. Neurophysiol. 43:596-597.

MURRAY, M.P., Seireg, A., & Scholz, R.C. 1967. Center of gravity, center of pressure, and supportive forces during human activities. J. Appl. Physiol. 23(6):831-838.

SMITH, O.J.M. 1962. Nonlinear computations in the human controller. IRE. Trans. Bio-Med. Electron. 9:125-128.

WILDE, R.W., & Westcott, J.H. 1963. The characteristics of the human operator engaged in a tracking task. Automatica 1:5-19.

Mathematical Model and Lagrangian Analysis for the Dynamics of the Human Body in the Crawlstroke

M.A. Bourgeois and L.A. Lewillie
Université Libre de Bruxelles, Bruxelles, Belgium

Various authors have tried to define forces and torques necessary for propulsion in the crawlstroke based upon Newtonian mechanics (Seireg and Baz, 1971; Seireg, Baz, and Pattel, 1971; Belokovsky and Kuznetsov, 1976). Other studies based on the electromyographic activity have tried to show qualitatively the muscular behavior produced by the swimmer (Lewillie, 1971; Lewillie, 1973; Tokuyama, Okomoto, and Kumamoto, 1976).

The mathematical model presented in this study utilizes the synthesis of both types of investigations. This synthesis is made possible by the investigation methods developed in our laboratory and based on the Lagrangian dynamics. Indeed this approach seems to be the most appropriate to the dynamic study of human motions (Hatze, 1977). This mathematical model shows qualitatively as well as quantitatively:

1. The variation of the generalized forces applied to a system with 2° of freedom, composed of two solid homogeneous cylinders, which are, as shown in Figure 1, on one side the propulsive arm in crawl and on the other side the rest of the human body;

2. The variation in muscular torques applied on the arm in its underwater trajectory, as well as of the muscular torques necessary to maintain the trunk in the horizontal position in crawl swimming;

3. The forces and hydrodynamic torques undergone by the swimmer at various speeds.

The results obtained by this method, however, are based on a mathematical simulation of swimming and not on experimental data.

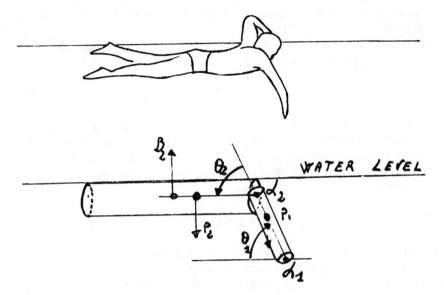

Figure 1 — Model for arm and trunk rotation in sagittal plane.

Mathematical Model

The general equation of the dynamics of the system is stated by the Lagrangian relation:

$$\frac{d}{dt}\left(\frac{\delta T}{\delta \dot{\theta}_i}\right) - \frac{\delta T}{\delta \theta_i} + \frac{\delta V}{\delta \theta_i} = \Gamma_i \tag{1}$$

in which T and V are, respectively, the kinetic energy and the potential energy of the system. The angle θ_i, in generalized coordinates, are shown in Figure 1. Γ_i is the generalized torque applied to each body and it appears clearly that it can be established on the basis of the kinematic parameters of the model. The simulation of the angular displacements θ_i is established on the basis of a Fourier's series, i.e.:

for $0 \leqslant \frac{t}{T} \leqslant \phi$, \hfill (2)

$$\theta_i(t) = \xi_0 + (\xi_1 - \xi_0)\left[\frac{t}{T} - \frac{\phi}{\pi} \sin\left(\frac{\pi t}{\phi T}\right)\right]$$

For $\phi < \frac{t}{T} \leqslant 1$,

$$\theta_i(t) = \xi_0 + (\xi_1 - \xi_0)\left[\frac{t}{T} + \left(\frac{1 - \phi}{\pi}\right) \sin\left\{\frac{\pi\left(\frac{t}{T} - \phi\right)}{1 - \phi}\right\}\right]$$

in which ϕ is the relative value of (t) for $\ddot{\theta}(t) = 0$ in relation to the total time (T) of the motion. ξ_0 and ξ_1 are the θ_i values for $t = 0$ and $t = T$, respectively.

The generalized torques Γ_i of Equation (1) can be considered as the sum of all the torques applied about the ith body from which

$$\Gamma_i = \Sigma \; \ell \ell_{Fi} \qquad (3)$$

The torques applied about each body are derived from the following sources: hydrodynamic forces, buoyancy forces, acceleration of the pivot point considered as the origin (α_1), and muscular activity.

The drag forces are determined by the general equation of fluid dynamics:

$$D_i = \frac{1}{2} \; CD_i \; \varrho \; S_i \; V_i^2 \qquad (4)$$

in which CD is a viscous drag coefficient defined by the Reynold's numbers corresponding to each body on the basis of the experimental results of Welsh (1953). V is the velocity of the push center of the mobilized body, ϱ is the water density, and S is the normal area of the displacement. The buoyancy force applied on each body is defined as function of the weight of the displaced water column during the human body immersion.

The moments resulting from these different forces are defined in relation to the motion α_i by

$$\ell \ell_{Fi} = F_i \cdot r_i \qquad (5)$$

in which r_i is the lever arm of the force F_i in relation to α_i.

The calculation of these different factors as well as of the energy transfers caused by the acceleration of the origin of the system α_1 allows us to mathematically isolate the torque resulting from the muscular activity on each considered body, i.e., the arm on one side and the trunk and the legs on the other side (here considered as only one body) in Equation (3).

Results and Discussion

The computer program established on the basis of this mathematical model allows the study of variations in the dynamic data applied on the human body for various values of angular displacements of each body in relation to time.

The basic characteristics of the motion studied in the program are the following:

1. The hand is considered as origin of the system.

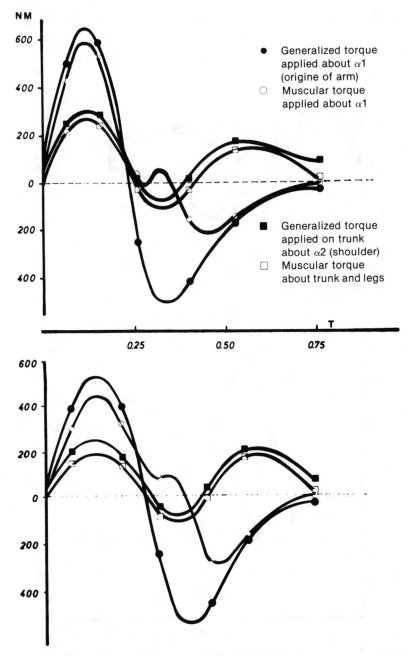

Figure 2—Generalized and muscular torques applied about α_1 and α_2 for $\phi = 0.35$ (a) and $\phi = 0.45$ (b) due to arm rotation in sagittal plane in water.

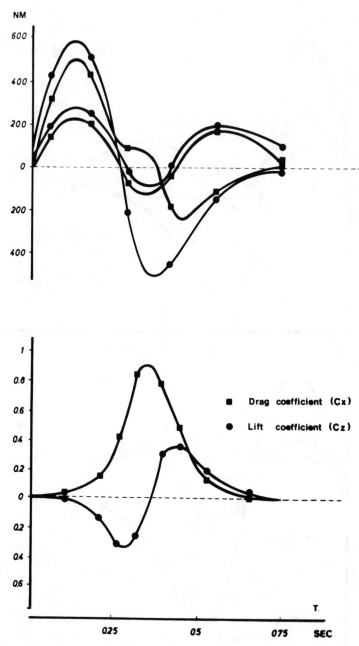

Figure 3 — Variation of drag and lift coefficients during the arm rotation in water for $\phi =$ 0.40.

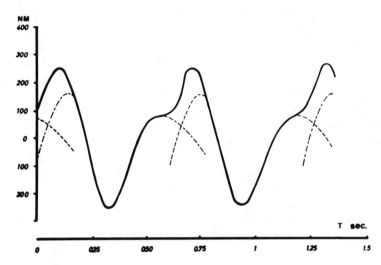

Figure 4—Generalized torque applied to the trunk and legs about α_2 when both arms are simultaneously in the water with a difference of phase of 160°.

2. The arm is straight in its underwater trajectory, which represents a pair approximation in the pressing and pulling phases; θ_1 fluctuates then from 0 to 180°.

3. The human body remains in the horizontal position, thus $\theta_2 = -\theta_1$.

4. The time of the underwater trajectory of the arm is 0.75 sec, which corresponds to an average speed of the shoulder displacement in a horizontal plane of 1.95 m/sec.

Figure 2 shows the time history of the generalized torques and the muscular torques applied on the arm and the rest of the human body for different values of ϕ, i.e., 0.35 T (a) and 0.45 T (b) during an underwater trajectory of the arm.

This shows the importance of the inertia at the end of the arm trajectory in its push phase.

As the horizontal speed of the arm is equal to zero when it comes out of the water at the end of the push phase, the dissipation of the accumulated kinetic energy at the beginning of the pull phase is affected by the muscles of the arm acting on the trunk (biceps brachii, pectoralis major).

The mathematical model shows well this phenomenon at the end of the motion. On another side these muscles exercise a positive motor torque in relation to the shoulder which allows the trunk to remain in a horizontal position.

Figure 3 shows the evolution of the drag coefficients (C_X) and lift coefficients (C_Z) during the underwater trajectory of the arm. It clearly appears that the drag coefficient C_X has a maximum of 0.89 when the arm is in a vertical position. Besides, the lift coefficient is negative during the pressing and pulling phases, the minimum of which is equal to -0.34 when the angle θ_1 is 57.4. This facilitates the arm motion during these two phases. That is why the muscular torque is inferior to the generalized torque at the beginning of the pulling phase of the arm and superior at the end of the pulling phase when the arm is in vertical position.

Figure 4 shows the evolution of the generalized torque applied to the human body when both arms are simultaneously in the water with a phase difference of 160°, when the second arm enters the water. This result is interesting from a methodological viewpoint as it gives an indication of the most efficient type of kicking to be used by the swimmer. This mathematical simulation shows the value of a two-beat kick with a powerful ascending phase of the legs which allows the body to stay in a horizontal position.

Conclusion

The developed mathematical model shows that inertia cannot be neglected in the crawlstroke. It also evidences the lift coefficient C_Z applied on the arm in its underwater trajectory.

Finally this simulation points out the importance of the two-beat kick in crawlstroke. Further studies will be necessary to improve this mathematical model, mainly in regard to the envisaged bodies and degrees of freedom in order to be able to input experimental kinematic data.

References

BELOKOVSKY, V.V., and Kuznetsov, V.V. 1976. Analysis of dynamic forces in crawlstroke swimming. In: P.V. Komi (ed.), Biomechanics V-B, pp. 235-242. University Park Press, Baltimore.

HATZE, H. 1977. A complete set of control equations for the human musculo-skeletal system. J. Biomechanics 10:799-805.

LEWILLIE, L.A. 1971. Graphic and electromyographic analysis of various styles of swimming. In: J. Vredenbregt and J. Wartenweiler (eds.), Biomechanics II, pp. 253-257. S. Karger, Basel.

LEWILLIE, L.A. 1973. Muscular activity in swimming. In: S. Cerquilini, A. Venerando, and J. Wartenweiler (eds.), Biomechanics III, pp. 440-445. S. Karger, Basel.

SEIREG, A., and Baz, A. 1971. A mathematical model for swimming mechanics. In: L. Lewillie and J.P. Clarys (eds.), Biomechanics in Swimming, pp. 81-103. Université Libre de Bruxelles.

SEIREG, A., Baz, A., and Pattel, D. 1971. Supportive forces as the human body during underwater activities. J. Biomechanics 4:23-30.

TOKUYAMA, H., Okamoto, T., and Kumamoto, M. 1976. Electromyographic study of swimming in infants and children. In: P.V. Komi (ed.), Biomechanics V-B, pp. 215-221. University Park Press, Baltimore.

WELSH, C.J. 1953. The drag of finite-length cylinders determined from flight tests at high Reynolds numbers for a Mach number range from 0.5 to 1.3. NACA TN 2941, Langley Aeronautical Laboratory, Washington.

Modelling of the Human Thorax

J. Cornelis, B. Van Gheluwe, and M. Nijssen
Vrije Universiteit Brussel, Brussels, Belgium

In many applications, knowledge of the geometry of the human thorax and exact localization of specific reference points on the thorax surface may be used to improve the quality of results and the accuracy of quantification, e.g., electrode positions in ECG-analysis (Cornelis, 1980; Horacek, 1974). This article describes a relatively simple and reliable technique for spatial reconstruction of cinematographically recorded body marks and for mathematical modelling of a body part (*in casu*, the thorax) as an analytical function (a linear combination of bicubic B splines, LCB). The methodology already outlined by Cornelis, Van Gheluwe, and Nijssen (1978) was refined in this study. The article includes: 1) a description of the influence of the positions of body marks (arbitrarily chosen), 2) the relationships between the LCB-coefficients and characteristic measurements on the thorax which may be used as absolute constraints for the determination of the LCB-coefficients, and 3) a presentation of the accuracy and examples of recent applications.

Procedure

Reconstruction of the Three Dimensional Coordinates of an Arbitrary Point on the Thorax Surface

Characteristic points are marked on the thorax surface (e.g., anatomical reference points, electrode positions). Cinematographical data of these points are collected by surrounding each test subject by several (6 or 8) cameras without special precautions of alignment. After simultaneous activation of these cameras and development of the films, the image coordinates of each body mark are collected.

The spatial coordinates (x_1, x_2, x_3) of each point on the thorax surface, visible by at least 2 cameras, are reconstructed by computer calculations. The mathematics of the reconstruction method are discussed by Van Gheluwe (1978) and the general set-up is described by Cornelis et al. (1978).

Description of the Thorax Surface by an Analytical Function

Reconstruction of the three-dimensional surface of the body under test is performed using the spatial coordinates (x_1, x_2, x_3) of the characteristic body marks and eventual anthropometric measures which can be used as constraints (sagittal and transverse diameters, thorax perimeters or horizontal arc lengths at certain heights). The aim of the procedure is the synthesis of a bicubic B spline function which adequately describes the thorax surface between the height of the iliac crest and acromion process (without arms).

A vertical axis (Z) is constructed through the center of the biacromial line and the transformation $(x_1, x_2, x_3) \rightarrow (r, \theta, z)$ is performed (see Figure 1a). The problem which must be solved now can be formulated as follows (see Figure 1b): given the values r_i (i = 1,...,q) of a set of points (θ_i, z_i) arbitrarily situated in R and belonging to the thorax surface r (θ, z) which is supposed to be 'smooth', find a function s (θ, z) defined in R which is a satisfying approximation to r (θ, z).

To solve this problem we use a 'global smoothing' method (Cornelis et al., 1978; Cornelis, 1980). In the R-domain (see Figure 2a) r (θ, z) is approximated by a linear combination of bicubic (order = 4) B splines $N_{4,i}$ (θ) $N_{4,j}$ (z) in which the variables θ and z are separated (see Figure 2b). The observation equations are

$$r_e = \sum_{i=-k}^{N} \sum_{j=-k}^{M} c_{ij} N_{4,i} (\theta_e) \cdot N_{4,j} (z_e) \, (e = 1, ..., q) \tag{1}$$

or r = A.c.

The over determined set of equations (1) has to be solved using a least squares method eventually taking into account several absolute constraints on the solution vector c formulated as

$$B.c = g \tag{2}$$

The two sets of equations (1) and (2) introduce two kinds of problems which should be taken care of by the solution method. Since the points (θ_e, z_e) are situated arbitrarily in the domain R, matrix A often happens to be rank-deficient (Hayes and Halliday, 1974). It is possible that formulation of different constraints causes unexpected contradictions or redundancies in Equation (2).

For the one dimensional smoothing problem, an *a priori* detection of eventual rank deficiency of the smoothing matrix from the relative posi-

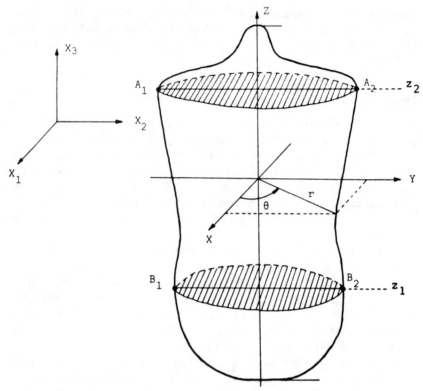

Figure 1a — Transformation from Cartesian coordinates (x_1, x_2, x_3) to cylindrical coordinates (r, θ, z) (A_1 and A_2 represent both acromion processes, B_1 and B_2 the iliac crests).

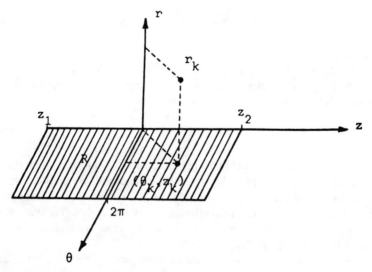

Figure 1b — Definition of the R-domain $\{0 \leq \theta < 2\Pi; z_1 \leq z \leq z_2\}$.

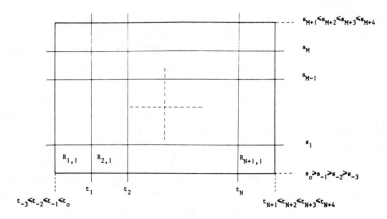

Figure 2a — The R-domain divided by knotlines.

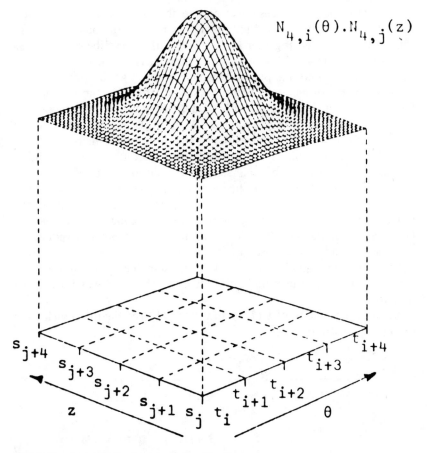

$$N_{4,i}(\theta) \cdot N_{4,j}(z)$$

Figure 2b — An example of an elementary B-spline.

tions of the data points and the spline knots is possible (Schoenberg-Witney condition, SWC). The synthesis of a generalization of the SWC to two-dimensional problems seems very difficult. In the special case of observation, points situated on $(M + 4)$ lines parallel with the θ-axis, a sufficient condition for matrix A having maximal rank $(M + 4)(N + 4)$ is the existence of an injection f: $\{Q_{ij}\} \to \{P_\ell\}$ of the collection of the areas $\{Q_{ij} (t_i < \theta < t_{i+4}, s_j < z < s_{j+4}); i = -3 \to N, j = -3 \to M\}$ on the collection of all observation points $\{P_\ell (\theta_\ell, Z_\ell); \ell = 1 \to q\}$ in such a way that $f(Q_{ij}) \epsilon Q_{ij}$ and that each of the $M + 4$ lines parallel to the θ-axis contains $(N + 4)$ of the observation points $f(Q_{ij})$. This means that to each support Q_{ij} of a B-spline $N_{4,i} (\theta) N_{4,j} (z)$ a *specific* data point must be associated which lies in this support and that the $(N + 4)$ points obtained by varying i and holding j constant have an identical z-coordinate. The proof of this theorem is given by Cornelis (1980).

The situation required for the validity of the theorem stated is often approximately met (e.g., when potential maps in ECG are measured it is very easy to place electrodes on successive horizontal layers, whereas it is practically impossible to position them in increments of constant angles). The theorem is used together with an algorithm described by Dierckx (1975) to provide good knotline positioning. It is, however, not always possible to find knotline positions which assure a good approximation accuracy of the observation points $r_\ell (\ell = 1, q)$ by the "smoothing surface" and which eliminate the rank of deficiency of A completely.

If the Grammian of the columns of A, $A^\tau A$, is singular, the solution vector c of (1), which minimizes the Euclidian norm of the residual vector, $r - Ac$, is not unique. If the supplementary constraint is introduced so that $\| c \|$ must be minimal, a unique solution—the minimal least square solution (MLSS)—is obtained which is given by Xr whereby X is the pseudo inverse of the rectangular matrix A. The MLSS satisfying Equation (2) is obtained through application of the method described by Hayes and Halliday (1974) which also takes care of possible redundancies and contradictions in Equation (2).

The constraint equations (2) are defined using linear relations between anthropometric measures and coefficients of the LCB which describes the thorax surface.

At first, periodicity of s (θ, z) and $\dfrac{\partial s(\theta, z)}{\partial \theta}$ (see Figure 1) yields the following constraint equations:

$$\sum_{i=-2}^{-1} \frac{c_{i,j} - c_{i-1,j}}{t_{i+3} - t_i} N_{3,i}(o) = \sum_{i=N-1}^{N} \frac{c_{i,j} - c_{i-1,j}}{t_{i+3} - t_i} N_{3,i} (2\pi) \qquad (3)$$

$$\sum_{i=-3}^{-1} c_{i,j} N_{4,i}(o) = \sum_{i=N-2}^{N} c_{i,j} N_{4,i} (2\pi) \qquad (j = -3,...,M) \qquad (4)$$

These equations can be simplified if t_{-3}, t_{-2}, t_0 coincide with $\theta = 0$ and t_{N+1}, t_{N+2}, t_{N+3}, t_{N+4} with $\theta = 2\pi$ (see Figure 2).

Secondly, the sagittal or transverse diameters (d) at heights z' are known as

$$s(\theta_1, z') + s(\theta_1 + \pi, z') = d \tag{5}$$

Also some thorax perimeters or arc lengths at specified heights (z') are used in this way.

An approximation of the measured value for these latter parameters is given by:

$$\sum_{k=L-4}^{L-1} N_{4,k}(z') \cdot \left[\sum_{\ell=i-3}^{i} c_{\ell k} \frac{t_{\ell+4} - t_\ell}{4} \right. \tag{6}$$

$$(\ell < p-3)$$

$$\sum_{\ell=i-3}^{i} c_{\ell k} \frac{t_{\ell+4} - t_\ell}{4} \sum_{r=0}^{3} \frac{(\theta_1 - t_{\ell+r}) N_{4-r, \ell+r}(\theta_1)}{(t_{\ell+4} - t_{\ell+r})}$$

$$+ \sum_{\ell=p-3}^{p} c_{\ell k} \frac{t_{\ell+4} - t_\ell}{4} \sum_{r=0}^{3} \frac{(\theta_2 - t_{\ell+r}) N_{4-r, \ell+r}(\theta_2)}{(t_{\ell+4} - t_{\ell+r})}$$

$$+ \sum_{\ell=i+1}^{p-4} c_{\ell k} \frac{t_{\ell+4} - t_\ell}{4} \quad \Bigg]$$

$$i < p-4$$

The validity of the approximation is demonstrated in Table 1.

Accuracy of the Reconstruction Method

The accuracy of the three-dimensional reconstruction of (x_1, x_2, x_3) coordinates is given by Cornelis et al. (1978) for a cylinder with known dimensions. Table 2 summarizes the results discussed by Cornelis (1980) for thoraxes of 42 test subjects.

The performances of the smoothing procedure are illustrated in Figure 3. The conclusion which can be deduced from this figure and from several tests on analytical geometric surfaces is that a number of internal knotlines, $N = 5$ and $M = 3$, leads to an RMS deviation ($\Delta 2$) between observation points (r) and smoothed thorax surface equal to 0.6 cm. The

Table 1

Differences between Measured and Reconstructed Perimeters for 10 Subjects

Perimeter-height	Mean of measured perimeter (cm)	Standard deviation (cm)	Mean of reconstructed perimeter (cm)
Mesosternal	91.64	4.274	90.72
Nipple	89.98	4.160	89.17
Infrasternal	87.00	4.061	86.39
Waist	79.23	3.647	77.74
Epigastricum	82.41	3.966	81.24

Table 2

**Absolute Accuracy of the Coordinates and Lengths in
Three-Dimensional Reconstruction using Two Cameras for Each Point**

x1 (cm)	x2 (cm)	x3 (cm)	Sagittal and transverse diameters (cm)	Height differences (cm)
0.56	0.67	0.37	1.57	0.83

relative RMS deviation ($\Delta 3$) is 0.04 and the relative accuracy (RN) defined as the mean difference between measured and reconstructed diameters is 1.3 cm. The choice of adequate absolute constraints [Equation (2)] at heights z_1 and z_2 (see Figure 1) is indispensable to obtain trustworthy borders of the thorax surface. Coincidence was imposed with known one-dimensional spline functions at heights z_1 and z_2 derived from a standard thorax model (Horacek, 1974), which was shaped for each individual according to measured perimeters and diameter values at z_1 and z_2. A good result is also obtained by introducing a large number of observation points for Equation 1 near the borders.

Results

The methods described here were used in the study of the inverse problem (Cornelis, 1980) in ECG-analysis. This study requires the thorax to be a closed body surface. A typical example of thorax reconstruction is illustrated in Figure 4.

The spatial accuracy of the reconstruction is about .5 cm (compare Table 2).

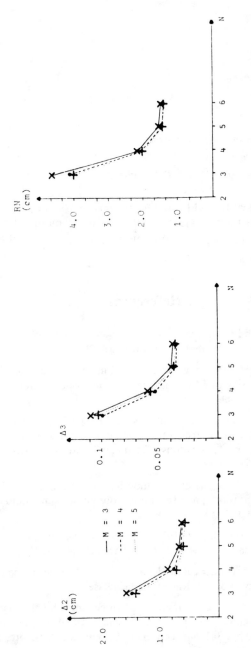

Figure 3 – Performance of the smoothing procedure, with Δ2 being the RMS deviation, Δ3 the relative RMS deviation, and RN the mean difference between measured and reconstructed diameters.

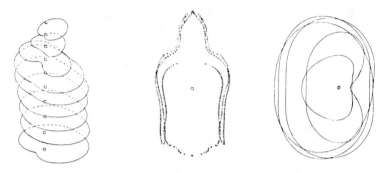

Figure 4—An example of thorax reconstruction.

Conclusion

This method proved to yield satisfactory results for the modelling of an individual human thorax. Because of its general nature, however, the method may be used for the reconstruction of the surface of other body segments.

References

BUZINGER, P., and Golub, G. 1965. Linear least square solutions by Householder transformations. Numerische Mathematik 269-276.

CORNELIS, J. 1980. Studie van de verwerking van cardiologische signalen (Study of the processing of ECG potentials). Unpublished doctoral dissertation, Vrije Universiteit Brussel.

CORNELIS J., Van Gheluwe, B., and Nijssen, M. 1978. A photographic method for the tridimensional reconstruction of the human thorax. Proceedings of the NATO Symposium on Applications of Human Biostereometrics (Soc. Photo. Opt. Instr. Eng.) 166:294-300.

DIERCKX, P. 1975. An algorithm for smoothing, differentiation and integration of experimental data using spline functions. Journal of Computational and Applied Mathematics 3:165-184.

HAYES, J.G., and Halliday, J. 1974. The least-squares fitting of cubic spline surfaces to general data sets. J. Inst. Maths Applics. 14:89-103.

HORACEK, B. 1974. Numerical model of an inhomogeneous human torso. Advances in Cardiology. Vol. 10. Karger Verlag, Basel.

VAN GHELUWE, B. 1978. Computerized three-dimensional cinematography for any arbitrary camera set-up. In: E. Asmussen and K. Jørgensen (eds.), Biomechanics VI-A, pp. 343-348. University Park Press, Baltimore, MD.

B.
Simulation

Two-dimensional Simulation of Human Movements during Take-off and Flight Phases

A.J. Spaepen, V.V. Stijnen,
E.J. Willems, and M. Van Leemputte
Katholieke Universiteit Leuven, Heverlee, Belgium

Our research in the last few years was aimed at a quantitative mechanical analysis of human motion using mathematical models. In a first step, these models are evaluated and optimized for individual subjects. The second step of the study deals with the development of motion equations for two-dimensional simulations, both for take-off and flight phases. It is shown that the mathematical problem in both cases may be reduced to the solution of a set of three differential equations of first or second order. This reduction is independent of the number of segments in the model.

Finally, model and simulations were evaluated for three gymnastic movements. Data from film analysis and force recordings were compared to results from simulated movements.

Methods

Evaluation and Optimization of the Model

We started from the Hanavan-model (Hanavan, 1964) to build a mathematical model of the body, consisting of 15 segments with homogeneous mass distribution. Then the model was slightly modified by linking the hands to the forearms. Also the trunk was divided in two equal segments, thorax and abdomen. The mechanical parameters of the new segments were calculated from those of the original model.

Thereupon the model was experimentally evaluated in the following way (Stijnen, 1980). A total of 34 male subjects, all students in physical education, underwent a series of tests. In the first series the position of

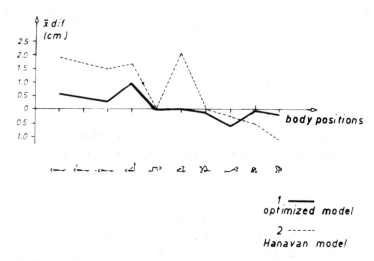

Figure 1 — Distances between the measured and calculated position of the gravity line for 10 different body positions.

the gravity line was derived from force recordings of a Kistler force-plate for 10 different stable body positions. These positions were simultaneously recorded on 35 mm Kodak film. After the tests, the position of the gravity line was calculated from film coordinates. The mean differences between the test values and the calculated ones are given in Figure 1 for the different body positions. The mean difference was 12 mm.

Based on a pendulum movement, the moment of inertia relative to the axis of the pendulum was measured in a second test series with the same subjects (see Figure 2). The test was performed three times for three different body positions. Each time the subject was on a table which rotated around a horizontal axis. Two springs attached the extremities of the table to the surroundings, so that a pendulum movement was possible. The force in one of the springs was recorded. The moment of inertia was deduced from the period and the logarithmic decrement of the pendulum movement. It was also calculated from film analysis of pictures taken simultaneously. Results from pendulum measurements and film analysis are compared in Figure 3. It is seen that the moment of inertia of the modified Hanavan-model for a straight position corresponds very well to its measured value, but that for more inclined positions the error can be as high as 1.7 kg • m².

To reduce the discrepancies between the experimental data and values obtained for the model, a procedure was devised for an optimum choice of mechanical parameters of the model. The relative mass of forearms and hands, upper arms, trunk and head, thighs, and lower legs were calculated such that the discrepancies became a minimum.

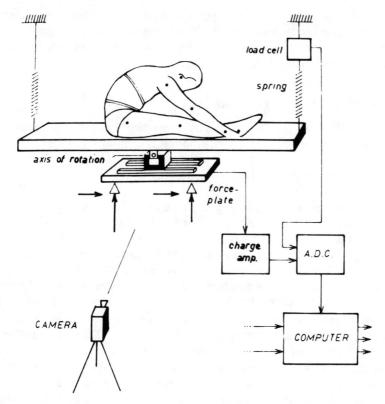

Figure 2—Experimental setup for the measurement of the moment of inertia.

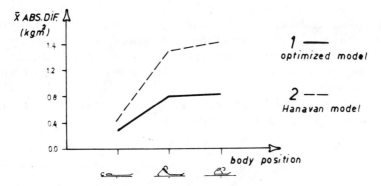

Figure 3—Differences between the measured and calculated moments of inertia relative to a lateral axis through the center of gravity.

Results for differences of the position of the gravity line were reduced from 12 to 6.6 mm (see Figure 1). There was also a slight improvement for the moment of inertia (see Figure 3). The mean error of this parameter was reduced from 1.2 to 1.0 kg • m².

Derivation of Motion Equations

The problem of simulating human movements can be described as follows. Suppose that for a given motion the following parameters are known:

1. for the starting position:
   a. the orientation of each segment;
   b. the position of each segmental center of gravity;
   c. the angular momentum of the whole body;
   d. the momentum of the whole body;
   e. the link between the model and its surroundings (no link for flight phases and a set of springs for take-offs).
2. during the complete movement: a function for each of the 13 joints which gives the angle of the joint at each moment as a function of time.

It was then asked to calculate from those data the trajectory and orientation of each segment during the whole motion. The solution for flight phases is based on the following three properties of the system:

1. the velocity and angular velocity of one segment are linearly related to the velocity and angular velocity of every other segment.
2. the constant angular moment of a system of N rigid bodies relative to a lateral axis through the center of gravity, is given by the following expression:

$$L = \sum_{i=1}^{N} I_i \, \omega_i + \sum_{i=1}^{N} m_i \cdot (x_i \cdot \dot{y}_i - \dot{x}_i \cdot y_i) \qquad (1)$$

with I_i the moment of inertia of a segment relative to its center of gravity; m_i the segmental mass; ω_i the angular velocity; x_i and y_i the coordinates of the mass center of the segment relative to the mass center of the whole body, and $\dot{x}_i$ and $\dot{y}_i$ the components of the velocity of the mass center of the segment relative to the mass center of the whole body. Consequently, the angular momentum is a linear combination of the velocity and angular velocity of the segments.

3. the constant total momentum of a system of N rigid bodies depends linearly on the velocity of the segments:

$$M \cdot \dot{x}_{cg} = \sum_{i=1}^{N} m_i \, \dot{x}_i \qquad (2)$$

$$M \cdot \dot{y}_{cg} = \sum_{i=1}^{N} m_i \cdot \dot{y}_i \qquad (3)$$

Using these properties, it can be shown that the unknown velocity and angular velocity of 13 segments may be mathematically eliminated from three motion equations:

$$M \cdot x_{cg} = C_1 = f_1 (\dot{x}_1) \qquad (4)$$

$$M \cdot \dot{y}_{cg} = C_2 = f_2 (\dot{y}_2) \qquad (5)$$

$$L = C_3 = f_3 (\dot{x}_1, \dot{y}_1, \omega_1) \qquad (6)$$

In this way the simulation is reduced to the solution of a set of three ordinary differential equations of first order. This number of equations is independent of the number of body segments in the model. In the case of the simulation of take-offs, the solution method is similar for second order parameters such as acceleration and angular acceleration.

The simulation is now reduced to a set of 3 ordinary differential equations of second order, with three unknowns, $\ddot{x}_1$, $\ddot{y}_1$ and α_1.

$$F_X = f_4 (\ddot{x}_1) \qquad (7)$$

$$F_y - G = f_5 (\ddot{y}_1) \qquad (8)$$

$$M_{cg} = \frac{dL}{dt} = f_6 (\ddot{x}_1, \ddot{y}_1, \alpha_1) \qquad (9)$$

Evaluation of the Simulation

Simulations were evaluated for three gymnastic movements: backward handspring, forward roll, and backward somersault. The movements were filmed at 200 frames/sec (Locam 16 mm). Take-off forces were measured simultaneously on a Kistler force-plate and sampled at 1 kHz with a 12-bit ADC. Starting positions and joint movements were derived from film analysis. Displacements and changes of orientation of all segments were then simulated and compared to the values obtained from film analysis.

Results

Because the conclusions for the three movements were similar, only the results for the backward handspring are discussed. The simulated movement is shown in Figure 4. The absolute angle of the thorax is shown in Figure 5 as a function of time. The mean difference between simulated and registered values does not exceed 7°. The comparison of the results of the three movements leads to the conclusion that the accuracy of the calculated rotational characteristics is directly related to the precision

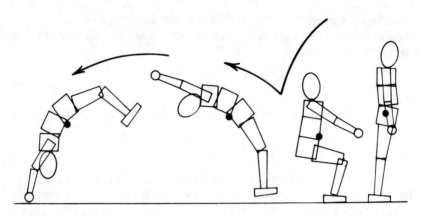

Figure 4—Simulated backward handspring.

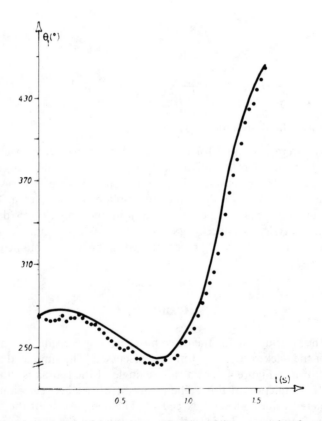

Figure 5—Absolute angle of the thorax as a function of time. .... = data from film analysis; — = simulated results.

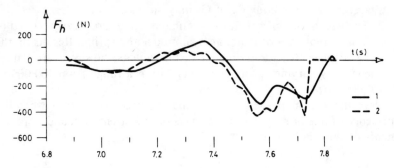

Figure 6 — Horizontal take-off forces as a function of time. 1 = measured values; 2 = simulated values.

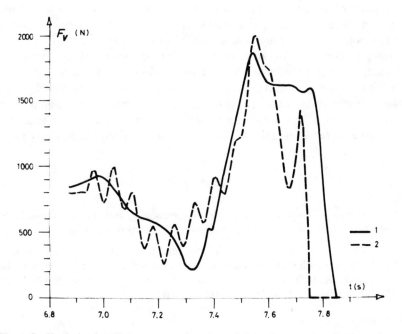

Figure 7 — Vertical take-off forces as a function of time. 1 = measured values; 2 = simulated values.

with which the moment of inertia and the position of the center of gravity of an individual subject is approached by the model.

Horizontal and vertical forces are evaluated in Figures 6 and 7. Two main differences occur:

1. The simulated take-off is ended 0.1 sec early. This effect is probably caused by inaccuracies in the second derivatives of the data for angular displacements in the joints and by the imperfect spring based link between the model and its surroundings. As a result, differences in

displacements of the body C. of G. may attain 0.2 m.

2. The oscillatory effects in the simulated values are clearly related to the natural frequency of mathematical model and spring. Results on the level of displacements and angular displacements, however, are hardly influenced by varying spring constants within a range from 100,000 to 2,000,000 N/m.

Nevertheless, the present methods may be used successfully when the influence of one or more varying factors on the orientation and displacements of the whole body is studied.

Conclusion

Although inevitable inaccuracies in data from film analysis and deficiencies in the construction of a mathematical model may clearly influence the results of a computer simulation of human movements, it is shown that results may be obtained from a simple set of only three differential equations with an accuracy that is proportional to the accuracy of the mechanical properties (C. of G., moment of inertia) of the model. When the interpretation of take-off forces is needed, however, additional effort will be necessary to obtain improved data for joint movements and to describe more adequately the link between model and supporting base.

References

HANAVAN, E. 1964. A personalised mathematical model of the human body. TR AMRL TDR 64 102, Wright Patterson Air Force Base, Ohio.

STIJNEN, V.V., Spaepen, A.J., and Willems, E.J. 1980. Models and methods for the determination of the center of gravity of the human body from film. In: A. Morecki, K. Fidelus, K. Kedzior and A. Wit (eds.), Biomechanics VII-A, pp. 558-564, University Park Press, Baltimore, MD.

Characteristics of a Multi-Body Model
for Human Level Walking

Y. Tagawa
Kurume Institute of Technology, Mukaino, Japan

T. Yamashita
Kyushu Institute of Technology, Tobata, Japan

Quantitative knowledge of the functions of the elements and of the mechanisms in human walking is required for the design of a bipedal walking machine and an artificial leg, and also for the therapy and surgery to treat a disabled lower extremity. Walking characteristics can be obtained by experimental methods under some limitations; i.e., some variables cannot be measured directly, or walking conditions may be distorted due to measurements, etc. On the other hand, the simulation approach is an effective tool for removing the limitations for the study of coordinated movements.

Various models have been used for simulation of walking (Chow & Jacobson, 1971; Townsend & Seireg, 1972; Gubina et al., 1974; Yamashita et al., 1976). In general, they use a simplified model, such as one rigid body having legs with or without mass, to examine the characteristics or the control strategies of the human while walking. As for the characteristics, the following have been examined: floor reactions, moments acting at each joint, movements of the center of gravity of body, etc.

In this study, a multi-body model consisting of 10 elements was introduced to simulate the characteristics of a human walking on a level surface and to examine the functions of the element. The rotations of the upper torso were analyzed in relation to stabilizing the unstable motion during the single-support phase during which the system was approximated by an inverted pendulum. The main characteristics showed a good agreement with experimental results obtained by other researchers, which are cited in Yamashita and Tagawa (1978).

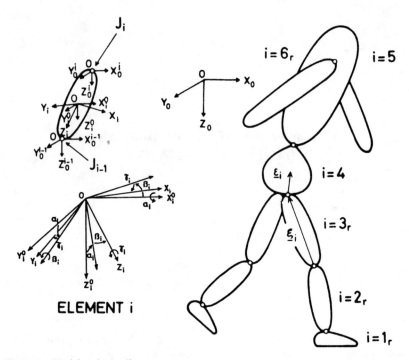

Figure 1—Model and coordinate systems.

Model and Physical Parameters

The model in this study had 10 rigid elements, which represented the upper limbs, the upper and lower torso, the thighs, the shanks, and the feet as shown in Figure 1. Adjacent elements were coupled by ideal joints. The body, which was supported by either one or both legs at the same time, depending on the supporting condition, was allowed to move in three-dimensional space. The motion of each element was controlled by the moments m_{xi}, m_{yi}, and m_{zi} applied at the i-th joint (J_i) which were all defined in the coordinates X_i, Y_i, and Z_i for the i-th element.

The parameter values of the model have been estimated based on the anthropometric data by properly approximating the shapes of the elements as a hemisphere, a column, truncated cones, or truncated elliptical cones. The normalized values of the parameters in the right leg system are shown in Table 1: the lengths are divided by the leg length ($\ell = 0.90m$); the inertial terms are normalized by $\ell^2 \Sigma M_i$ ($\Sigma M_i = 70.44$ kg).

Table 1

Normalized Values of Physical Parameters

		Arm	Torso Upper	Torso Lower	Thigh	Shank	Foot
M_i [-]		0.05	0.366	0.176	0.107	0.053	0.019
ϵ_{ir} [-]	X	0	0.063	−0.063	0	0	0.18
	Y	0	0.17	−0.126	0	0	0
	Z	0.638	−0.344	−0.121	−0.5	−0.411	0.089
ϵ_{ir} [-]	X	0	0.063	0	0	0	0.055
	Y	0	0	−0.126	0	0	0
	Z	0.277	−0.307	−0.033	−0.256	−0.18	0.067
I_i [-]	X	0.0016	0.0545	0.0049	0.0089	0.0025	0.0001
	Y	0.0016	0.0526	0.0016	0.0089	0.0025	0.00024
	Z	0.00003	0.0047	0.0045	0.00035	0.00007	0.00017

Procedure

An outline of the process for the analysis using the locomotion model is shown in Figure 2. Equations of motion for the locomotion model were derived by applying the principle of mechanics in a conventional way. In general, the equations for a walking model having many degrees of freedom are not only nonlinear but also very complicated. Therefore, linearization techniques were used in this simulation.

The number of the unknown variables in the equations exceeded the number of the equations. Therefore, some variables were prescribed based on experimental results, which can be accurately measured, to make the equations a closed system. By the prescriptions of the movements of the hip joint and of the foot angles at heel contact and toe-off (see Figure 3), the angular motions at the leg joints during the stance phase were uniquely determined through geometrical relationships. The angular variables of the leg in the sagittal plane were approximated in the second order form to get better characteristics. The amplitude of the vertical oscillation of the hip joint was formulated so that it was dependent on the step length, but independent of the cycle time of walking. When the angular motion at the leg joints during the stance phase was numerically calculated using the geometrical relationship however, there were possibilities that the angular velocity might become infinite. To overcome this difficulty, the vertical displacement of the hip joint was

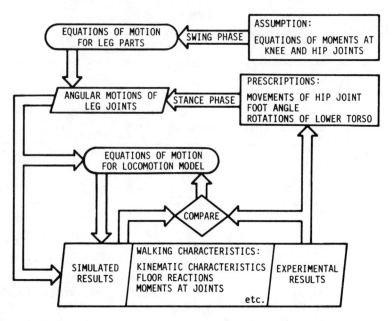

Figure 2—Outline of the process for analysis using the locomotion model.

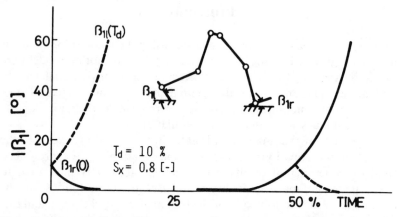

Figure 3—Foot angles during the stance phase. The origin of the time scale was defined as the instant when the right foot (subscript: r) was placed on the ground.

forced down by a predetermined small amount ΔL.

On the other hand, the angular motion of the leg during the swing phase can be derived by two methods: one is to prescribe the motion of the leg system based on experimental results, the other is to solve the equations of motion for the parts of the leg in order to satisfy the continuity conditions of the angular positions and velocities. In the latter case,

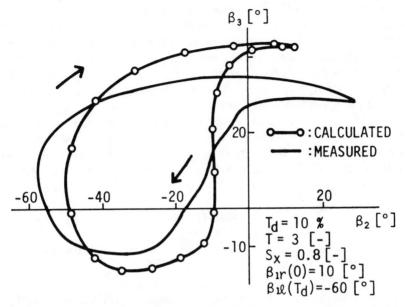

Figure 4—Angular motions of shank (β_2) and thigh (β_3) during one walking cycle. The measured value was cited from Eberhart et al. (1954).

the moments of the leg joints during the swing phase were assumed to be a function of the first order of time. The angular motions of the leg joints using the latter method are shown in Figure 4 with the corresponding measurements (Eberhart et al., 1954).

Finally, the equations of motion for the upper torso were easily solved to satisfy the boundary conditions of the continuity of the angular positions and velocities, because nonlinear interactions of the unknown variables between the sagittal and frontal planes vanished by the linearization. The analysis was repeated for various values of the following parameters: step length, S_x; walking cycle time, T; double-leg support phase T_d; rotations of lower torso, α_4 and γ_4; movement of hip joint, Y_H; and angular displacements of foot at heel contact and toe-off, $\beta_{1r}(0)$ and $\beta_1(T_d)$. The simulated results were compared with the walking characteristics obtained experimentally for walking on a level surface.

Results

The steady characteristics of human walking have been obtained with the knowledge of the functions of the elements in the human body. The prescription of the motion of the hip joint made the analysis much easier.

The main results were as follows:

1. Both the floor reaction and the moments at the leg joints showed

good agreement with the experimental results.

2. The rotation of the upper torso in the sagittal plane changed with the walking speed and the foot angle at toe-off but not with the step length. The amplitude of the rotation was greater in slow walking than in fast walking, and this fact suggests that the body is stabilized dynamically in fast walking and statically in slow walking. A strong push at toe-off forced the upper torso backward.

3. The movements of the upper limbs and the rotations of the lower torso contributed to making the variation of the vertical reaction force small. When you walk faster you must utilize the motions of the upper extremities and lower torso in order to achieve a smooth walking pattern.

References

CHOW, C.K., & Jacobson, D.H. 1971. Studies of human locomotion via optimal programming. Math. Biosci. 10:(3/4):239-306.

EBERHART, H.D., Inman, V.T., & Blesler, B. 1954. The principal elements in human locomotion. Human limbs and their substitutes. Hafner, pp. 437-471.

GUBINA, F., Hemami, H., & McGhee, R.B. 1974. On the dynamic stability of biped locomotion. IEEE Trans. Biomed. Eng. 21(2):102-108.

TOWNSEND, H.A., & Seireg, A. 1972. The systhesis of bipedal locomotion. J. Biomech. 5:71-83.

YAMASHITA, T., Kabashima, K., & Sakatani, Y. 1976. Control of macro-model to simulate human level walking. Reprints of Second CISM-IFToMM Symp. on Theory and Practice of Robots and Manipulators, pp. 183-192.

YAMASHITA, T., & Tagawa, Y. 1978. Two-body and massless leg model for human level walking. Proc. Int'l Conference on Cybernetics and Society, Tokyo, pp. 44-48.

Design of a Seven-Link-Biped
By State Feedback Control

Tsutomu Mita, Toru Yamaguchi, and Toshio Kashiwase
Chiba University, Japan

There are many studies (for example, Hemami & Farnsworth, 1977; Kato et al., 1981) on the realization of biped locomotion. There are no studies, however, except Kato's (1981) in which the biped successfully walks. The difficulty of the realization is caused by the fact that the biped is dynamically unstable without control.

The most effective controlling method for biped locomotion is considered to be the state feedback, which has been studied intensively in modern control theory since the biped is a multivariable dynamic system having complex interactions between all variables.

The controlling strategy derived by Kato et al. (1981) uses a tracking control of the center of mass but the stabilizing action seems to be very slow. In our studies, although a simple support compensates for the sagittal dynamics under existing circumstances, the biped called CHIBA-WALKER 1 (CW-1) walks any arbitrary number of steps and it takes only 1 sec to walk one step. Such high speed biped locomotion should be used as a walking machine for persons handicapped in the lower half of the body.

Mechanical Part of CHIBA-WALKER 1

As shown in Figure 1, the biped CW-1 is composed of seven links (the hip, two upper legs, two lower legs, and two feet) and six DC motors which actuate the torques within the legs. Relative angles between these legs are measured by six potentiometers mounted beside the motors. Four touch-sensors attached under each foot send a signal which informs whether the foot is on the ground or not. The height of CW-1 is about 75 cm and it weighs about 15 kg.

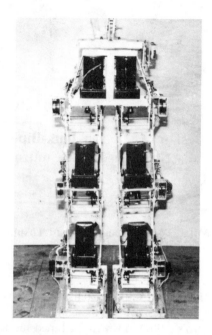

Figure 1—Front view of CHIBA-WALKER 1.

Attitude Control and Gait Control

State Feedback and Attitude Control

The coordinates which determine the motion of the biped are defined as shown in Figure 2, where p's are angles of the links with respect to the vertical and u's are input torques between the links.

The following non-linear differential equation is derived using Lagrange's equation of motion (Hemami & Farnsworth, 1977):

$$A(p)\ddot{p} + B(p)\dot{p}^2 + C\dot{p} + D(p) = Eu \tag{1}$$

$$: p = (p_1, .., p_5)^T, \; p^2 = (p_1^2, .., p_5^2)^T, \; u = (u_1, .., u_5)^T$$

$$A(p)(5 \times 5) = [L_{ij}\cos(p_i - p_j)], \; B(p)(5 \times 5) = [L_{ij}\sin(p_i - p_j)],$$
$$D(p)(5 \times 1) = [M_i\sin p_i]$$

$$E = \begin{bmatrix} 1 & -1 & & & \\ & 1 & -1 & & \\ & & 1 & 1 & \\ & & & -1 & 1 \\ & & & & -1 \end{bmatrix}$$

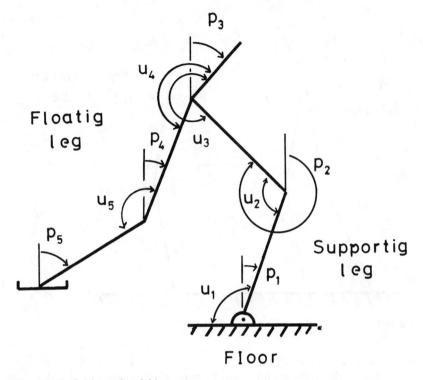

Figure 2 – Definitions of variables.

where $C(5 \times 5)$ is a matrix which represents coefficients of friction and T denotes the transposition. Several unknown parameters included in $A(p)$, $B(p)$, C, and $D(p)$ are determined by appropriate experiments (Yamaguchi, 1981), but the detail is omitted here. The relation between p's and relative angles (q's) which are measured by potentiometers is described by:

$$\begin{bmatrix} p_1 \\ p_2 \\ p_3 \\ p_4 \\ p_5 \end{bmatrix} = \begin{bmatrix} 1 & & & & \\ 1 & 1 & & & \\ 1 & 1 & 1 & & \\ 1 & 1 & 1 & -1 & \\ 1 & 1 & 1 & -1 & -1 \end{bmatrix} \begin{bmatrix} q_1 \\ q_2 \\ q_3 \\ q_4 \\ q_5 \end{bmatrix} \qquad [p = (E^{-1})^T q] \qquad (2)$$

At the outset for the attitude control of this biped it was assumed that the foot of the supporting leg is constrained to the floor (see Figure 2).

If $\bar{p}$ is given as a constant equilibrium which specifies some desired attitude of the biped, $\bar{p}$ must satisfy the following steady-state equation:

$$D(\bar{p}) = E\bar{u} \qquad (3)$$

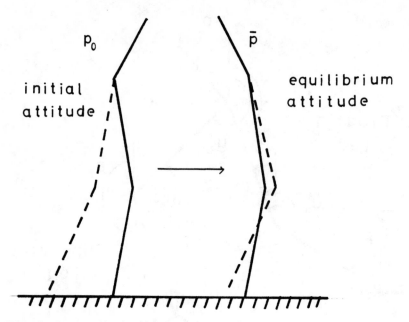

Figure 3—One example of attitude control.

where $\bar{u}$ is a biasing input vector which is used for maintaining the desired attitude. This attitude must be maintained in spite of the fact that small disturbances exist. Therefore, an appropriate control is required. For this small perturbation, Δp and Δu were introduced such that they satisfy

$$p = \bar{p} + \Delta p, \quad u = \bar{u} + \Delta u \tag{4}$$

Substituting (3) and (4) into (1) yields the following linearized equation:

$$A(\bar{p})\Delta\ddot{p} + C\Delta\dot{p} + G(\bar{p})\Delta p = E\Delta u \tag{5}$$

$$: G(\bar{p}) = \text{diag}(M_1\cos\bar{p}_1, M_2\cos\bar{p}_2, M_3\cos\bar{p}_3, M_4\cos\bar{p}_4, M_5\cos\bar{p}_5)$$

provided higher powers of Δp and $\Delta\dot{p}$ are neglected. Equation (5) can be transformed to a state space representation

$$\Delta\dot{x} = A\,\Delta x + B\,\Delta u$$
$$: \Delta x = (\Delta p,..., \Delta p_5, \Delta\dot{p}_1,..., \Delta\dot{p}_5)^T$$

$$A = \begin{bmatrix} 0 & I_5 \\ -A(\bar{p})^{-1}G(\bar{p}), & -A(\bar{p})^{-1}C \end{bmatrix} \qquad B = \begin{bmatrix} 0 \\ A(\bar{p})^{-1}E \end{bmatrix} \tag{6}$$

Table 1

Equilibrium, Eigenvalues, and Weighting Matrices

Equilibrium	$\bar{p}$	10	-10	-30	-15	25	(deg.)
	$\bar{q}$	10	-20	-20	-15	-40	(deg.)
Eigenvalues of A		0.135, 0.852, 2.82, -0.249, $-3.30 \pm j$ 3.25					
		-7.36, -28.7, -28.9, -31.1					
Weighting matrices		Q = diag (5000, 5000, 10000, 10000, 10000)					
		R = diag (1, 1, 1, 1, 1)					

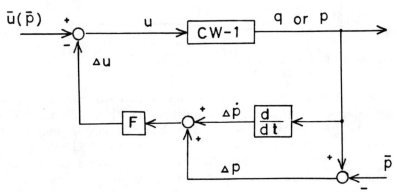

Figure 4—Construction of control system.

where Δx is the state vector of the biped. The values of the parameters of A and B vary according to the given equilibrium p. Since the biped is dynamically unstable without control, some of the eigenvalues of A have positive real parts. When the attitude of equilibrium is given as illustrated in Figure 3, for example, the eigenvalues of CW-1 become those shown in Table 1.

To maintain the equilibrium the state feedback was used.

$$\Delta u = -F\Delta x = -F(\Delta p, \Delta \dot{p})^{T} \qquad (7)$$

The construction of the control system is illustrated in Figure 4. The feedback coefficient matrix (F) is chosen so that the closed loop system

$$\Delta \dot{x} = (A - BF)\Delta x \qquad (8)$$

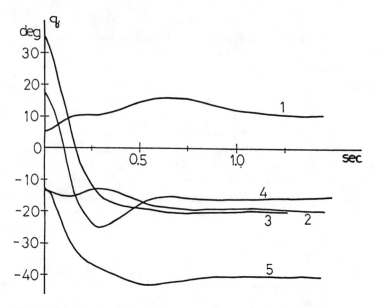

Figure 5 — Result of the simulation.

becomes asymptotically stable. This means that real parts of all the eigenvalues of A-BF are negative. In this study, F was determined by using the linear optimal control and the linear decoupling control. For the sake of simplicity, however, only the case where the linear optimal control was used is explained. In this control, F is chosen so that it minimizes the following integrated quadratic performance

$$J = \int_0^\infty (\Delta p^T Q \Delta p + \Delta u^T R \Delta u) \, dt \qquad (9)$$

where Q and R are weighting matrices which are semi-positive definite and positive definite, respectively.

It is derived by digital computer simulations which simulate the non-linear closed loop dynamics so that almost all attitudes can be approached from any initial attitude. One example of the simulation is given in Figure 5. In this simulation, the initial attitude is given by $p_0 = (5.6, -7.4, 27.7, 9.9, 23.4)$ [$q_0 = (5.6, -13.0, 35.1, 17.8, -13.5)$] and the equilibrium attitude is given by $\bar{p} = (10.0, -10.0, -30.0, -15.0, 25.0)$ [$\bar{q} = (10.0, -20.0, -20.0, -15.0, -40.0)$]. Both attitudes are illustrated in Figure 3, and the weighting matrices are also shown in Table 1.

Gait Control by State Feedback

For gait control it was assumed that the biped has an initial attitude il-

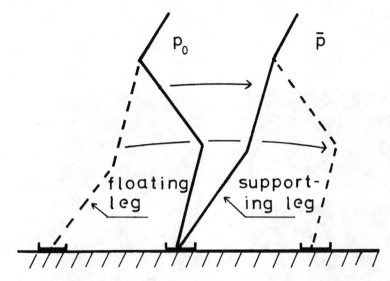

Figure 6—Gait control.

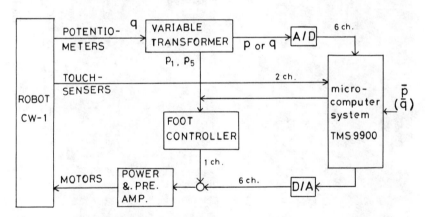

Figure 7—Controlling circuit.

lustrated in Figure 6 at some moment. If the biasing input is switched to ū which specifies the next equilibrium $\bar{\varphi}$ shown in the same figure as soon as the moment in which the swinging leg is identified, the biped could walk a step. This is the main idea of this study. This can be realized accurately only if the assumed constraint, such that the foot of the supporting leg is pinned to the floor, holds during the gait period. As a matter of fact, such a constraint is violated and the biped staggers soon after the center of mass passes through the head of the foot of the supporting leg. If the biasing input, however, is switched when the foot of the floating leg reaches the ground, the biped continues the gait while recovering its stability.

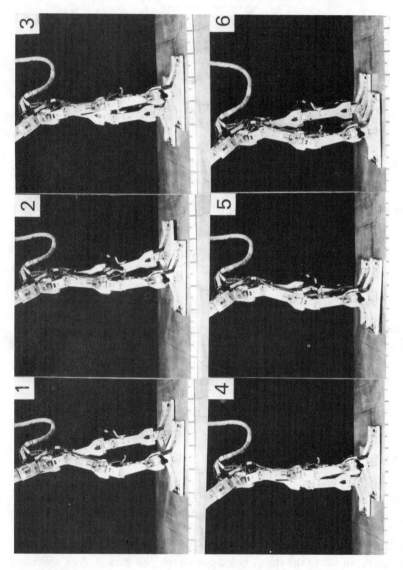

Figure 8 — Locomotion of CHIBA-WALKER 1 (each row shows one step).

A rather important problem is the following. It is a question whether the foot rises or not since only the next equilibrium is provided rather than the trajectories of the legs. Using digital computer simulations, however, it is shown that this is possible provided Q and R in Equation 9 are chosen properly. Another method in which several equilibriums are set over the gait period has also been studied and experimented with (Yamaguchi, 1981) but the detail is omitted.

Experimental Equipment and Results of Experiments

The control system for CW-1 is shown in Figure 7. The roles of the microcomputer system, TMS 9900 (Texas) are: 1) differentiating detected signals using the digital three-point-formula and reconstructing approximated value of $\Delta\dot{p}$, 2) multiplying Δp and $\Delta\dot{p}$ by F and synthesizing the controlling signal $u = -F(\Delta p, \Delta\dot{p})^T + \bar{u}$; 3) checking the signals sent from the touch-sensors and deciding the supporting leg, and 4) changing the order of numbers of p's.

The biasing input ($\bar{u}$) and the optimal feedback gain matrix (F) are calculated by a large scale computer in advance and stored in a floppy disk of the microcomputer system.

The sampling time of the micro-computer system is about 4 msec due to the times of the multiplying operations. There are about 50 multiplications since the size of the gain matrix (F) is 5 by 10.

Since the motor which moves the foot of the swinging leg is not used for the attitude control at all, a minor analog servo-system controls this motor so that the foot comes parallel with that of the supporting leg. The actuating motors are all 15w DC servo motors.

By this controlling method and control system, CW-1 successfully walks any pre-determined number of steps at a frequency of one second per step. One example of the locomotion is shown in Figure 8.

Conclusions

Using the state feedback control studied in modern control theory, the biped locomotion CW-1 successfully walks. Under existing circumstances CW-1 has simple support in order to achieve ideal locomotion. Such a support, however, should be removed in continuing studies.

References

HEMAMI, H., & Farnsworth, R.L. 1977. Postural and gait stability of a planar five link biped by simulation. IEEE Trans. on AC, 22(3):452-458.

KATO, I., et al. 1981. Studies on biped locomotion. Preprint of 25th Symposium on System and Control, pp. 141-144. (in Japanese)

YAMAGUCHI, T. 1981. Control of biped locomotion using modern control theory. Masters Thesis submitted to Dept. of Elec. Eng., Chiba Univ.

YAMAGUCHI, T., Kashiwase, T., Yamazaki, T., & Mita, T. 1981. Control of biped locomotion using modern control theory. Preprint of 10th Symposium on Control Theory, pp. 229-232. (in Japanese)

C.
Miscellaneous

A Finite Element Analysis of
Tibial Component Design in Total Knee Arthroplasty

Ken-ichi Murase, Susumu Tsukahara, and Susumu Saito
Fukushima Medical College, Fukushima, Japan

Roy D. Crowninshield and Douglas R. Pedersen
University of Iowa, Iowa City, Iowa, U.S.A.

Total joint replacements have become one of the most successful procedures for orthopedic surgeons, since Charnley (1976) improved total hips. Compared with total hips the rate of failure of total knees is high. The dominant cause of mechanical failures comes from the loosening of the tibial component. To prevent the loosening, many designs of prostheses have been developed throughout the world (Charnley, 1976).

A comparative study of different geometries of tibial components and the influences of changing the material properties and the size of the prosthesis was conducted using two-dimensional and axisymmetric three-dimensional finite element methods (Chang et al., 1981). Model variations included central fixation post type and tray type both with and without metal support.

Under the unbalanced loading condition, the largest magnitude peak stresses were predicted within the various components of the structures. Maximum stresses in both bone cement and cancellous bone were reduced by using the metal support. The central fixation post of the prosthesis played an important role in decreasing the stresses in the cancellous bone under the unbalanced loading condition. The change of size of the tibial component had little effect on stress distribution.

Methods and Materials

Geometric data were derived from X-ray tomography of the tibia. There were four models produced for analysis. Model 1 and model 2 were the central fixation post type and model 3 and model 4 were the tray type. In model 2 and model 4 the lower part of the tibial component was changed from high density polyethylene (H.D.P.) to metal as metal support (see Figure 1).

Before analysis, convergency of predicted structural displacement was examined and it was found to occur with about 300 elements. The models used in this research had about 450 triangular elements (see Figure 2).

To investigate the effect by changing the tibial component plateau three sizes of models were made (see Figure 3). Type L covered the entire area of bone, type M covered all the cancellous bone area, and type S was slightly smaller than type M.

The following three loading conditions were studied using two-dimensional analysis and two conditions for three-dimensional analysis:

2-D Analysis	3-D Analysis
1. Symmetric loading	1. Symmetric loading
2. Medial loading only	2. One side loading
3. Lateral loading only	

In all cases, the resultant force was equal to five times body weight; 3000 Newtons and the direction of load was vertical and perpendicular to the plateau surface (Morrison, 1970).

The model was divided into 10 different material regions (Goldstein et al., 1980; Pugh et al., 1973). The values of modulus of elasticities are included in Table 1. For representing cancellous bone, six moduli were used. The value in the proximal part was higher than that in the distal part. In two-dimensional analysis, the value at the medial side was higher than that at the lateral side of the tibial plateau. These values for cancellous bone were higher than those used by others (Hayes et al., 1978; Santavicca, 1979; Vichnin et al., 1979) so the analysis was repeated with one lower value of 0.3 GPa. for all cancellous bone regions. But the results showed the same tendency of stress distribution.

Results and Discussion

The analysis by a finite element method depends upon the geometry of the model, loading conditions, and material properties. When this method is utilized for human body analysis, it requires significant simplification and idealization of structure as input data. Despite these limitations of a finite element method, it is a quite useful analysis for providing a technique to predict quickly, without clinical trials or experimental methods, the changes in stress distribution resulting from variations in design of a prosthesis.

The primary cause of mechanical failure is the loosening of the tibial

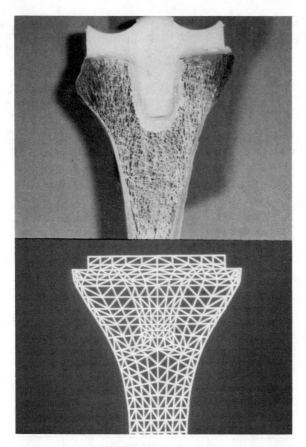

Figure 1a – A dry bone tibia with a central fixation post tibial component and axisymmetric finite element model.

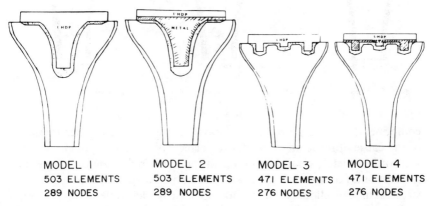

MODEL 1
503 ELEMENTS
289 NODES

MODEL 2
503 ELEMENTS
289 NODES

MODEL 3
471 ELEMENTS
276 NODES

MODEL 4
471 ELEMENTS
276 NODES

Figure 1b – Four models used in this investigation.

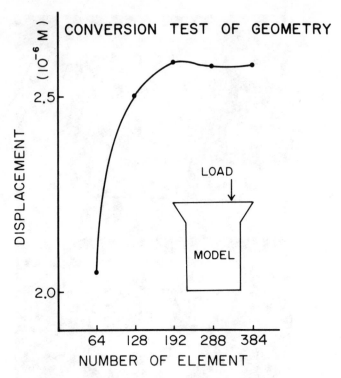

Figure 2—Convergence of predicted vertical displacements at the loading site with an increasing number of elements.

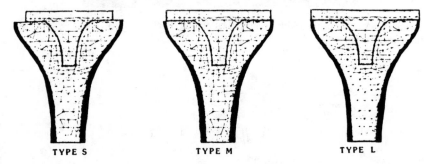

Figure 3—Models for differences in the size of tibial component plateau.

component, but the prevention of loosening requires an understanding of the cause of loosening. The desired effects of tibial component design are not completely clear, however. Although the reduction of stresses in a prosthetic component is clearly desirable, a general reduction of stresses in cancellous bone is not so clearly desirable. Overload or underload have an influence on bone remodeling, which may contribute

Table 1

The Values of Modulus of Elasticities

Material	Poisson's ratio	Mod. of elasticity GPa
UHDP	0.30	1.2
Metal	0.30	200.0
Bone cement	0.23	2.0 (0.5, 1.0, 4.0, 8.0)
Cortical bone	0.30	18.0
Cancellous bone 1	0.30	0.3
Cancellous bone 2	0.30	0.8 (0.3)
Cancellous bone 3	0.30	1.5 (0.3)
Cancellous bone 4	0.30	2.3 (0.3)
Cancellous bone 5	0.30	3.3 (0.3)
Cancellous bone 6	0.30	4.0 (0.3)

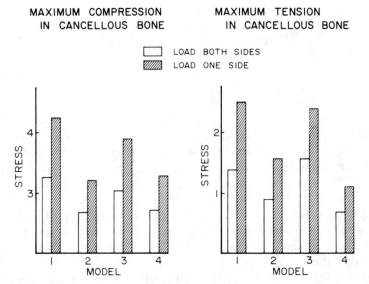

Figure 4—Maximum stresses in cancellous bone under symmetric loading conditions. Results of three-dimensional analysis.

to the loosening of the tibial component. But it is believed that in order to more effectively prevent loosening, large magnitude stresses should be reduced by the type of design of the tibial component. Analysis using this method is useful in providing information regarding the tendencies of the stress distribution associated with various prosthetic designs.

Under symmetric loading conditions, maximum stresses in cancellous bone were found beneath the loading site of tibial plateau in all models except model 2. In model 2, it occurred at the region around the tip of the

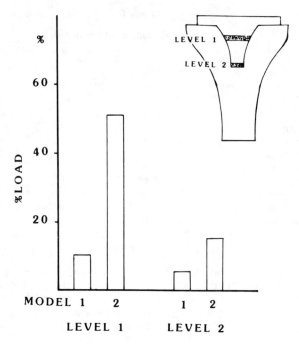

Figure 5 – Percent load transmitted by the post. The metal central fixation post has greater capability of transmitting load compared with the H.D.P. post.

post. By using the metal support, predicted maximum stresses in both bone cement and cancellous bone were reduced by about 40% (see Figure 4). The load transmitted by the post was calculated to compare the effect of H.D.P. and the metal post (see Figure 5). The results showed that the metal central fixation post could transmit about 3 times the load of the H.D.P. post. The effect of transmitting and diffusing the load by a metal post reduced the maximum stresses beneath the loading site and the corresponding maximum stresses in cancellous bone were evident around the tip of the post in model 2.

An unbalanced load, applied unilaterally to the tibial component, caused the highest stresses within both the bone cement and the cancellous bone according to the results of a two-dimensional and three-dimensional analysis. Therefore, it is very important to obtain a good alignment of the lower limb at the time of the operation in order to avoid high stresses. There were large differences between the results of the two-dimensional and three-dimensional analyses. In the results of three-dimensional analysis, stresses were increased about two times according to the concentrated value of the load. But the results of two-dimensional analysis showed the important role of the post of the prosthesis (see Figure 6). The maximum compressive stress in cancellous bone by the tray type was about 50% higher than that by the post type under the

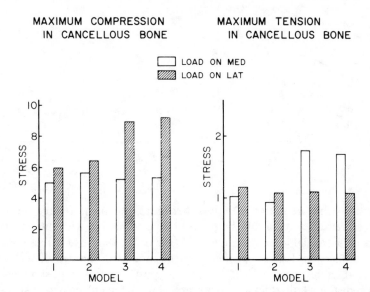

Figure 6 — The maximum stresses in cancellous bone under unbalanced loading conditions. Results of two-dimensional analysis.

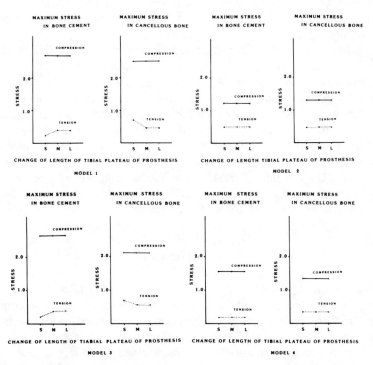

Figure 7 — Variations of maximum stresses in the bone cement and cancellous bone produced by altering the size of the prosthesis.

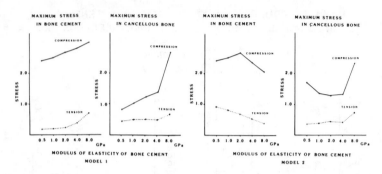

Figure 8 — The results of changing the modulus of elasticity of the bone cement.

lateral loading condition. On the other hand, the maximum tensile stress was increased about 100% by the tray type compared with the post type under the medial loading conditions. One of the reasons for these different results was that the geometry of the models in the two-dimensional method emphasized the effect of the post, since the post extended through the whole thickness from anterior surface to posterior surface. The other reason may have been due to the material properties of cancellous bone such that in the two-dimensional analysis the medial region of cancellous bone had a higher elasticity than the lateral, but in the three-dimensional analysis, the elasticity was symmetric throughout the cancellous bone.

Variations in the size of the tibial component plateau had little effect on the magnitude of stresses (see Figure 7). There were no changes in the maximum compressive stresses in the bone cement and the cancellous bone by enlarging the size. In experimental research of cement strain, Walker et al. (1981) reported similar results in that large surface components did not reduce bone cement strain significantly.

Variations of the predicted stresses in bone cement and cancellous bone according to the change of modulus of bone cement are shown in Figure 8. The maximum stresses were increased in the bone cement and the tendency of changes in the stresses in cancellous bone depended on the models when the modulus of the cement was increased.

Conclusion

This analysis suggested that stresses in the tibia after total knee replacement could be significantly affected by the design of the tibial component and its material properties. Stresses were reduced by using a metal support, and the central fixation post played an important role under the unbalanced loading condition. The most effective design of those analyzed in this study in reducing the stress levels within the bone cement and surrounding cancellous bone appeared to be the central fixation post with a metal support. Tibial component sizes had little effect on stress distribution.

References

CHANG, T.S., Crowninshield, R.D., and Arora, J.S. 1981. A semi-analytic axisymmetric code ASISYM. Technical Report No. 82, Material Division, College of Engineering, University of Iowa.

CHARNLEY, J. 1976. Total knee replacement. Report of a conference held in London, September, 1976. Lancet, May 8.

GOLDSTEIN, S., Wilson, D.L., Sonstegard, D.A., and Matthews, L.S. 1980. An experimental determination of numerical values for elastic modulus and ultimate strength of human tibial trabecular bone as a function of location. Trans. Ortho. Res. Soc. 5:58.

HAYES, W.C., Swenson, L.W., and Schurman, D.J. 1978. Axisymmetric finite element analysis of the lateral tibial plateau. J. Biomech. 11:21-33.

MORRISON, J.B. 1970. The mechanics of the knee joint in relation to normal walking. J. Biomech. 3:50-61.

PUGH, J.W., Rose, R.M., and Radin, E.L. 1973. Elastic and viscoelastic properties of trabecular bone. J. Biomech. 6:475-486.

SANTAVICCA, E.A. 1979. The analysis and design of tibial components for knee replacement. Unpublished thesis.

VICHNIN, H.H., Hayes, W.C., and Lotke, P.A. 1979. Parametric finite element studies of tibial component fixation in the total condylar knee prostheses. Trans. Ortho. Res. Soc. 4:99.

WALKER, P.S., Thatcher, J., Ewald, F.C., and Milden, J. 1981. Variables affecting the cement stresses and the tilting of tibial components. Trans. Ortho. Res. Soc. 6:158.

Hereditary and Environmental Determination of Biomechanical Characteristics in Human Motion Ontogenesis

V.K. Balsevich, A.G. Karpejev, and E.E. Martin
Institute of Physical Culture, Omsk, USSR

The different aspects of exogeneous and endogeneous determination of human motion were investigated by many authors (Balsevich, 1976; Bailey et al., 1978; Malina, 1978; and others). But only recent studies (Gollnick et al., 1972; Zatsiorskiy and Sergienko, 1975; Forsberg et al., 1976; Komi et al., 1976; Murase et al., 1979) had represented the fundamental data about the genetic factors of human motion determination. Meanwhile it is not clear up to this time whether these factors change their influence in ontogenesis or not. At the same time some authors had obtained interesting results about the age peculiarities of genetic determination of neurophysiological parameters of human activity (Ushakov, 1977) and about the possibility of differentiations in the process of animal muscle fibers' evolution under the influence of the fixed conditions for physical activity (Dehl and Aas, 1979).

The methodology of twins investigations is the most popular in the studies of human motion determinations (Zatsiorskiy and Sergienko, 1975; Komi et al., 1976; Murase et al., 1979). Our study was designed to evaluate the heretability estimate in monozygous (MZ) and dizygous (DZ) twins of both sexes with regard to the ontogenesis of biomechanical, morphological, and physical fitness characteristics utilized in human motion.

Methods

Fifty-one MZ and 78 DZ pairs of both sexes ranging in age from 7 to 17 years were used as subjects. Each of the MZ and DZ groups of 12-13 year old subjects consisted of 9 male and 9 female pairs.

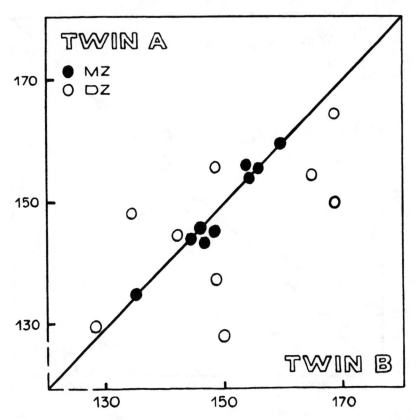

Figure 1 — Intrapair comparison of height parameters of identical (MZ) and nonidentical (DZ) male twins 12 to 13 years of age.

The biomechanical parameters of motion were studied on the pattern of a ball throwing movement. The main features were the time characteristics of different phases of coordinated arm and leg movements and their rhythm. These data were obtained by using the force platform and special transducers for acceleration measurements. Morphological and physical fitness characteristics were obtained by traditional procedures of anthropological measurements and performance testing.

Results

The intrapair variances for the MZ pairs in morphological parameters were smaller than those for DZ males and females in all investigated age groups. The least concordance of morphological characteristics was observed for boys and girls 12 to 13 years of age (see Figure 1).

The comparison of maximal step frequency in the running test of MZ

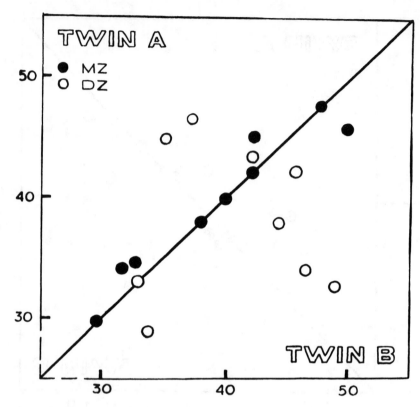

Figure 2 — Intrapair comparison of maximal frequency of steps in running test for MZ and DZ male twin pairs 12 to 13 years of age.

and DZ twins indicated somewhat greater influence of heredity, especially in female pairs (see Figure 2).

In the joint mobility test the intrapair variances differed significantly between MZ and DZ pairs of either sex in all age groups except for male subjects 12 to 13 years of age.

In 30 m sprint the environmental determination of running speed was observed. The same regularity was observed for the results of the long jump and the ball throwing tests.

Biomechanical parameters of the ball throwing had not demonstrated any differences between the intrapair variances of the MZ and DZ twins of 12 to 13 years of age in either sex (see Figure 3). The computation of comparison tests for 28 biomechanical characteristics did not reveal any differences between MZ and DZ male and female pairs of the mentioned age group. The tendency for hereditary determination was observed only for the following parameters: the time of arm movement in the ball throwing test and the time interval of acceleration of the throwing arm. The remaining biomechanical parameters of arm and leg motions, in-

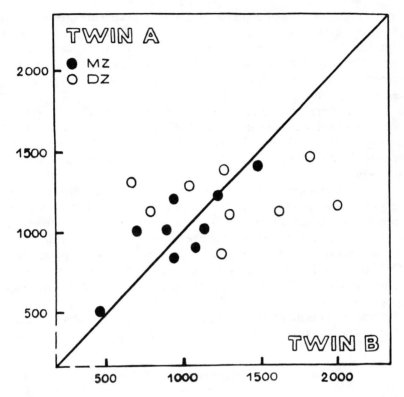

Figure 3 — Intrapair comparison of biomechanical parameters of ball throwing for DZ and MZ male twins 12 to 13 years of age.

cluding the rhythmical characteristics, produced no significant difference between the two twin types of either sex.

Discussion

The comparison of the results of this investigation with the data of earlier studies confirms the opinion about the complicated character of hereditary and environmental influences on the development of human motor function in ontogenesis. First of all it seems reasonable to emphasize that this influence is not the same for different components of human motorics and morphology as well as for different age groups of males and females.

The unambiguous determinations of physiological, anthropometrical, and biomechanical aspects of running performance (Murase et al., 1979) were not confirmed in respect to the ball throwing test. This can be explained by the structural differences of these two kinds of movement.

The coordination of motion in ball throwing is much more artificial than the sprint running. At the same time heredity was found to have a greater influence on maximal frequency of steps in running as tested in this study. The reason may lie in the natural biomechanical structure of this exercise as reported for the natural coordination of the sprint running test used by Murase et al. (1979).

The joint mobility determination can be connected with the peculiarities of morphological development. Probably, these peculiarities are of the same nature as the muscle fiber types distribution. The genetic components for distribution of muscle fiber types were identified by Komi et al. (1976).

It was demonstrated in neurophysiological studies that different periods of ontogenesis differ in the character of hereditary and environmental influences on the age evolution of different functions (Ushakov, 1977). For example, the age of about 13 years was characterized as a period of less influence of heredity on the physical parameters. The same peculiarity of neurodynamic parameters was indicated for males and females 12 years of age. The results of the present study support these data.

It may be concluded that some aspects of a given problem became clearer now, but others still remain obscure and require further study.

References

BAILEY, D.A., Malina, R.M., and Rasmussen, R.L. 1978. The influence of exercise, physical activity, and performance on the dynamics of human growth. In: F. Falkner and G.M. Sanner (eds.), Human growth 2 (Postnatal growth), pp. 475-505. Plenum Press, New York.

BALSEVICH, V.K. 1976. Biological rhythms in the development of human locomotions in ontogeneses. In: P.V. Komi (ed.), Biomechanics V-B, pp. 141-145. University Park Press, Baltimore.

DAHL, H.A., and Aas, O.G. 1979. The effect of activity on the normal development of rat muscle fibre types. In: K. Fidelus and A. Morecki (eds.), VIIth International Congress of Biomechanics (abstracts), pp. 96-97. Warsaw.

FORSBERG, A., Tesch, P., Sjödin, B., Thorstensson, A., and Karlsson, J. 1976. Skeletal muscle fibers and athletic performance. In: P.V. Komi (ed.), Biomechanic V-A, pp. 112-117. University Park Press, Baltimore.

GOLLNICK, P.B., Armstrong, R.B., Saubert, C.W., IV, Piehl, K., and Saltin, B. 1972. Enzyme activity and fiber composition in skeletal muscle of untrained and trained men. J. Appl. Physiol. 33:312-319.

KOMI, P.V., Viitasalo, J.T., Havu, M., Thorstensson, A., and Karlsson, J. 1976. Physiological and structural performance capacity: effect of heredity. In: P.V. Komi (ed.), Biomechanics V-A, pp. 118-123. University Park Press, Baltimore.

MALINA, R.M. 1978. Secular changes in growth, maturation, and physical performance. In: R.S. Hutton (ed.), Exercise and Sport Sciences Reviews. 6:203-255. Franklin Institute Press, Philadelphia.

MURASE, Y., Hoshikawa, T., Amano, Y., Ikegami, Y., and Matsui, H. 1979. Biomechanical analysis of sprint running in twins. In: K. Fidelus and A. Morecki (eds.), VIIth International Congress of Biomechanics (abstracts), pp. 179-180. Warsaw.

USHAKOV, G.K. 1977. The peculiarities of twins development. "Medicina," Moscow. (in Russian)

ZATSIORSKIY, V.M., and Sergienko, L.P. 1975. The influence of heredity and environment on the development of human physical fitness. Theor. and Prac. of Phys. Cult. 6:22-28. (in Russian)

V.
INSTRUMEN-
TATION AND
METHODOLOGY

Keynote Lectures

Methodological Aspects of
Sport Shoe and Sport Surface Analysis

Benno M. Nigg and Simon Luethi
University of Calgary, Canada

Jachen Denoth and Alex Stacoff
Swiss Fed. Institute of Technology, Zurich, Switzerland

Movement is becoming less and less dominant in our daily lives. The number of people who sit during their daily work is increasing. The trip from home to work is usually made by car, bus, etc. Movement is an important part of human life, however—movement is life. This may be a major reason for the enormous increase in running or jogging that has occurred in several countries in the last few years. The present popularity of the jogging movement and the increasing number of people running has led sport shoe manufacturers and athletes to consider the question: "What are relevant criteria for judging the quality of running shoes?" The sport shoe manufacturers want to know how running shoes should be constructed. The athlete wants to know what kind of running shoes one should buy. Thus, manufacturers and athletes alike need criteria for several aspects (e.g., price, durability, "quality," protection). Analogous developments can be noted with sport fields. Enormous numbers of gymnasiums and outdoor sport fields have been built, and this development has led sports facility contractors and users to ask the question: "What are relevant criteria for judging the quality of sport floors?" To evaluate such criteria, tests for sport shoes (Cavanagh et al., 1980) and sport floors (Kolitzus, 1976) were developed and used.

The goal of this article is to: 1) present some general considerations concerning "tests" from a biomechanical point of view, 2) discuss specific considerations and methods in connection with running shoes and sport floors, and 3) describe a possible approach to biomechanical analysis of sport shoes and/or sport floors.

General Considerations Concerning Biomechanical Tests

Tests can be formed with or without human subjects. Tests of both types can be used for single tests or for long term testing. The tests without human subjects — *the material tests* — are normally easy to conduct but may be affected by two problems:

1. It is possible that external conditions (boundary conditions) are not correctly specified. Errors in frequency, range, position, and mass are possible examples. The usually acting effective mass (Denoth, 1980) may be 5 kg, for instance, but the test is performed with a mass of 1 kg. The actual forces on the floor may be 5000 N but the tests are carried out with forces of 500 N.

2. It is possible that the test may not provide a valid criterion on which to base the evaluation. If, for instance, the durability of sport shoes is examined and such shoes usually break at the sole, a test of the upper part of the shoe would not be a valid test with respect to durability.

Neither example is cited from existing test procedures but are hypothetical in order to illustrate the nature of the two problems.

The boundary conditions described in 1 can be determined by a *single test on human subjects*. It is not only important, however, that the boundary conditions are correct but that they are complete. The tester, therefore, has to know what the independent variables of measurement are.

$$x = f(x_1, x_2, \ldots, x_n) \tag{1}$$

If, therefore, x is to be measured, the boundary conditions for all the independent variables $x_1, x_2, \ldots x_n$ must be considered as important.

However, single tests on human subjects will never answer the questions of the validity of a test. The validity of a measured parameter can only be determined by *long term studies with human subjects*. They can provide information concerning the problems that are relevant. They can also show whether the parameters analyzed describe the problem that needs to be solved.

Biomechanical testing should, therefore, start with measurements with human subjects. These single tests and long term studies provide a more complete basis for material tests. All testing that does not start from this base runs the risk of using erroneous boundary conditions, of being irrelevant, and/or of being invalid.

The difficulties of developing a test can be illustrated with an example for testing shock absorption (see Figure 1). Two tests were performed with seven running shoes systematically varied according to the hardness of the heel. One was carried out with subjects running at 5.5 ± 1 m/sec over a force platform. The first contact on the floor was with the heel and the corresponding passive peak force (Nigg et al., 1981) was

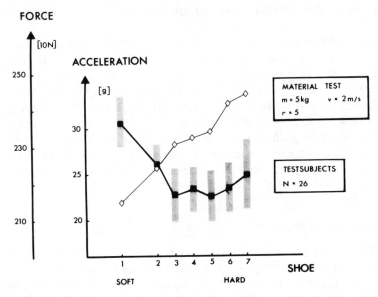

Figure 1 — Results of reaction force measurements with human subjects (running at 5.5 ± 1 m/sec) and acceleration measurements (material test with a shot r = 5 cm, m = 5 kg, v = 2 m/sec) for seven running shoes systematically varied according to the hardness of the heel.

measured. The material test was carried out with a shot dropping onto the heel part of the same shoes (m = 5 kg, v = 2 m/sec). The acceleration was measured with an accelerometer. The result shows that the characteristics of the two measurements were different. Except for the shoe with the softest heel, there was little difference in the passive peak force while there was a steady increase in the accelerations as the heels became progressively harder. By using nothing more than these two tests, one cannot determine which result should be considered as relevant and therefore used as a criterion.

Specific Remarks on Sport Shoes and Sport Floors

Biomechanical analyses of sport shoes and sport floors have been conducted by several authors. Bonstingl et al. (1975), Cavanagh (1978), Hort (1979), Rheinstein et al. (1978) and Bates (1981) measured forces, torques, and/or centers of pressure in connection with sport shoes. Krahl (1978) and Nigg and Luethi (1980) described cinematographical methods to analyze sport shoes. Sport shoes and injuries or pain were described by Wietfield and Thiel (1978) and Segesser and Nigg (1980). The use of foot orthotic devices and/or methods to quantify the influence of such devices on human movement were reported by Nigg et al. (1977; 1978; 1981), Bates et al. (1977) and Segesser et al. (1978). Cavanagh (1978) and

Cavanagh et al. (1980) presented applied tests used to rank running shoes.

External and some internal forces and/or torques associated with the use of different sport floors were measured by Nigg et al. (1974), Browers and Martin (1974; 1975), Baumann and Stucke (1980), Nigg (1980), Nigg and Denoth (1980) and Denoth and Nigg (1981). Accelerations recorded at the shoe were reported by Prokop (1976). The correlation between the construction of a sport floor and performances were described by McMahon and Green (1978). Segesser (1976) described an apparent correlation between the loading of the human body on different surfaces and the occurrence of pain. Kolitzus (1976) developed norms for use in the testing of sport floors.

Loading as a Criterion

Implicit or explicit in most of the studies cited is the basic idea that there is a connection between human health, on the one hand, and sport shoes and/or sport floors on the other. From a biomechanical point of view the loading of the human body — one aspect associated with human health — seems to be important. The relevance of the load placed on the human body can be shown by a survey made with tennis players. Of the 2481 cases analyzed, some of the players reported pain in relation to their tennis activity. More than 25% reported pain in the lower back, in the knees, and/or in the foot. The percentage of people with pain (see Figure 2) was related to the type of floor they were used to playing on. The cases where the origin of pain could be located elsewhere (e.g., previous ski injury) were eliminated from the percentage with pain. Under the prerequisite that pain is a consequence of overloading an element of the human body, this result is a justification that loading of the human body is a relevant criterion for the quality of a tennis floor. Analogous studies could, and should, be performed for other cases, where footwear and playing surfaces are to be evaluated (e.g., jogging, sport shoes).

Different Approaches

If it is shown that loading is a relevant criterion, two basic considerations can be made: 1) where are the limits of loading? and 2) how can loading be reduced?

Limits of Loading. In order to determine the limits of loading, critical values for human tissues determined by Yamada (1973) can be used. He measured, for instance, that the critical pressure for cartilage was 500 N/cm^2 (a critical value is one beyond which structural damage will occur to the tissue). With an estimated congruent area of the knee joint of

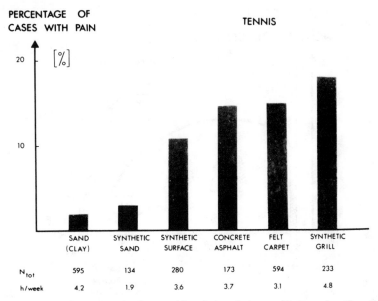

Figure 2 — Percentage of tennis players with pain in relation to the type of surface (2481 cases).

~4 cm² the critical force would be 2000 N, a value that is exceeded in many human movements. The example shows that these figures for 'material constants' of the human body should be accepted with caution (Viidik, 1980). In addition, it demonstrates that general limits cannot be determined with the knowledge currently available (Baumann and Stucke, 1980). As a consequence of this: there is not enough information available to permit the development of norms or standards.

Reduce Loading. The results in Figure 2 and the findings of several other studies (Hess and Hort, 1973; Segesser, 1976; Prokop, 1976) indicate that in many sport activities, the loading of some elements of the human body is too great and should be reduced. If this conclusion is accepted, it is obviously important to determine in which element or elements of the human body the loading should be reduced. In the studies cited, this question is usually addressed only in very general terms and/or in terms of plausibility. Systematic long term studies, however, should be conducted to determine the most endangered elements for specific sport activities. The research then should concentrate on the load reduction of this special element of the human body.

If, for instance, a long term study reveals that the cartilage of the knee joint is endangered in a particular sport activity, a method should be used that determines the forces in the knee joint. If the relevant forces

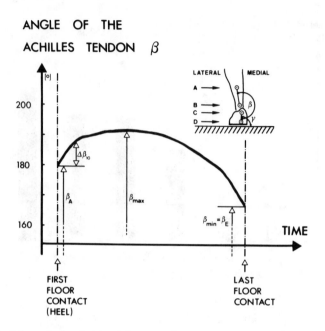

Figure 3—Schematic description of the normalized $\beta(t)$ function.

are high frequency forces (passive forces) the approach described earlier (Denoth, 1980) using the effective mass m* can be applied:

$$F_{knee} = (m^* - m_{tibia}) \cdot \ddot{x}_{tibia}$$

If the forces in the active phase (second part of the force time curve) are relevant, the methods described by Paul (1965), Baumann and Stucke (1980), and others can be applied.

If it were found that the ligaments of the foot are endangered, methods should be used that determine forces in the corresponding ligaments (Procter et al., 1981) or methods that use substitute parameters. Such a method with substitute parameters is illustrated in Figure 3 (Nigg et al., 1980). The following will describe this in more detail. Four points are marked on the posterior side of the human foot and leg. Two on the heel representing the position of the calcaneus in the standing position (if the subject is barefoot) and two on the posterior side of the lower leg, representing the position of the Achilles tendon in standing position. If the subjects are wearing shoes, the two heelpoints are fixed on the heel of the shoe so that a line joining them would form a 90° angle with the floor in the unloaded situation. Instead of measuring the internal forces in the ligaments the external parameters $\beta(t)$ and $\gamma(t)$ are used. The difference between the results from measurements on the heel and on the shoe was

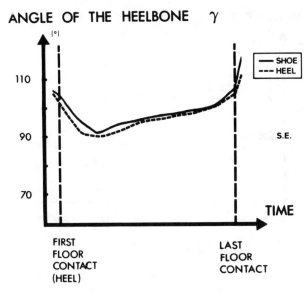

ANGLE OF THE HEELBONE γ

Figure 4—Differences in the angle of the heelbone measured on the shoe or on the heel.

investigated with a few subjects (N = 3) with small holes in the shoe. The results showed that in the dynamic situation (running with first contact with the heel) the functions $\beta(t)$ and $\gamma(t)$ for both the heel and the shoe had the same shape. Systematic shifts are possible for individuals (see Figure 4) but in the movements analyzed they were smaller than 2°. Averaged curves for the Achilles tendon angle in running barefoot and with running shoes (v = 4 ± 1 m/sec) are shown in Figure 5. The result shows that pronation and supination are greater in running with shoes than in running barefoot. This demonstrates that the mechanics of the foot during floor contact can be influenced by the construction of the running shoe. Results for the angle of the calcaneus ("angle of pronation" $\Delta\gamma_{10}$) for different running shoes are shown in Figure 6. The basic shoe was used without and with medial support applied in four different positions from anterior to posterior. The illustrated parameter indicates the change of the position of the 'heelbone' in the first tenth of the floor contact period which corresponds to ~ 30 msec. This time interval corresponds to the duration of the passive forces in landing. It is interesting to compare the results of running with shoes to those without.

The results presented suggest that this method can be applied for sport shoe analysis. The method is relatively simple and shows impressive results. Some limitations should be noted, however:

1. It is plausible that the ligaments of the foot are less loaded by small 'bend in angles', but there is no proof that this assumption is correct. Proof of such might be obtained using analytical or descriptive methods.

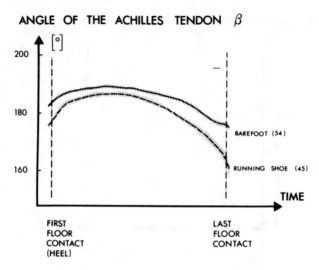

Figure 5—Differences in the angle of the Achilles tendon in running barefoot or with running shoes.

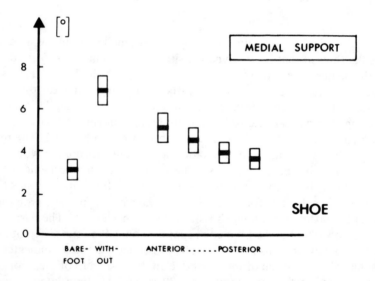

Figure 6—Change of the angle of the calcaneus in the first tenth of floor contact ("Angle of pronation" $\Delta\gamma_{10}$) for barefoot and footwear with different positions of the medial support.

2. It is plausible that the parameters $\beta(t)$ and $\gamma(t)$ are related to the loading of the ligaments of the foot. There is no proof, however, that these parameters are relevant in the analysis of loading the ligaments of the foot.

Conclusion

A review of the literature concerning the biomechanical analysis of sport shoes and/or sport floors leads to the conclusion that a considerable number of different methods to measure kinetic and kinematic parameters have been developed. They are impressive from the point of view of the technical possibilities (Nicol, 1977; Cavanagh, 1978), the description and illustration of movements or forces (Krahl, 1978; Aebersold, 1980), and the number of analyzed cases and other aspects.

Some of the studies attempt to draw conclusions concerning the practical selection of sport shoes and/or sport floors. It is the authors' conviction that such studies should contain the following parts:

1. Determination of the criterion to be used (e.g., price, loading, durability).

2. Proof that this criterion is relevant.

3. Determination of the independent parameters that describe the parameter of interest (e.g., loading of the knee joint $= f(x_1, \ldots, x_n)$).

4. Measurements of the independent parameters $x_1, \ldots, x_n$.

If biomechanical measurements are replaced by material tests, the following additional points should be fulfilled:

5. Adjustment of the boundary conditions.

6. Proof that the material tests are valid.

Material tests that do not cover the six points cited should—from a scientific and biomechanical point of view—not be used as a basis for decision of the practical use of a sport shoe and/or a sport floor.

References

AEBERSOLD, P., Stuessi, E., and Debrunner, H.U. 1980. Computerunterstuetzte Ganganalyse (Gait analysis with the help of computers). Orthop. Praxis 10:836-840.

BATES, B.T., James, S.L., Osternig, L.P., and Mason, B. 1977. The use of foot orthotic devices to modify lower extremity mechanics. Paper presented at the Am. Soc. of Biom. Convention, Oct.

BATES, B.T., James, S.L., Osternig, L.P., Sawhill, J.A., and Hamill, J. 1981. Effects of running shoes on ground reaction forces. In: A Morecki, K. Fidelus, K. Kedzior and A. Wit (eds.), Biomechanics VII. University Park Press, Baltimore. (in press)

BAUMANN, W., and Stucke, H. 1980. Sportspezifische Belastungen aus der Sicht der Biomechanik (Loading in sport from a biomechanical point of view). In: H. Cotta, H. Krahl, and K. Steinbrueck, Die Belastungstoleranz des Bewegungsapparates, pp. 55-64. Georg Thieme, Stuttgart.

BONSTINGL, R.W., Morehouse, C.A., and Niebel, B.W. 1975. Torques developed by different types of shoes on various playing surfaces. Medicine and Science in Sports 2:127-131.

BROWERS, K.D., and Martin, R.B. 1974. Impact absorption, new and old Astro-Turf at West Virginia University. Medicine and Science in Sports 3:217-221.

BROWERS, K.D., and Martin, R.B. 1975. Cleat-surface friction on new and old Astroturf. Medicine and Science in Sports 2:132-135.

CAVANAGH, P.R. 1978. A technique for averaging centres of pressure paths from a force platform. J. Biomechanics 11:487-491.

CAVANAGH, P.R. 1978. Testing procedure. Runner's World Oct.:70-80.

CAVANAGH, P.R., Hinrichs, R.N., and Williams, K.R. 1980. Testing procedure for the Runner's World Shoe Survey. Runner's World Oct.:38-49.

DENOTH, J. 1980. Forschungsmethoden—Methoden zur Bestimmung von Belastungen (Research methods to determine load on the human body). In: B.M. Nigg and J. Denoth (eds.), Sportplatzbelaege, pp. 41-69. Juris Verlag, Zurich.

DENOTH, J., and Nigg, B.M. 1981. Influence of different floors on load of lower extremities in man. In: A. Morecki, K. Fidelus, K. Kedzior and A. Wit (eds.), Biomechanics VII-B, pp. 100-105. University Park Press, Baltimore.

HESS, H., and Hort, W. 1973. Erhoehte Verletzungsgefahr beim Leichtathlehtik training auf Kunststoffboeden (Increased danger of injuries on artificial track and field sport surfaces). Sportarzt und Sportmedizin 12:282-285.

HORT, W. 1979. Des Sportschuh auf modernen Kunststoffbelaegen (The sport shoe on modern sport floors). Leistungssport 3:206-209.

KOLITZUS, H. 1976. DIN 18065 Sportplaetze: Teil 6, Kunststofflaechen, FN-Bau, Berlin.

KRAHL, H. 1978. Kinematographische Untersuchungen zur Frage der Fuss-gelenkbelastung und Schuhversorgung des Sportlers (Cinematographical study concerning the loading of ankle joint in connection with sport shoes). Orth. Praxis 11:821-824.

McMAHON, T.A., and Greene, P.R. 1978. Fast running tracks. Scientific American 239:112-121.

NICOL, K. 1977. Druckverteilung unter dem Fuss bei sportlichen Abspruengen und Landungen im Hinblick auf eine Reduzierung von Sportverletzungen (Pressure distribution under the foot during take off and landing in order to reduce sport injuries). Leistungssport 3:220-227.

NIGG, B.M. 1980. Belastung und Beschwerden beim Tennisspielen (Load and pain in tennis). In: B.M. Nigg and J. Denoth (eds.), Sportplatzbelaege, pp. 70-77. Juris Verlag, Zurich.

NIGG, B.M. 1980. Biomechanische Ueberlegungen zur Belastung des Bewegung-sapparates (Biomechanical considerations concerning the loading of the human body). In: H. Cotta, H. Krahl, and K. Steinbrueck (eds.), Die Belastungstoleranz des Bewegungsapparates, pp. 44-54. Georg Thieme, Stuttgart.

NIGG, B.M., and Denoth, J. 1980. Sportplatzbelaege (Sportfloors). Juris Verlag, Zurich.

NIGG, B.M., Denoth, J., and Neukomm, P.A. 1981. Quantifying the load on the human body, problems and some possible solutions. In: A. Morecki, K. Fidelus, K. Kedzior and A. Wit (eds.), Biomechanics VII-B, pp. 88-105. University Park Press, Baltimore.

NIGG, B.M., Eberle, G., Frey, D., Luethi, S., Segesser, B., and Weber, B. 1978. Gait analysis and sportshoe construction. In: E. Asmussen and K. Jorgensen (eds.), Biomechanics VI-A, pp, 303-309. University Park Press, Baltimore.

NIGG, B.M., Eberle, G., Frey, D., and Segesser, B. 1977. Biomechanische Analyse von Fussinsuffizienzen (Biomechanical analysis of the insufficient foot). Medizinisch-Orthop. Technik 6:178-180.

NIGG, B.M., and Luethi, S. 1980. Bewegungsanalysen beim Laufschuh (Movement analysis and running shoes). Sportwissenschaft 3:309-320.

NIGG, B.M., Leuthi, S., Segesser, B., Stacoff, A., Guidon, H., and Schneider, A. 1981. Sportschuhkorrekturen. Ein biomechanischer Vegleich von drei verschiedenen Sportschuhkorrekturen (A biomechanical comparison of three different sportshoe corrections). Orthopaedie 119. Enke Verlag, Stuttgart. (in press)

NIGG, B.M., Neukomm, P.A., and Unold, E. 1974. Biomechanik und Sport. Orthopaede 3:140-142. Springer Verlag, Berlin.

PAUL, J.P. 1965. Bioengineering studies of the forces transmitted by joints. In: R.M. Kenedy (ed.), Biomechanics and Related Bioengineering Topics, pp. 369-380. Pergamon Press, Oxford.

PROCTER, P., Berme, N., and Paul, J.P. 1981. Ankle Biomechanics. In: A. Morecki, K. Fidelus, K. Kedzior and A. Wit (eds.), Biomechanics VII-A, pp. 52-56. University Park Press, Baltimore.

PROKOP, L. 1976. Sportmedizinische Probleme der Kunststoffbelaege (Sportmedical problems of the artificial floorsurfaces). Sportstaettenbau und Baederanlagen 4:1175-1181.

RHEINSTEIN, D.J., Morehouse, C.A., and Niebel, B.W. 1978. Effects on traction of outsole composition and hardnesses of basketball shoes and three types of playing surfaces. Medicine and Science in Sports 4:282-288.

SEGESSER, B. 1976. Die Belastung des Bewegungsapparates auf Kunststoffboeden (Loading of the human body on artificial surfaces). Sportstaettenbau und Baederanlagen 4:1183-1194.

SEGESSER, B., and Nigg, B.M. 1980. Insertionstendinosen am Schienbein, Achillodynie und Ueberlastungsfolgen am Fuss—Aetiologie Biomechanik, therapeutische Moeglichkeiten (Tibial insertion tendinoses, Achillodynia and damage due to overuse of the foot—etiology, biomechanics, therapy). Orthopaede 9:207-214.

SEGESSER, B., Ruepp, R., and Nigg, B.M. 1978. Indikation, Technik und Fehlermoeglichkeiten einer Sportschuhkorrektur (Indication, technique and error possibilities of sportshoe correction). Orthopaedische Praxis 11:834-837.

VIIDIK, A. 1980. Elastomechanik biologischer jewebe. In: H. Cotta, H. Krahl, and K. Steinbrueck (eds.), Die Belastungstoleranz des Bewegungsapparates, pp. 124-136. Georg Thieme, Stuttgart.

WIETFIELD, K., and Thiel, A. 1978. Veraenderungen am Sportschuh bei Verletzungen und Schaeden des Bewegungsapparates (Changes in the sportshoe as a consequence of injuries). Orthopaedische Praxis 11:838-840.

YAMADA, H. 1973. Strength of Biological Materials. R.E. Krieger Publishing Company, Huntington, New York.

Some Possibilities and Problems in Collecting Data for the Biomechanical Analysis of Sport Techniques in Flatwater Canoe Events

Gert Marhold and Hartmut Herrmann
Deutsche Hochschule für Körperkultur, Leipzig, DDR

The general aim in canoe racing events consists, as in many other sports, of minimizing the time required to complete the race. The efforts of an athlete are, therefore, concentrated upon maintaining a maximal average velocity, irrespective of specific tactical points of view.

One must consider the rather complicated structure of competitive performances with the individually conditioned aspects, i.e., the general and specific mental and physical, as well as material-technical elements and the multitude of relationships in order to assess the intensity and efficiency of the athlete's efforts.

Consistent mastery of a sport technique is recognized within a total structure as an important factor whenever high competitive results are to be achieved. As everyone knows, sport technique is the typical way and method to solve an athletic motor task, i.e., independent of unusual and individual traits (Dyatchkov, 1973; Hochmuth, 1975; Marhold, 1978).

The actual motor task, from a biomechanical viewpoint, can be formulated in one way while the endurance component is taken into consideration so that an optimal part of the total performance can be mobilized during every single cycle and also utilized and transformed as effectively as possible via the available mechanisms and the water medium. The technique applied is expressed within a specific system of simultaneous and consecutive partial movements which can be presented objectively from a biomechanical standpoint by an ensemble of mechanical curves and data.

The application of adequate measuring procedures in canoe racing events must first take into account the specific conditions (water medium, temperature deviations, action space up to 1000 m, individual or crew, boat type kayak or Canadian). Further, basic technical measuring requirements must be satisfied (reliable signal transmission in terms of phase and amplitude, and accurate calibration).

In addition to these requirements, technological conditions must be maintained (e.g., high confidence in applying them, reasonable amount of time). Finally, general measuring principles have to be considered as far as possible (like absence of reaction, i.e., no additional energy supply or inhibition or changes in the motor process that is to be analyzed).

This structure of conditions for the selection of measuring procedures seems obvious at first. There are, however, considerable technological limitations and problems when gathering data which are demonstrated in this article.

First, there is the velocity factor. The introductory remarks and the characterized specific motor task emphasized the central position of this mechanical factor and its corresponding informational value for theory and practice. Finally, the effect of all the partial biomechanical actions must be reduced to the mechanical criterion of a high average velocity with as small a range of variability within the cycles as possible. Technical capabilities for measuring boat speeds are rather limited under special conditions. The advantages provided by non-contact photometric measuring procedures (Hochmuth, 1975; Gutewort & Blumentritt, 1978; Matsui, 1978; Terauds, 1979, 1980; Plagenhoef, 1979; Struble et al., 1981) could not be utilized because of the temporal and economic problems of continuous recording over a 1000 m course. The application of radar speedometers (Straszynska, 1973) has been excluded because of negative results obtained with systems that are also moving. Neither can speedometric procedures with thread connections (Donskoi, 1961; Marhold, 1974; Gutewort & Sust, 1976) be applied over such long distances. After having evaluated and analyzed special transmitter systems (Emtchuk & Zharev, 1970; Schneider et al., 1978), a speedometric impulse-technical system was used that was attached to the boat according to the requirements mentioned previously and in principle very similar to the apparatus described by Schneider (1980).

The system consisted of an impeller (see Figure 1) that was fitted with two permanent magnetic pieces and was able to rotate on an axis lengthwise to the boat. It was mounted on the steering fin (kayak) or course-stabilizing fin (Canadian) (Münch, 1973).

As a result of the rotation caused by the resistance of the water to the boat's relative motion and by the impeller's form, the permanent magnets were moved along a stationary induction-coil. The frequency of the voltage impulses from the transmitter was standardized for the values to be measured. Therefore, the voltage impulses were transformed into a continuous-analog D.C.-signal by means of a digital-analog transducer, thus making the appropriate voltage amplitude proportional to the speed of rotation.

Figure 1 – Impeller mounted on the fin of the boat.

The speedometers were calibrated in a test-basin for the construction of ships similar to that described by Clarys et al. (1972). Calibration curves were established specific to the transmitter. It was expected that this interference with the boat's flow relationship would cause changes in the resistive forces. This, however, was not the case with the single kayak. Additional problems were present, however, because of the varying frequency and unexpected intensity of the steering activities. Steering deflections occasionally resulted in an unbalancing of the measuring potentiometer within the bridge circuit (see Figure 2).

Therefore, steering deflections had to be recorded continuously by an arrangement of velocity transducers fixed on the submerged part of the kayak. By doing so, the effects of steering deflections on the signals could be eliminated. The effects of steering deflections upon the velocity signal are shown in Figure 3.

An increase in resistance of 2% occurred as a result of the measuring fin on the Canadian-single as compared with the kayak-single within a velocity range of 3 to 4 msec^{-1}.

Further signal processing and additional loads inevitably connected therewith which are more or less obvious are not considered, nor are the consequences related to them explained here. Instead, other transducers that were applied to analyze this sport technique are described. Based upon the special biomechanical tasks previously mentioned, force is required to utilize an optimal portion of the performance as effectively as

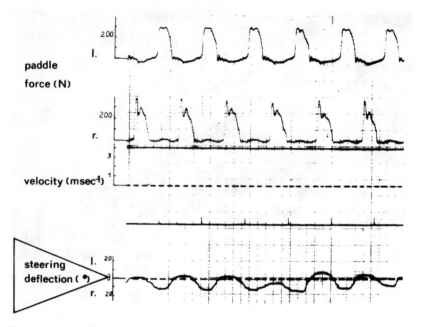

paddle
force (N)

velocity (msec⁻¹)

steering
deflection (°)

Figure 2 — Paddle forces (N) for left and right sides and steering deflections due to an un-balancing of the measuring poteniometer.

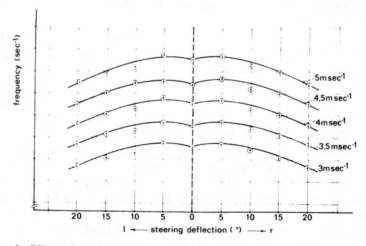

Figure 3 — Effects of steering deflections on velocity.

possible for propulsion during any cycle. We were particularly interested in those forces that act upon the blade, e.g., the pressure or the pulling force at the handles and the pressure at the blade of the paddle. These are usually measured as the result of the bending of the shaft or are calculated from the leverage. Various components of the reactive force in the

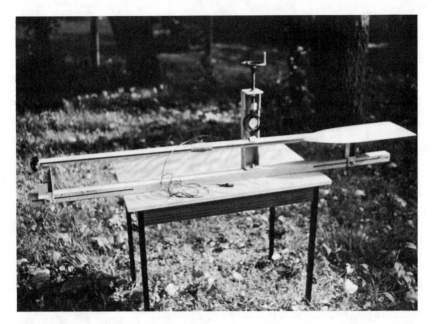

Figure 4 — Calibration of an instrumented paddle.

propulsive direction come into play depending on the position of the blade in the water. Strain gauge measurements were used following procedures utilized in rowing (Sarytchev et al., 1969; Ishiko, 1970; Asami et al., 1978; Schneider et al., 1978; Schneider, 1980) and in canoe racing events (Tonev et al., 1973; Joachimsthaler, 1967; Usoskin & Ganshenko, 1977). Strain-gauge wires were attached to the shaft of the paddle by means of a two-component adhesive and connected with a half-bridge. Regular 4.5 V batteries were used as a D.C. power supply. The bridge supply voltage was stabilized by means of a monolithically integrated circuit. The measuring paddles were calibrated by using a special auxiliary instrument, thus achieving an approximate point-like support at the level of the applied force. The paddle was loaded stepwise at the point of the pulling force. A number of simplifications were achieved by this force measuring procedure, the effects of which cannot easily be assessed technically. Specifically, the simplifications include 1) a point of action of pressure or pulling force at the paddle shaft that was considered to be constant and 2) a constant point of force action at the center of the paddle blade.

The extent of the first simplification can only be calculated theoretically, but the extent of the latter is obvious. It is known from kinematics analyses of the paddle action in the Canadian-single that about 100 msec are required until the given situation has been reached. Deviations must be expected in the case of the completely submerged paddle blade be-

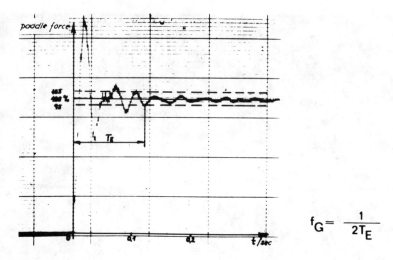

$$f_G = \frac{1}{2T_E}$$

Figure 5 — Dynamic behavior of the transducers showing the characteristic function for a unit step load.

cause of varying depths of entry and of different motions of the water.

If measured values of individual athletes are to be interpreted, these influences may be considered as systematic errors. Caution is necessary, however, when interindividual comparisons are made.

The dynamic behavior of the transducer was also investigated. This was done experimentally by assessing the characteristic function for a unit step load (see Figure 5).

Calculations were made to find the upper limit of frequency (F_G) from the transient time (T_E), i.e., that period of time after which the final value of a unit load approached 95%. It came to 4.2 sec $^{-1}$ which was thought to be sufficient. In addition to those forces acting on the paddle, we were also interested in those forces exerted along the longitudinal direction of the boat via the kinematic chain and the effect of the velocity of the boat created by the interaction of resistive and inertial forces.

The transfer mechanism of the forces of the human body between the paddle and the boat is mobile and can be deformed. The direct measurement of forces at the footboard has the advantage as previously mentioned of measuring the real force of propulsion. Furthermore, reactive forces are caused by body motions in relation to the boat which can be evaluated at the same time. In order to measure these forces, a bending transom was applied as a signal producing unit which had strain-gauges mounted on it and was supported on both sides by the foot board. Roller bearings were built-in between the footboard and the bottom of the boat allowing motion along the boat's longitudinal axis as shown in Figures 6 and 7.

Calibration was achieved by means of a hoop force-measuring shackle

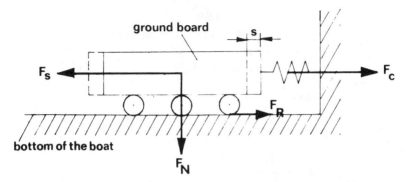

Figure 6—Roller bearing placed between the footboard and the bottom of the boat.

Figure 7—Signal producing unit linked to the footboard in the bottom of the boat.

and a strut between the bending transom and built-in counterfort. Because of the dependency of roller friction forces, force calibration curves were established for different normal forces. The dynamic behavior of the transducer was experimentally analyzed as was previously shown for the paddle. The limiting frequencies were found to depend on the actual displacement as shown in Table 1.

Therefore, these signal producing systems were thus available to measure and record important biomechanical values. The specialist will certainly see from this presentation that the instrumentation described

Table 1

Dynamic Behavior of the Transducer

Weight force F_G (N)	Transient time T_E (sec)	Limiting frequency f_G (sec^{-1})
900	0.18	2.8
800	0.15	3.3
700	0.11	4.5
600	0.06	8.3

here may have an effect on overall performance. Therefore, it was necessary to check whether the motor process that was to be analyzed originally was happening at all under the given circumstances. In addition to the mass normally moved consisting of, for example, an 80 kg athlete and a 12 or 16 kg boat (minimal weights according to regulation for the kayak-single or Canadian-single, respectively), there was a 1 kg increase in paddle weight and about 2.5 kg was added as a result of the measuring equipment (instruments, current supply, additional attachments, and cable connections).

Additional mass was required because of the necessity to make the signals visible within the boat, to record them, store them, or to transmit them telemetrically to appropriate sites on land. An additional 1 to 3 kg would be added depending on the technical design if, for instance, a 4-channel tape recorder was utilized in the boat. Because of these technical and technological problems, a telemetric system was utilized. Further mass increases were thereby reduced to 0.5 kg.

At the same time, it created a practical opportunity to perform a functional test of data collection from the shore. A change in technique would be required of the athlete because of the total additional mass of 3 to 4 kg as compared with the normal displacement creating increased water resistance and inertial forces that would be noticeably different from the normal conditions. It could be deduced from theoretical considerations that the boat velocity during the so-called recovery phase (nonpropulsive phase) would show greater decelerations, and, therefore, greater accelerations would have to be achieved by greater strength efforts in order to maintain the same average boat velocity, the same amplitudes of movement of the paddle, and constant stroke frequency. When the strength effort remains constant, recovery times must be reduced for the same motion amplitude or the stroke frequency and average velocity will decrease. Obviously changes can be expected in any case as a result of the increases in the experimental values shown in Figure 8. These were recorded during tests with specific masses added, as

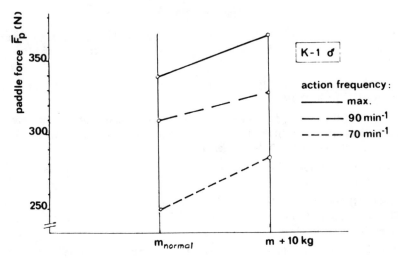

Figure 8 — Changes in paddle force at different stroke frequencies with an additional 10 kg load.

compared with the mass under normal conditions, and demonstrate quite plainly the relationship.

It becomes quite clear that serious reactions may occur when the measuring systems described are applied in singles. These effects can basically be avoided or essentially reduced by: 1) miniaturizing the measuring systems, 2) reducing the mass of the current system of attachments and connections, and/or 3) reducing the mass of the boat shell.

No solutions were immediately available for the first two possibilities. The third solution was successfully achieved by the transition to plastic shells.

In principle, it is possible to stay below the minimal mass of kayak-singles (12 kg) or Canadian-singles (16 kg) as stipulated in the competitive regulations by reducing the mass of the processed polyester and by reducing the amount of color supplements. Therefore, the boats were only of normal weight with the added measuring equipment. Consequently, it was possible to analyze the techniques of canoe racers without essential reactions to changes and to compare the individual biomechanical values.

In summary, it should be stressed again that an attempt was made to demonstrate with a special example the versatility required when selecting, adjusting, and applying biomechanical procedures. Technological and methodological aspects are of greater importance if the remarkably improved technical measuring instruments are utilized which are being provided by scientific-technological progress.

References

ASAMI, et al. 1978. Biomechanical analysis of rowing skill. In: Amsussen and Jørgensen (eds.), Biomechanics VI-B, pp. 109-114. University Park Press, Baltimore.

AUTORENKOLLEKTIV. 1980. Training von A bis Z (Training from A to Z.) Sportverlag Berlin DDR.

CLARYS, J.P., Jiskott, J., & Lewillie, L. 1972. A cinematographic, electromyographic and resistance study of waterpolo and competition frontcrawl. In: Biomechanics III, pp. 446-452. S. Karger, Basel.

DONSKOI, D.D. 1961. Biomechanik der Körperübungen (Biomechanics of body exercises.) Sportverlag Berlin DDR.

DYATCHKOV, V.Wm. 1973. Die Vervollkomnung der Technik der Sportler (The perfection of techniques of sportsmen.) In: Theor. u. Prax. d. KK. Sportverlag Berlin DDR 22, Beiheft 1.

EMTCHUK, J.F., & Zharev, N.V. 1970. Die Steuerung der speziellen Vorbereitung des Rudereres (The special preparation of rudders for oarsmen.) Ubersetzg. a.d. Russ. DHfK Leipzig DDR Ü. 3643. Moskva: Fizkult. i sport.

GÖTTE, K. 1974. Taschenbuch Betriebsmeßtechnik (Handbook of activity measuring techniques.) VEB Verlag Technik Berlin DDR.

GUTEWORT, W., & Sust, M. 1976. Digital speedometry at high sampling rate. In: P.V. Komi (ed.) Biomechanics V-B, pp. 441-448. University Park Press, Baltimore.

GUTEWORT, W., & Blumentritt, S. 1978. Impulslichtphotogrammetrie-Möglichkeiten und Grenzen (Light impulse photogrammetry-possibilities and limitations.) In: Marhold (ed.) Biomechanische Untersuchungsmethoden in Sport. pp. 33-56. DHfK Leipzig DDR.

HOCHMUTH, G. 1975. Biomechanik sportlicher Bewegungen (Biomechanics of sport movements.) Sportverlag Berline DDR. (3. Auflage)

ISHIKO, T. 1970. Biomechanics of rowing. In: Vredenbregt/Wartenweiler (eds.) Biomechanics II, pp. 249-252. V.S. Karger, Basel.

JOACHIMSTHALER, F. 1967. Das Studium der Zusammenhänge zwischen der Kraft des Durchzugs und der äußeren Form der Bewegung beim Paddeln (The study of relationship between the drag force and other forms of motion during paddling.) In: Sbornik vedecke rady UV. Verlag Sport und Touristik Prag, pp. 117-142.

MARHOLD, G. 1974. Biomechanical Analysis of the Shot-put. In: Nelson and Morehouse (Eds.) Biomechanics IV, pp. 175-179. University Park Press, Baltimore.

MARHOLD, G. 1978. Biomechanische Merkmale der Entwicklung sportlicher Techniken (Biomechanical characteristics in the development of sport techniques.) In: Theor. u. Prax. d. KK. Sportverlag Berlin DDR 27(9):691-697.

MATSUI, H. 1978. Automatic analyzing system of cinematography. In: Marhold (Ed.), Biomechanische Untersuchungsmethoden in sport, pp. 17-21. DHfK Leipzig DDR.

MÜNCH, H. 1973. Wissenschaftlich-technische Betrachtung des Systemteils Gerät in Kanu-Rennsport unter der Sicht der Erschließung möglicher Leistungsreserven (Scientific and technical observations of part systems instruments in canoe racing from the viewpoint revealing possible performance reserves.) Unveröffentlichtes Manuskript DHfK Leipzig DDR.

PLAGENHOEF, S. 1979. Biomechanische Analyse des Kanu-Rennsports (Biomechanical analysis of canoe racing events.) In: Res. Quart. 50(3):443-459.

SARYTCHEV, S., Krasnopevzev, G., & Lazareva, A. 1969. Tensometrische Untersuchungen im Rudern (Investigations of rowing.) In: Sammelband "Grblya na baydarkach i kanoe" Fiz. kult. i sport Moskau.

SCHNEIDER, E. 1980. Leistungsanalyse bei Rudermannschaften (Analysis of performance at rowing championships.) In: Aus der Wissenschaft für die Praxis Bd. 6 Limpert-Verlag Bad Homburg BRD.

SCHNEIDER, E., Angst, F., & Brandt, J.D. 1978. Biomechanics in rowing. In: Asmussen and Jörgensen (Eds.) Biomechanics VI-B, pp. 115-119. University Park Press, Baltimore.

STRASZYNSKA, D. 1973. Die Anwendung eines Radargeschwindigkeitsmeßgerätes im Sprinttraining (The use of a radar speedometer in sprint training.) In: Leicht-athletik Warschau 18(7):26, Beilage.

STRUBLE, K.R., Erdman, A., & Stoner, L.J. 1981. Cmputer-aided cinematic analysis of rowing. In: A. Morecki and K. Fidelus (Eds.) Biomechanics VII, University Park Press, Baltimore.

TERAUDS, J. 1979. Science in biomechanics cinematograph. In: Proceedings of the Clinic on High Speed Biomechanics Cinematograph. Academic publishers, Del Mar, CA.

TERAUDS, J. 1980. Achievements and prospects for the use of cinemethods at major international sport events. In: Scientific World Congress, Sport in Modern Society, Tbilisi (USSR).

TONEV, T., Tsvetkov, A., & Boytchev, K. 1973. Untersuchungen der Charakteristika des Schlagzyklus im Einer-Kanu bei der Arbeit im Ruderbecken und unter normalen Bedingungen (Investigations of the characteristics of the stroke cycle in single canoe while working in a rowing basin and under normal conditions.) In: Fragen der Körperkultur Sofia 10:596-601.

USOSKIN, E.G., & Ganshenko, J.V. 1977. Kräfte die beim Paddelboot und Kanurudern auf das Ruder einwirken (Forces on the canoe and the canoe paddle affecting the paddler.) In: Grebnoy sport-Ezegodnik Moskau Fizkult. pp. 59-62.

A.
Instrumentation

The CoSTEL Kinematics Monitoring System: Performance and Use in Human Movement Measurements

Aurelio Cappozzo
Universita degli Studi, Rome, Italy

Tommaso Leo
Universita degli Studi, Ancona, Italy

Velio Macellari
Istituto Superiore di Sanita, Rome, Italy

Stereophotogrammetry has proved to be the most effective technique for the analysis of gross body movements. Optoelectronic devices have been developed as a substitute for the conventional photographic cameras permitting the direct feeding of the target landmark position information to a digital computer and thus automation of the measurement (Winter et al., 1972; Jarret et al., 1974; Lindholm, 1974; Brügger & Milner, 1978; Leo & Macellari, 1981). Several such systems are currently used in laboratories of biomechanics. There still are serious difficulties, however, in bringing these systems into a clinical environment or other field situations. This is mainly due to the imperfect accomplishment of the following features: accuracy, reliability, and simplicity of usage.

The accuracy of these systems is assessed in terms of their ability to permit the reconstruction of the displacement, velocity, and acceleration functions of a moving target landmark. This implies that attention be paid to the statistical characteristics of the random error of measurement in addition to sampling frequency.

Reliability of the system essentially refers to the invariance of its performance with respect to contingent circumstances such as measurement field illumination and movements performed by the test subject.

Usage simplicity is considered in terms of measurement management, which should be thoroughly automatized, and in terms of calibration procedures, which should be simple and required only when the geometry of the stereo system is modified.

In this article, the biomechanically oriented features of an automatic kinematic monitoring system (CoSTEL) are discussed according to the criteria already outlined. The working principle of CoSTEL has been previously reported (Leo & Macellari, 1981).

The CoSTEL

CoSTEL is the name given to an optoelectronic remote sensing device designed for the measurement of the linear coordinates of projections of a target light emitting point onto three non-coplanar image space axes. These projections are obtained by means of toroidal lenses (anamorphic objectives) and the relevant coordinates measured by CCD linear array sensors placed in the focal planes of the objectives. Each transducer reading identifies a plane on which the target point and the objective nodal axis lie. Three of such planes univocally identify the spatial position of the target point in the laboratory system of reference.

The resolution of each transducer depends on the distance between two adjacent photoelements in the CCD sensor, on the principal distance of the lens, and on the distance between the measurement field and the objective nodal axis. The stereometric unit of CoSTEL mounts three CCD-143 Fairchild arrays with 2048 photoelements. The angular resolution of each transducer is 13.8×10^{-3}. This entails a linear resolution on each coordinate in the object space of 1.2 mm for a target point placed at 5 m from the objectives. For the arrangement of the three transducers shown in Figure 1, the measurement field was 2.2 m wide, 2.2 m high, and had a depth of 0.6 m.

CoSTEL is used with Telefunken CQX-19 infrared radiation emitting markers. These are telecontrolled. Since only one marker can be tracked by the transducers at a time, the eight markers, which are considered necessary for one-side whole body movement analysis, were fired sequentially and cyclically. The maximal frequency with which the position of each marker is sampled, up to 1 kHz as far as the CCD sensor is concerned, is limited by the amount of marker radiating energy that reaches one CCD photoelement. For the set-up shown in Figure 1, the relevant reliability of the system is unaffected by sampling frequencies less than 100 Hz. This is an adequate value for whole body movement analyses. The good matching between the marker radiation frequency and the CCD maximal sensitivity frequency, in addition to an automatic adjustment of the transducer threshold level as a function of the background light level, allows for a high insensitivity of CoSTEL to spurious radiation sources.

Two CoSTEL units, one for each side of the subject are controlled by a microcomputer through a Master Unit (see Figure 1). For gait analysis

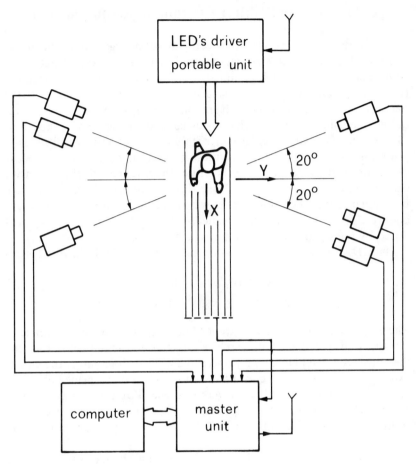

Figure 1 — Experimental set-up.

measurements, information relative to stride temporal factors, arising from an instrumented walk path, are fed into the Master Unit. The computer thoroughly manages the measurement consistent with relevant basic information given through the keyboard. These include the number of markers to be activated, the sampling frequency, and the specific transducers being activated during the particular measurement. An analytical stereophotogrammetric model is also implemented in the microcomputer. It yields the 3-D instantaneous coordinates of the target markers in the laboratory frame. These data are thereafter stored in a permanent file (floppy disc). They may be retrieved and processed in the same or an other computer, and the results displayed according to the practical requirements.

Calibration and Object Coordinate Reconstruction

Given the measured image space coordinates of a target point, the afore-mentioned photogrammetric model used yielded relevant object space coordinates. This model comprised an image distortion correction (I.D.C.) operator for each transducer, and a 3-D projective (3-D.P.) operator. The identification of the parameters of these two operators required the calibration of each transducer and of the stereometric unit (S.U.). The remarkable performance stability of CoSTEL permitted an a priori and single identification of these parameters for a given geometrical set-up of the three transducers within the S.U. and of the S.U. with respect to the measurement field. The three transducers of the S.U. were positioned so that their optical axes were aimed at the object space origin and were positioned in the X-Y plane (see Figure 1). The inaccuracy produced by these alignments was not critical.

The I.D.C. operator for the j-th transducer had the following form:

$$x_j = x_j' + (p_{1j}(x_j' - x_{oj})(k_j - k_{oj})^2 \tag{1}$$
$$+ p_{2j}(k_j - k_{oj}) + p_{3j}(x_j' - x_{oj})); \ j = 1,2,3$$

where x_j was the corrected image coordinate; x_j' the transducer output; k_j was the undistorted coordinate of the target point radial projection onto the objective focal plane, with respect to an axis orthogonal to the sensor axis; x_{oj} and k_{oj} were the coordinates of the focal plane principal point. All aforementioned coordinates were referenced to a measurement axis system. The coordinate, k_j, was estimated from the relevant object space coordinates through an obvious 2-D inverse projective operator (Wolf, 1974).

The 3-D.P. operator that linked the object and image spaces had the following form:

$$(x_j c_{1j} + c_{2j})X + (x_j c_{3j} + c_{4j})Y + (x_j c_{5j} + c_{6j})Z + x_j c_{7j} + 1 = 0; \ j = 1,2,3 \tag{2}$$

where X, Y, and Z were the target point object space coordinates; x_1, x_2, and x_3 the relevant corrected image coordinates.

The I.D.C. operator parameters were assessed, for each transducer at a time, in the following way. A planar calibration grid, with 0.2 m mesh, was constructed with overall dimensions of 2×2 m. Active markers were placed on the grid intersections for a total of 121 different positions. Relevant measured image coordinates (x_1', x_2', x_3') were obtained from each transducer placed at a 5 m distance from the grid—this was consistent with the experimental set-up shown in Figure 1—and with its optical axis orthogonal to the grid and aimed at its central intersection. The parameters p_{ij} were thereafter identified from Equation 1 using a least squares procedure. The coordinate, x_j, was made equal to the relevant undistorted coordinate calculated, as with k_j, using the 2-D inverse projective operator already mentioned. After correction, the standard deviation of the coordinate residuals throughout the entire viewing field was found at 0.9 LSB (which corresponds to a reconstructed grid coordinate standard deviation at 1.1 mm).

The S.U. calibration was carried out as follows. A three-dimensional grid, with 0.2 m mesh, was constructed. It had overall dimensions 0.8 × 0.8 × 0.4 m, and active markers were placed on each intersection (calibration points). This calibration object was placed at the center of the measurement field. The symmetrical axes of the calibration object defined the object space system of reference. The following operations were carried out which led to the estimation of the 3-D.P. operator parameters:

1. The calibration point measured image coordinates (x_1', x_2', x_3'), together with the corresponding object coordinates, were fed in the I.D.C. operator.

2. The corrected image coordinates of the calibration points (x_1, x_2, x_3), together with the relevant object coordinates (X, Y, Z) were used, within a least squares procedure, for the parameter identification of the 3-D.P. operator c_{ij}. Since this operator was linear with respect to its parameters, the estimated values of these latter parameters were optimal in the global sense.

The object coordinate reconstruction of a given target point was carried out using the following procedures:

1. The transducer outputs were fed in the 3-D.P. operator and provisional values of the object coordinates obtained. These latter object coordinates permitted the calculation of k_j in the I.D.C. operator.

2. The transducer outputs were then fed in the I.D.C. operator, together with the provisional object coordinates, and relevant corrected image coordinates obtained. It should be emphasized that the accuracy with which k_j in Equation 1 was estimated had no important effect on the effectiveness of the I.C.D. operator.

3. These latter coordinates permitted, through the 3-D.P. operator, the calculation of the relevant object coordinates.

Since the active markers were fired sequentially, the relevant coordinate samples needed a posterior synchronization. This was done using interpolating splines that were resampled at evenly spaced instants in time, equal for all markers.

Measurement Error

The aforementioned planar grid was placed in various positions within the measurement field and its intersection point coordinates recon-

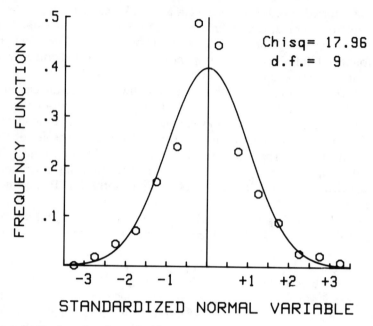

Figure 2 — Random error frequency histogram (900 samples, s.d. = 0.39 mm) with normal distribution.

structed. The relevant maximal errors for the X, Y, and Z coordinates were found to be 2.9, 7.2, and 2.6 mm, respectively and occurred at the very periphery of the field.

The statistical properties of the random error were assessed. This was done by replicating a number of times the observation of the same trajectory. An active marker, mounted on a rotating arm, was made to move along circular trajectories, with radius equal to 0.5 m, and lying on the X-Z plane. The trajectory center was placed in nine different positions so that the entire measurement field was covered. The angular velocity was constant at approximately 1 rev/sec. The CoSTEL sampling frequency was 100 Hz. Nine ensembles of ten replicate observations were obtained for each marker coordinate. The mean value across each ensemble was calculated. The difference between this mean value and the surveyed relevant coordinate was within system resolution, throughout the entire measuring field. Ensembles of ten sample stochastic sequences (random error) were obtained as the differences between the relevant coordinate observations and ensemble averages. These stochastic sequences were tested for ergodicity, lack of correlation, and normality.

The standard deviations of the stochastic sequences were calculated as ensemble averages and time averages. The former standard deviations did not show trends with respect to time nor with respect to the trajectory position within the measurement field. Since the aforementioned statistical properties did not undergo significant variations when they were assessed over single sample records or over the several samples of each ensemble, the random error could be assumed ergodic throughout the entire measurement field. The relevant standard deviation was 0.39, 0.94, and 0.34 mm for the X, Y, and Z coordinates, respectively.

The covariance matrix was calculated for one sample sequence (autocovariance), for two sample sequences within the same ensamble, and for two sample sequences belonging to two different ensembles (crosscovariances). The autocovariance matrices were dominantly diagonal and the crosscovariance matrices were, with good approximation, zero matrices. A lack of statistical correlation between any two samples of the random error could thus be assumed.

The type of frequency distribution of the random error was also investigated through the chi-square test. For this test, all sequences of each ensemble were used. This could be done because of ergodicity of the random variable. For a normal probability distribution the test gave a significance level at around 1% for all ensembles. The distributions were remarkably symmetrical and, on the average, 94% of the errors fell within plus or minus two standard deviations, and 72% within plus or minus one standard deviation (see Figure 2).

The previous statistical tests were repeated with the rotating arm moving at 2.5 rev/sec. Results were consistent with those already discussed. The invariance of the random error statistical properties to experimental conditions, that can be inferred from this discussion, permitted the use of noise reduction and differentiating techniques independent of the particular signal being processed.

Acknowledgment

This work was partly a research project sponsored by the Centro di Ingegneria Biomedica, Universita degli Studi, Roma, Italy.

References

BRÜGGER, W., & Milner, M. 1978. Computer-aided tracking of body motions using c.c.d.-image sensor. Med. Biol. Engng. Comp. 16:207-210.

JARRET, M.O., Andrews, B.J., & Paul, J.P. 1974. Quantitative analysis of locomotion using television. ISPO World Congress, Montreaux, Switzerland.

LEO, T., & Macellari, V. 1981. On line microcomputer system for gait analysis data acquisition based on commercially available optoelectronic devices. In: A. Morecki and K. Fidelus (eds.), Biomechanics VII-B, pp. 163-169. National Scientific Press, Warsaw, and University Park Press, Baltimore.

LINDHOLM, L.E. 1974. An optoelectronic instrument for remote on-line movement monitoring. In: R.C. Nelson and C.A. Morehouse (Eds.), Biomechanics IV, pp. 510-512. University Park Press, Baltimore.

WINTER, D.A., Greenlaw, R.K., & Hobson, D.A. 1972. Television-computer analysis of kinematics of human gait. Comp. Biomed. Res. 5:498-504.

WOLF, P.R. 1974. Elements of Photogrammetry. McGraw-Hill, New York.

Evaluation of Total Knee Arthroplasty Patients Using the VA Rancho Gait Analyzer

James C. Otis, Dennis F. Fabian, Albert H. Burstein, and John N. Insall

Hospital for Special Surgery, New York, N.Y., U.S.A.

The objectives of this study were to compare the preoperative and postoperative gait performances of a group of total knee arthroplasty (TKR) patients to each other and also to the performances of a normal population. Previous work has shown that it is possible to define a relationship between single stance time (SST) and velocity during normal gait (Perry et al., 1976; Otis & Burstein, 1981). It was a specific aim of this study to examine the velocity and SST measurements for preoperative and postoperative TKR patients to determine if there existed a relationship between the variables which define and differentiate the gait of patients who require total knee arthroplasty from the gait of normal subjects, and also to determine whether or not this relationship for the postoperative patient approximates that of the normal subjects. In addition, the variables of SST and velocity were examined to determine whether or not they could be used to independently characterize abnormal gait function in this patient population.

Single stance time was chosen as a measurement variable because it reflects: 1) the patient's ability to support load on each leg, 2) the symmetry of gait when comparing the two sides, and 3) the degree of normal bipedal gait rhythm when examined in relationship to velocity. In addition, free walking velocity, which is an indicator of overall performance and efficiency of gait, was also examined.

Methods

Right and left SSTs and average velocity were measured using the VA-Rancho Gait Analyzer which was developed by the Pathokinesiology Service at the Rancho Los Amigos Hospital for the Veterans Administration.[1] The instrumentation consists of foot switches, a manual or remote start/stop controller, a waist pack recorder, and a calculator. The foot switches are worn as insoles in the individual's street shoes and indicate when the foot is weight bearing. At the end of each run the recorder is connected to the calculator and right and left single stance time as a percentage or normal (%NSST) and the average velocity for 6 m are calculated for each trial.

The calculator is programmed with a calibrating function which defines the relationship between absolute single stance time (SST) and velocity. For each measured velocity the calculator internally determines the expected SST for a normal individual, and uses that value to normalize the measured SST for the right and left sides following each trial. With the help of the Rancho group the Gait Analyzer was recalibrated based upon a normal study of 67 females and 69 males at this institution (Otis & Burstein, 1981). This provided separate data bases for females and males, respectively, which were used in this study.

TKR patients (M = 23) with unilateral or bilateral osteoarthritis of the knee joint were evaluated preoperatively and postoperatively. The population included 17 females and 6 males ranging from 49 to 85 years with a mean age of 68 years. Seven cases were bilateral and 16 cases were unilateral. The preoperative alignment was noted with 12 knees in valgus, 12 in varus, and 6 in a neutral position.

Patients were tested preoperatively and six months postoperatively. On both occasions the patients were instructed to walk at a free velocity. Five consecutive trials were conducted and the left and right %NSSTs and velocities were recorded. The mean values for measured velocity, left %NSST, and right %NSST for the five runs were used for the determination of the %NSSTs based upon the performance of a normal population from the Hospital for Special Surgery. In addition, the Hospital for Special Surgery Knee Rating (Insall et al., 1976) was completed at each test session and correlated with the Gait Analyzer scores.

Each patient in the group underwent TKR utilizing either the Total Condylar (TCP I) or Total Condylar (TCP II) prosthesis with patellar components. One orthopedist performed all of the surgery.

Results

Preoperatively, the unilateral patients demonstrated a relationship between SST and velocity that was significantly different from normal for both the affected and the unaffected sides. This difference is illustrated in Figure 1, which shows the linear regressions of percent normal single stance time (%NSST) versus velocity with slopes of .537 and .427 %/cm/sec for the affected and nonaffected sides, respectively. The difference between these slopes was not significant; however, the mean

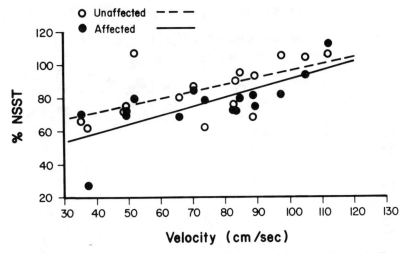

Figure 1 — Preoperative %NSST vs velocity for the nonaffected (dashed line) and affected (solid line) sides of 16 unilateral patients. The slopes are .427 and .537%/cm/sec, respectively.

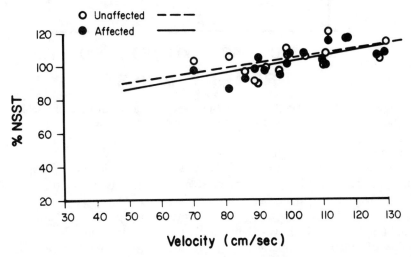

Figure 2 — Postoperative %NSST vs velocity for the nonaffected (dashed line) and affected (solid line) sides of 16 unilateral patients. The slopes are .290 and .328%/cm/sec, respectively.

%NSSTs were both significantly different than 100% (p < .01) and significantly different from each other (p < .05). Postoperatively, the slopes of .328 and .290 for the affected and nonaffected sides, respectively, demonstrated relationships between %NSST and velocity (see Figure 2) that were significantly different from the normal relationship; however, these slopes were not significantly different from each other. In ad-

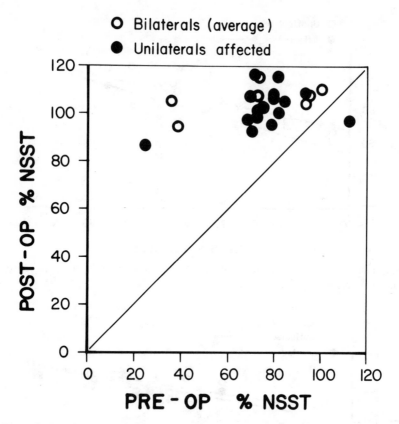

Figure 3—Postoperative %NSST vs preoperative %NSST of the affected sides of 16 unilateral patients and the average of both sides for the seven bilateral patients.

dition, the %NSST's had increased to normal levels and were not different from each other. A comparison of postoperative velocity to preoperative velocity for this group shows a significant increase (p < .01) from 73.9 cm/sec to 101.0 cm/sec.

The seven bilateral patients demonstrated a preoperative %NSST of 73.7 (± 26.2)% which was significantly (p < .01) less than the normal value of 100%. Postoperatively, a significant (p < .001) increase to 107.5 (± 6.9)% was demonstrated. In addition, a significant (p < .001) increase in velocity from 65.5 (± 26.0) to 112.5 (± 13.3) cm/sec was demonstrated. Postoperatively, the bilaterals could not be differentiated from the unilaterals based upon %NSST and velocity measurements.

Discussion

Through the use of single limb support time and velocity data it has been possible to document in a quantitative manner a significant improvement in gait performance following total knee arthroplasty. It is interesting to note the preoperative scores for the unilaterals. These patients demonstrate a significant reduction in single limb support time not only of the affected extremity, but also of the unaffected side. These results indicate that the unilateral preoperative patients compensate in their gait pattern by increasing their period of double support to minimize the asymmetry of their gait.

The measurements obtained from The Hospital for Special Surgery Knee Score Sheet were examined to see if any correlation existed between the quantitative results obtained with the Gait Analyzer and the subjective scores given to patients in the clinic. The subjective scores were compared to the velocity and %NSST measurements obtained from each patient. Although the score sheet showed a mean improvement of 48% for the TKR population, a significant correlation could not be found between the degree of improvement as measured by the score sheet and the improvement in %NSST and velocity measured with the Gait Analyzer. Of all the categories on the score sheet, only the pain score was found to have a statistically significant correlation with velocity and %NSST.

Although the improvement measured with the Gait Analyzer correlated only with the pain scores in the Hospital for Special Surgery Knee Rating, the ability to provide functionally significant measures of gait performance in TKR patients by using stance time parameters is evident from the plot of postoperative %NSST versus preoperative %NSST shown in Figure 3. This figure demonstrates that all the patients, except for one who began with a normal SST, increased their %NSST, thus indicating that velocity dependent SST is a useful measurement when applied to this clinical situation.

Acknowledgment

The authors are grateful to the Veterans Administration Prosthetics and Research Center and to the staff of the Pathokinesiology Service of the Rancho Los Amigos Hospital for their cooperation and the Clark Foundation for their financial support.

Footnote

1. Veterans Administration Contract No. V101 (134) p-244.

References

INSALL, J.N., Ranawat, C.S., Aglietti, T., & Shine, J. 1976. A comparison of four models of total knee replacement prosthesis. J. Bone and Joint Surgery 58A.

OTIS, J.C., & Burstein, A.H. 1981. Evaluation of the VA-Rancho Gait Analyzer, Mark I. BPR 10-35.

PERRY, J., Antonelli, D.J., & Bontrager, E.L. 1976. VA-Rancho Gait Analyzer Final Project Report. Pathokinesiology Service Report No. 4, Rancho Los Amigos Hospital, Downey, CA.

Pressure Distribution Measurements by High Precision Piezoelectric Ceramic Force Transducers

E.M. Hennig, P.R. Cavanagh, and N.H. Macmillan
The Pennsylvania State University,
University Park, Pennsylvania, U.S.A.

In many studies of gait, it is important to determine 1) the magnitude, direction, and point of application of the resultant force acting on the foot and 2) the way in which this force is spatially distributed over the plantar surface of the foot. Lord (1981) has recently reviewed many of the attempts made in the last century to measure these quantities. The present paper describes a new piezoelectric device for studying the distribution of the vertical contact stresses generated between the plantar surface of the foot and the shoe insole with greater spatial and temporal resolution than ever before; and the companion paper (Cavanagh et al., 1981) reports on the application of the device to the study of walking.

Methods

The heart of the device is a 'pressure-sensitive shoe insole' consisting of a flexible array of 499 4.78 mm square, 1.2 mm thick lead zirconate titanate transducers embedded in a 3 to 4 mm thick layer of highly resilient silicone rubber that is impervious to moisture and electrically insulating. These transducers have silver electrodes diffusion bonded to their major surfaces and are laid out in a square pattern at a center-to-center spacing of 6 mm on a sheet of thin copper gauze cut to the shape of a U.S. size 10 right foot. The gauze serves as a common ground, and the separate connection to the upper electrode of each transducer is made via a thin annealed copper wire that runs through the silicone rubber to the edge of the array and thence up the leg of the subject to a small (25 cm × 18 cm × 15 cm, 2.9 kg) back pack. Figure 1 shows the array at an early stage of construction, with the copper gauze installed beneath the

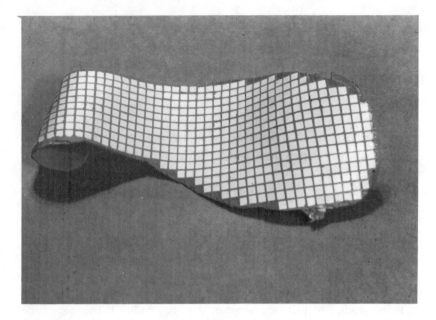

Figure 1—Prototype transducer array in the shape of a size 10 right insole at an early stage of construction.

transducers but neither the connections to the upper electrodes nor the outer covering of silicone rubber in place.

Each piezoelectric ceramic transducer forms part of a simple circuit (see Figure 2) in which it is connected to the inverting input of a high input impedance ($\sim 10^{12}\Omega$) field effect transistor operational amplifier used as a charge amplifier. In the configuration shown in Figure 2 (Hennig et al., 1981)—i.e., with a 3.3nF capacitor and a 3GΩ resistor connected in parallel between the same input and the amplifier output—the output voltage from the amplifier is virtually independent of the capacitances of the transducer and the cable linking this to the amplifier. As a result, it is possible to install the amplifier in a back pack rather than having to incorporate it into the shoe insole. This output voltage also is directly proportional to the charge generated by the transducer, with the constant of proportionality being determined by the capacitance of the feedback capacitor connected across the amplifier. Moreover, since the transducer is made from an appropriately oriented piece of an orthotropic material having a highly linear and hysteresis-free piezoelectric response (see subsequent explanation), it is in principle possible to mount it in such a manner that the charge it generates is 1) directly proportional to the vertical component of the stress generated by the foot against its upper surface and 2) independent of the medio-lateral and antero-posterior shear stresses generated concomitantly.

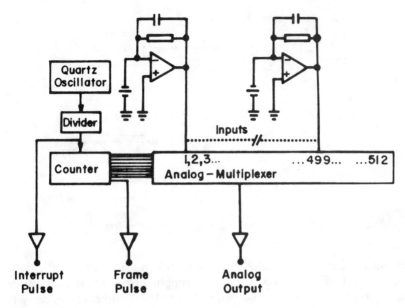

Figure 2 — The transducer-amplifier units incorporated into a multiplexing circuit that produces the outputs from them on a single line. Interrupt pulses and frame pulses are also generated.

The circuit shown in Figure 2 also has one other notable feature. This is that the resistor in parallel with the feedback capacitor provides each charge amplifier with an automatic reset. The time constant of 10 sec is long enough to permit loads generated by slow walking to be measured accurately, but short enough to prevent any extensive build-up of charge either by pyroelectric generation (if the temperature of the operating environment changes) or by integration of the amplifier input offset current (Hennig et al., 1981).

To further mitigate against this latter error, measurements are made not of the amplifier output voltage per se, but rather of the change in this voltage when the transducer is loaded. This requires that data acquisition begin with a scan of the entire array of 499 transducers immediately prior to the foot-strike of interest. To initiate this scan, the subject walks through a light beam so positioned a few cm above ground level that it is broken by the foot carrying the array only tens of milliseconds before this touches the ground (see Figure 3). The breaking of this beam is a necessary condition for the analog-to-digital converter (ADC) to sample the multiplexed signal (see Figure 2) — a process which can be carried out at frequencies of 100, 50, 25, or 12.5 kHz. These frequencies correspond to 'framing rates' — i.e., rates of scanning of the complete transducer array — of approximately 200, 100, 50, and 25 Hz, respectively.

Since the total current consumed by the 499 operational amplifiers and

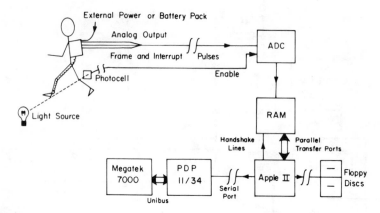

Figure 3 — Block diagram of the data processing and display system.

the various supporting electronics, all of which are carried in the back pack, is approximately 1.5 A, power is supplied from an external source via a cable trailed by the subject (see Figure 3). Two other cables carry 1) the analog output from the multiplexer to the ADC and the 64k byte random access memory (RAM) and 2) an interrupt pulse which controls the sampling.

A typical walking foot-fall generates at the highest framing rate a series of some 50,000 to 60,000 bytes of data which are stored in the RAM. In order to save software overhead time and thereby ensure the highest possible rate of data throughput, this operation is performed completely independently of the Apple II microcomputer.

The incorporation of this computer into the data processing system provides great operational flexibility. In particular, once the transducer array has been calibrated, the system can be operated both in the laboratory, where data processing and display are performed by a PDP-11/34 computer and Megatek 7000 graphics display system (see Figure 3), and in clinics, hospitals, etc., that do not have such facilities. In this latter situation, the Apple II can transfer the data to a floppy disc via its own RAM buffer. In addition, it can be used to perform the difference calculations necessary to convert appropriate subsets of the data held in its RAM buffer to vertical contact stress distributions and to display these either graphically or alphanumerically on its own CRT terminal, albeit only slowly. Inside the laboratory, however, the PDP-11/34 and the Megatek 7000 provide for faster processing and considerably more sophisticated graphic display of the data than is possible with the Apple II alone.

Figure 4—Load frame used for calibrating individual transducers.

Results

The Performance of Individual Transducers

To determine the responses of individual transducers to different levels and rates of variation of uniaxial compressive stress, a variety of experiments were carried out on individual transducer-amplifier units.

Initially these units were tested manually—at low loading rates by squeezing the transducer under test between a spring-loaded anvil assembly and a Kistler Type 9233A quartz reference transducer, and at high loading rates by laying the transducer under test on top of the same reference transducer and tapping the assembly with a suitably padded hammer.

Subsequent testing was carried out under more reproducible loading conditions by means of the load frame shown in Figure 4. This consists of three parts: 1) the same quartz reference transducer with the addition of a small probe-type anvil; 2) two movable cross-heads linked by a coil spring; and 3) an electrically driven eccentric cam mounted below the lower cross-head. The specimen is carried on the upper cross-head and loaded against the anvil. The magnitude and rate of variation of this load are controlled by varying the stiffness of the spring and the speed of rotation of the eccentric cam.

The output voltages V_{out} and V_{ref} generated during these experiments

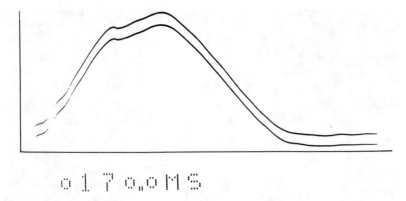

Figure 5—a: Time variation of the output voltage generated by a typical transducer-charge amplifier unit (V_{out}) and a quartz reference transducer (V_{ref}) when both were subjected to the same slowly varying stress. b: V_{out} plotted against V_{ref} under the same conditions as in a.

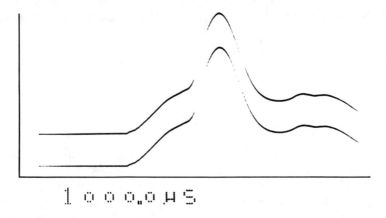

Figure 6—a: V_{out} and V_{ref} as functions of time when both transducers were subjected to the same rapidly varying stress. b: V_{out} plotted against V_{ref} under the same conditions as in a.

by the transducer-amplifier unit under test and the reference transducer assembly, respectively, were recorded by means of a Nicolet dual beam digital storage oscilloscope.This provided a convenient means of obtaining plots both of the variation of V_{out} and V_{ref} with respect to time and of the variation of V_{out} with respect to V_{ref} in forms in which they could be photographed to obtain permanent records.

　　Figures 5a and b are just such presentations of the results of a manual loading experiment in which the total loading time was 170 msec and the peak stress 800 kPa, and Figures 6a and b are the corresponding plots for a hammer blow that produced an approximately 1 msec duration stress pulse of comparable magnitude. It is apparent from these figures and from the results of similar experiments involving peak stresses up to 1500 kPa and loading times up to 700 msec that the lead zirconate titanate transducers can match the dynamic characteristics, the less than 2% linearity and less than 1% hysteresis, and the better than 0.5 kPa resolution of the reference transducer. A more complete account is given by Hennig et al. (1981).

In addition, the load frame was also used to monitor the effects of temperature on the performance of individual lead zirconate titanate transducers. These experiments showed that the piezoelectric coefficients varied by less than 1.5% between 10° and 40°C.

The Performance of the Completed Insole

Subsequent experiments with the load frame showed that neither the linearity nor the hysteresis of the transducers changed when they were incorporated into the insole. However, the maximum deviation of the relative sensitivity about its mean value was thereby increased from the ± 5% of this mean that is characteristic of the as-received material to ± 20%. This unacceptably high variation was corrected by using the load frame to measure a separate calibration factor for each transducer in the completed insole, and then storing these factors in the Apple II and PDP-11/34 computers for use in calculating the vertical contact stresses from the measured differences in output voltage.

Summary

A new piezoelectric device for measuring the distribution of the vertical component of the stresses generated between the plantar surface of the foot and the insole of a shoe while running or walking has been developed to the prototype stage. Tests of the individual transducers used in the device, each of which has an area of approximately 23 mm^2, show that these are capable of following 1 Hz to 1 kHz compressive stress pulses reaching peak stresses as high as 1500 kPa with an accuracy of a few percent. Since such stresses and rates of stressing are typical of a wide variety of bodily movements, it is apparent that the technique should have applications in areas of biomechanics far removed from the studies of gait for which it was developed. It might also be noted that, since piezoelectric ceramics typically have a Young's modulus in the range 10^{11} to 10^{12} Nm^{-2}, the present transducers can for all practical purposes be regarded as incompressible. Consequently, by controlling the amount and the compliance of the matrix material used to embed them, it is possible to create transducer arrays that are at once sufficiently flexible and sufficiently incompressible that they have little tendency to perturb the gait pattern under study.

Acknowledgment

This work was supported by U.S. Public Health Service Grant No. 1-R01-AM26410-01-AFY.

References

CAVANAGH, P.R., Hennig, E.M., Bunch, R.P., and Macmillan, N.H. 1982. A new device for the measurement of pressure distribution inside the shoe. In: H. Matsui and K. Kobayashi (eds.), Biomechanics VIII-B, pp. 1089-1096. University Park Press, Baltimore.

HENNIG, E.M., Cavanagh, P.R., Albert, H.T., and Macmillan, N.H. 1981. A piezoelectric method of measuring the vertical contact stress beneath the human foot. J. Biomed. Engng. 4:213-222.

LORD, M. 1981. Foot pressure measurement: a review of methodology. J. Biomed. Engng. 3:91-99.

A New Device for the Measurement of Pressure Distribution Inside the Shoe

P.R. Cavanagh, E.M. Hennig,
R.P. Bunch, and N.H. Macmillan
The Pennsylvania State University,
University Park, Pennsylvania, U.S.A.

Although there have been many investigations of the interaction between the foot and the ground, there has been no comprehensive study of the corresponding interaction between the foot and the shoe (Lord, 1981). Yet this latter interface is certainly the more important one, since many foot pathologies are the result of the development of abnormal forces at locations inside the shoe. With this in mind, the present authors have recently developed 1) an array of 499 piezoelectric ceramic transducers embedded in silicone rubber that can be fitted into a size 10 shoe in place of the sock liner and 2) the supporting data acquisition, processing and display system needed to handle the approximately 100,000 8-bit words of data generated per second by the array when it is operated at its highest temporal resolution. The various items of hardware and software that make up the device are described briefly in a companion paper in this volume (Hennig et al., 1981) and in more detail elsewhere (Cavanagh and Ae, 1980; Hennig et al., 1982).

With the development of this prototype device essentially complete, the authors have in recent months begun a pilot study of the vertical contact stress distributions (or 'pressure distributions') generated by normal (i.e., symptom-free) male subjects walking in a running shoe (Bunch, 1981). The present article includes some preliminary results from this study.

Methods

To collect the data, the subject — who wore the transducer array in his right shoe and its associated electronics in a small back pack (see Figures 1 and 2) — walked in a straight line from a predetermined point until he passed through three light beams. From his time of transit between the first and the last of these beams, which were set at hip height some 5 m apart, a check was made that the subject's average speed fell within the range $(1.5 \pm 0.1 \text{ msec}^{-1})$ deemed acceptable. By breaking the center beam with his right foot, the subject caused the analog-to-digital converter (ADC) to begin sampling the serial analog signal synthesized by the multiplexer from the output of the individual transducer-amplifier units. This sampling procedure was carried out at 100 kHz, at which frequency the complete array of transducers was scanned about once every 5 msec, corresponding to a 'framing rate' of approximately 200 Hz.

Under these conditions, a typical walking foot-fall of 500 msec duration generated 50 to 60k bytes of data that were transferred to a 64k byte random access memory (RAM) in real time. These data represented an initial 'reference scan' of the output voltages from all 499 transducer-amplifier units some tens of milliseconds prior to foot-strike, the several further such scans obtained between the breaking of the center light beam and the moment of foot-strike, and the 100 or so subsequent such scans obtained during the course of the foot-fall.

After they were transferred serially from the RAM to a PDP-11/34 computer, these data were processed via a three-stage calculation. First, each of the individual data points (output voltages) recorded during the actual foot-fall was 'corrected' by subtracting from it the value of the output voltage obtained from the corresponding transducer-amplifier unit during the initial reference scan. This served to eliminate errors due to integration of the individual amplifier input offset currents by the amplifier circuitry (Hennig et al., 1982). Then, to eliminate errors due to the differences in gain of the individual transducer-amplifier units, each corrected output voltage was converted to a stress by multiplying it by the experimentally determined calibration factor for the appropriate transducer-amplifier unit. And thirdly, the complete vertical contact stress distribution at any chosen moment in time was derived by linearly interpolating between the last value of this stress obtained from each transducer-amplifier unit prior to the moment of interest and the first value obtained thereafter. This step represented an attempt to correct for the fact that, because the transducer array is scanned serially, no two stress measurements were made at precisely the same instant.

Finally, the vertical contact stress distributions obtained in this fashion were displayed on the screen of the Megatek 7000 graphics display unit in the form of perspective drawings of a three-dimensional plot of the stress acting on each element as an ordinate above an outline of the foot drawn upon the ground (Cavanagh and Ae, 1980). It also proved useful to make use of the ability of the Megatek 7000 to rotate and translate this plot so that it could be seen from different perspectives.

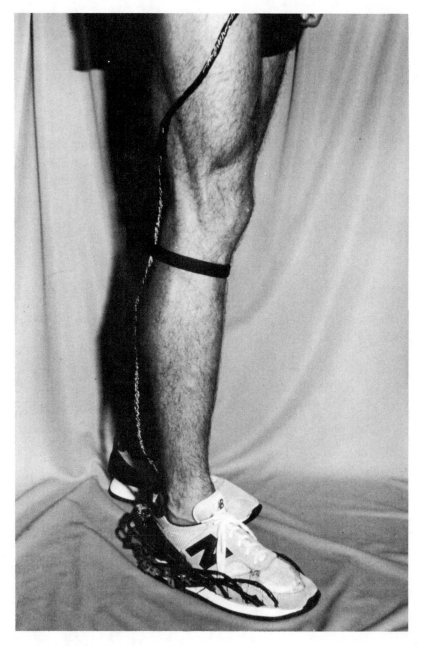

Figure 1—Subject wearing a shoe with the 499 element insole inside. Note the small diameter of the bundle of wires leading from the insole.

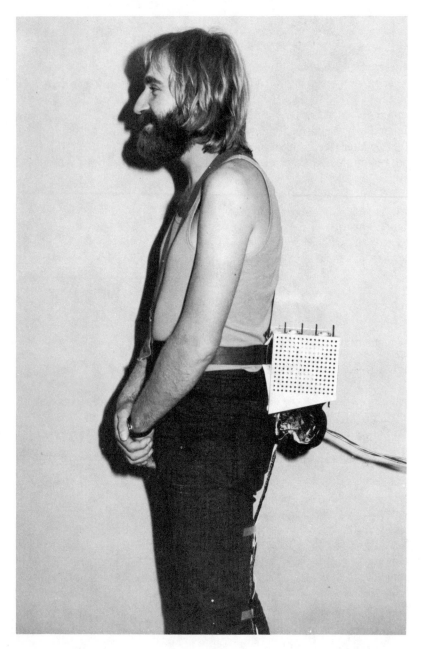

Figure 2—Subject wearing the backpack containing the amplification and multiplexing circuitry. Power, analog output, and enable lines are shown leading from the backpack.

Results

Figure 3 shows the way in which the vertical contact stress distribution generated by a normal subject varied with time after the first contact of the shoe with the ground. In Figure 3a, some 30 msec into the foot-fall, the total vertical load is distributed over the rear third of the foot in such a manner that the stress is higher on the lateral side than the medial and higher along the edges of the foot than along the mid-line. It is only some 70 msec later (see Figure 3b) that any contact stress appears in the fore-foot region. Note that while the stress on the lateral side of the foot is still similar to that 30 msec after first contact, there is now a dramatic increase in stress under the medial aspect of the rearfoot region. This pattern is certainly affected by the contours of the foot-bed in the shoe used. Figure 3b shows also that anterior movement of the applied load leaves the vertical contact stresses higher on the lateral side of the foot than the medial and accentuates the drop in stress between the edges and the mid-line of the foot. At 174 msec into the foot-fall (see Figure 3c) there has been a slight increase in the stresses underneath the forefoot region while the rearfoot stresses have decreased considerably; and 225 msec after first contact (see Figure 3d) the stress levels beneath the heel and forefoot are comparable and low enough to suggest that this is the time when the vertical component of the total ground reaction force is a minimum. Figure 3e shows that there is a time about 300 msec into the foot-fall when most of the vertical force is carried on the metatarsal heads and none on the rearfoot. Further transfer of weight in the anterior direction (see Figure 3f) serves to differentiate between the stresses acting on the first and second metatarsal heads and the great toe. Figure 3g shows that the stresses acting on the first metatarsal head and the great toe continue to build up for at least another 70 msec, and Figure 3h shows that the stresses acting on this toe remain high even 460 msec after first contact. By this time, however, the stresses under the metatarsal heads are declining, particularly along the lateral border of the foot. This last figure makes readily apparent the importance of the great toe at this late stage of the support phase.

Figure 4 shows four different projections of the stress distribution some 350 msec into the foot-fall when, just as in Figure 3f, most of the vertical load is carried on the metatarsal heads and the great toe. This figure shows clearly the way in which the Megatek 7000 graphics display unit can manipulate the image to provide additional insight. In particular, the 'valley' between the metatarsal heads and the toes that forms such a dramatic feature of Figure 4c is hardly visible in Figure 4b.

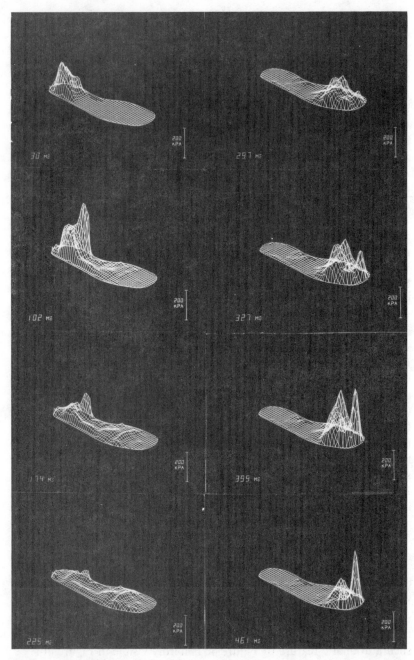

Figure 3—Evolution of the pressure distribution beneath the plantar surface of the foot during a walking foot-fall.

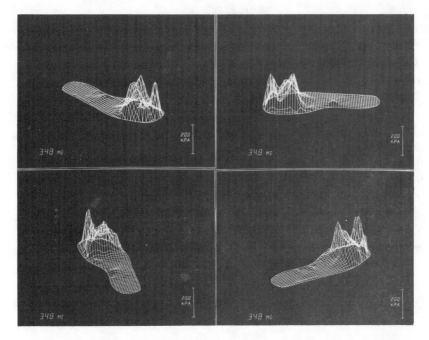

Figure 4 – Views of the same distribution of pressure from different perspectives.

Conclusion

This paper has described the first results obtained from a versatile and powerful system for the measurement of vertical contact stress distributions (pressure distributions). Note, however, that the use of this system is not limited to studies of human locomotion, for it has a high enough frequency response and sufficient temporal resolution to enable it to be employed successfully in far more demanding situations. Moreover, its ability to provide graphic feedback to orthopedists and others involved in the care of the human foot opens up a new avenue for the diagnosis and treatment of locomotor disorders. Experiments in this laboratory are currently exploring this application.

Acknowledgment

This work was supported by U.S. Public Health Service Grant No. 1-R01-AM26410-01-AFY.

References

BUNCH, R.P. 1981. In-Shoe Pressure Measurement During Walking. M.S. Thesis, The Pennsylvania State University.

CAVANAGH, P.R., and Ae, M. 1980. A Technique for the Display of Pressure Distributions Beneath the Foot. J. Biomech. 13:69-75.

HENNIG, E.M., Cavanagh, P.R., and Macmillan, N.H. 1981. Pressure Distribution Measurements by High Pressure Piezoelectric Ceramic Force Transducers. In: H. Matsui and K. Kobayashi (eds.), Biomechanics VIII-B, pp. 1081-1088. University Park Press, Baltimore.

HENNIG, E.M., Cavanagh, P.R., Albert, H.T., and Macmillan, N.H. 1982. A Piezoelectric Method of Measuring the Vertical Contact Stress Beneath the Human Foot. J. Biomed. Engng. 4:213-222.

LORD, M. 1981. Foot Pressure Measurement: A Review of Methodology. J. Biomed. Engng. 3:91-99.

Application of Microwaves to Continuous Measurement of the Velocity during Sprint Running

Yoshimitsu Inoue
Kobe University School of Medicine, Kobe, Japan

Masatoshi Shimada and Akira Tsujino
Osaka Kyoiku University, Osaka, Japan

Yukihiro Goto
Osaka City University, Osaka, Japan

The microwave technique has been applied to limited fields such as communication systems and long range radar. Recently, the development of economical microwave sensories has been realized, extending its application to many other fields (Miyai, 1977). For example, the microwave technique is now used for measuring automobile speed, preventing crashes, and measuring distances, since it is capable of detecting motion by analyzing the reflected waves from an object. But the technique has not yet been used to measure changes in velocity during human sprint running. Until now, the changes in velocity have been measured at regular intervals (Ikai et al., 1963; Tsujino and Goto, 1975) or at every pace (Honma et al., 1981) by different types of electronic instruments. These methods, however, are assumed to directly and/or indirectly obstruct the performance of a runner.

The present study was designed for microwave application to measure the speed of human runners, minimizing the possible obstructions to the runner. Using the microwave technique, a Doppler radar speedometer was experimentally developed to continuously measure the changes in velocity during sprint running. The characteristics of the Doppler shifts detected by this instrument were examined through experimentation. For the purpose of putting it to practical use, so as to rapidly obtain the velocity curve, the Doppler radar speedometer was connected to a

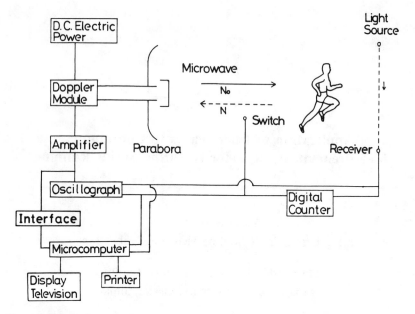

Figure 1 — Block diagram of Doppler radar speedometer and display instrument.

display instrument, composed of a microcomputer, cathode-ray tube, and printer.

Development of a Doppler Radar Speedometer using Microwaves

The Doppler radar speedometer developed depends on the Doppler effect of the microwave. Its block diagram is given in Figure 1. The Doppler effect caused by the changes in the distance between the wave source and the reflector is due to differences in the frequency between the transmitted wave and the reflected wave. As soon as D.C. voltage was allowed to pass through the Doppler module, the Gunn diode in the module oscillated and the microwave was transmitted through the parabolic antenna. The transmitted waves struck the runner and the reflected waves after bouncing off the runner were received by parabolic antenna. The frequency differences between the transmitted and received waves were detected by the Doppler module. The detected Doppler shifts were amplified and were illustrated on an electromagnetic oscillograph. Running velocity was obtained from the illustrated shifts. The details are as follows:

$$\triangle N = N_0 \sim N = \frac{2N_0}{C} V \qquad (1)$$

where ΔN = Doppler shift, N_0 = Frequency of transmitted wave, N = Frequency of received wave, C = propagation velocity of microwave, V = running velocity. N_0 = 10.525 GHz, and $C = 2.9979 \times 10^8$ m/sec, were substituted into the Doppler formula (1):

$$\Delta N = 70.22V \tag{2}$$

The electromagnetic oscillograph paper was usually driven at 100 cm/sec in the present experiment when the wave length of the Doppler shift on the paper was λ cm, ΔN equalled $\frac{100}{\lambda}$. This value was substituted into formula (2):

$$V = \frac{1.424}{\lambda} \text{ (m/sec)} \tag{3}$$

Substituting every wave length into formula (3), the changes in velocity were calculated.

Characteristics of the Doppler Shifts

Influences of Body Motion on the Doppler Shift

During sprint running, a certain degree of displacement and modification generally appears in the velocity from body part to body part. The regional differences in velocity are due to the reflected wave frequencies. In the present study, therefore, two model experiments were performed to elucidate the influence of displacement and modification in the human body on the Doppler shift. The transmitted wave was allowed to strike the aluminum reflection plates set at various positions on a disc rotating continuously, but the rotational speed of the disc was kept constant during all of the experiments.

Influences of the Two Different Parts Moving at Different Speeds in the Same Object on Doppler Shift Form. In this experiment, the distance from the rotational axis to the position of the plate was changed, 18.5 cm and 8.5 cm, respectively, and the two plates were simultaneously set at both positions. The wave form obtained simultaneously from the two plates set at different positions coincided with the composite wave form given by each plate. This result infers that when some parts of the body moved at different speeds, the composite wave form given by each part could be illustrated.

Influences of Reflection Plate Area on Doppler Shift Amplitude. The amplitude of the Doppler shift was examined in relation to the area of

Figure 2—Doppler radar speedometer and display instrument in position.

reflection plate. A high correlation coefficient was obtained between the amplitude of the Doppler shift and the reflection plate area, and the wider the area was, the larger the amplitude.

These two model experiments strongly suggested that the Doppler shift depends mainly on the size of the exposed area. Considering this suggestion, the Doppler shift detected on human locomotion would be influenced by the motion of the trunk having a larger surface area than the other parts.

Application of the Doppler Radar Speedometer for Continuous Measurement of Human Sprint Velocity

The beam range of the microwave transmitted from the speedometer was nearly 5° in width. Care had to be taken therefore, so that there was no other movement within this range when this instrument was used to measure the velocity of the runner (see Figure 2). Figure 3 shows the velocity curve of Subject H.A., who took 12.08 sec to run 100 m. It was observed that the velocity curve obtained from every wave was considerable as far as the fluctuation was concerned and the curve obtained

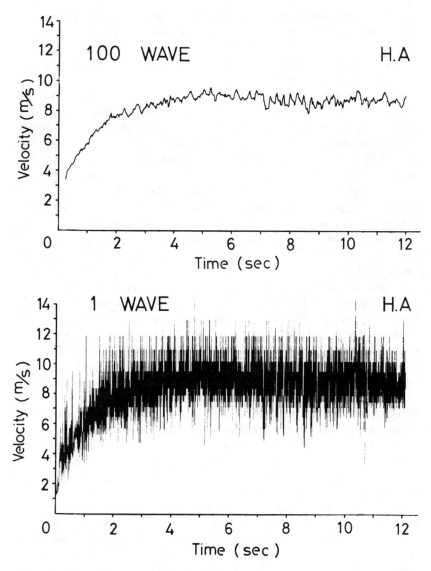

Figure 3—The velocity curve obtained with the Doppler radar speedometer, when subject took 12.08 sec to run 100 m.

from mean of every 100 waves was smooth. Consequently, the latter clarified the time when the velocity reached its peak and where it started to decrease progressively. In this figure, the velocity reached its peak at about 4.5 sec and began to decrease at about 9 sec after the start of the race, respectively. These results were nearly the same as those obtained with a photo cell (Ikai et al., 1963; Tsujino and Goto, 1975). The following three experiments were conducted to confirm if this speedometer was able to measure the continuous changes in velocity during human sprint running.

The velocity curve throughout one running cycle obtained by the speedometer was compared with the velocity given by a high-speed camera, especially at both the center of the trunk (the center was regarded as that point between the external auditory meatus and the great trochanter of the femur) and the body's center of gravity. The running form throughout one cycle at the 40-m point was filmed at a sampling rate of 200 frames/sec, and every second frame was analyzed with a film analyzer and the graph-pen system. The velocity levels measured by the speedometer were almost the same as those obtained from the trunk and the center of gravity. The level was decreased during the first half (break phase) and increased during the latter half (kick phase) of contact phase. This pattern was nearly the same as those shown by other methods (Goto and Tsujino, 1974). These results suggest that at least at the 40 m point, this speedometer was able to measure the velocity change from parts near the center of the trunk and/or the center of gravity of the body.

An additional experiment was performed to examine whether the speedometer could measure in detail the changes in velocity throughout the 100 m sprint, as could be measured at the 40 m point. In order to find the contact phase, the Doppler shifts and basogram were simultaneously illustrated on the oscillographic paper, through the speedometer and a wireless transmitter, respectively. From the start of the sprint until the 3rd to 4th step, the speed increased independently of the contact and stride phase. Subsequently, a decrease in speed during the first half and an increase during the latter half of the contact phase were repeated until the end of the run. From these results, it was clear that throughout the 100 m sprint the speedometer gave nearly the same pattern of change in velocity as the pattern observed in many other experiments by analyzing films (Goto and Tsujino, 1974).

The level of the running speed given by the speedometer was compared with the level calculated from the time required at each step obtained from the basogram and the length of each step measured in the usual way. The level of the two curves was almost the same throughout the sprint.

The results obtained from these experiments indicate that with the speedometer, it is possible to show the change in speed during each step throughout a 100 m sprint.

Development of a Display Instrument for Speed Measurements

In measuring changes in speed, the speedometer recorded more than

7000 Doppler shifts during a 100 m sprint. It takes a long time to actually measure a wave length and to calculate the change in velocity. This method, therefore, cannot rapidly illustrate the velocity curve just after running. For this purpose a display instrument, consisting of a microcomputer, a cathode-ray tube, and a printer which was experimentally developed, was used to rapidly show the velocity curve. The wave forms of the Doppler shifts were converted to square waves, and then, the number of the waves during each given time was counted by an interface (see Figure 1). These numbers were input into the microcomputer and the velocity curve of the sprint was displayed on the cathode-ray tube and the printer. The curve displayed on the printer was exactly the same as that obtained from the former method without the use of the display instrument.

Conclusion

In an attempt to continuously measure the change in velocity during a sprint and to minimize the possible effect on the performance of a runner, a kind of Doppler radar speedometer was experimentally developed. The practical characteristics of the Doppler shifts detected by this speedometer were examined. The speedometer was able to determine the changes in speed during each step throughout the 100 m sprint. Moreover, the speedometer was connected to a microcomputer to rapidly analyze the Doppler shift signals directly and accurately display the velocity curve of the sprint on the cathode-ray tube and printer. It can be concluded that this instrument can be advantageously applied in measuring the velocity curve of human beings during a sprint race.

References

GOTO, Y., and Tsujino, A. 1974. A kinesiological study of sprint running: The changes in velocity of the center of gravity within the body and movement of the limbs in each running cycle. Ann. Phys. Educ. Osaka City Univ. 9:53-67.

HONMA, K., Tsujino, A., Kazama, T., Goto, Y., and Matsushita, K. 1981. A basic study of learning method with sprint running of junior high school students: The relationships between step frequency and step length from the viewpoint of velocity in sprint running. Bul. Jpn. Curriculum Res. and Dev. 7:217-226.

IKAI, M., Shibayama, H., and Ishii, K. 1963. A kinesiological study of sprint running. Jpn. J. Phys. Educ. 7:59-70.

MIYAI, Y. 1977. Microwave sensory: Some instruments using microwave. Interface (June):55-62.

TSUJINO, A., and Goto, Y. 1975. An analysis of sprint running in 2- to 12-year-old children with respect to the running patterns. Memoirs of Osaka Kyoiku University 24:253-262.

Precise, Rapid, Automatic 3-D Position and Orientation Tracking of Multiple Moving Bodies

R.W. Mann, D. Rowell, G. Dalrymple, F. Conati,
A. Tetewsky, D. Ottenheimer, and E. Antonsson
Massachusetts Institute of Technology,
Cambridge, Massachusetts, U.S.A.

The acquisition of kinematic data from spatial arrays of moving elements are a common experimental requirement in machinery analysis and design and in studies in the natural, biological, and psychological sciences. The multiple bodies may be linked or independent. A particular case in biomechanics is the study of human movement where the element array is comprised of linked body segments.

We have developed a kinematic data-acquisition and dynamic analysis system of substantial generality, but focused on human movement. Criteria at the outset of the study included:

1. As normal and natural performance as possible for the particular human, minimizing the artificial aspects of the experimental environment and the burden on the human subject.

2. High precision of full three-dimensional kinematic data, including element translations and rotations relative to an absolute reference frame, at suitably high data-acquisition rates in a form suitable for subsequent dynamic or other analyses.

3. Automaticity, the total absence of human intervention in data acquisition and quantization in order to eliminate human subjectivity and error and reduce drudgery.

4. Very rapid processing of kinematic data so as to provide element position and rotation information in real-time, e.g., for use as input for control of an ongoing experiment. Efficient and quick processing of data off-line as in dynamic analyses using kinematic data.

5. Precise, automatic, and rapid mandated computer participation. Organize and link computer subroutines to achieve flexibility and versatility of system application. Provide extensive graphical access for pro-

gram assembly and supervision and for rapid visual evaluation of data veracity.

6. Couple kinematic analyses to automatic dynamic estimation of intersegmental forces and torques.

7. Acquire data from free-ranging, i.e., two- and three-dimensional human movement patterns within a large volume, not restricted to short linear activity. Capability of accommodating animal, pathological, and sport movement patterns.

8. Provide absolute calibration of system optical, kinematic and dynamic performance defining overall system reproduciblity, resolution, and accuracy.

Method: Opto-Electronics

A Selspot I (Selective Electronics AB, Partille, Sweden) was interfaced using Direct Memory Access to a Digital Equipment Corporation (Maynard, Massachusetts) PDP 11/60 minicomputer which executes the kinematic and dynamic routines. The 11/60 is in turn DECnetted to a PDP 11/40 which serves as a display processor and dynamic graphics terminal using a VT-11/VR-14.[1]

Selspot I consists of 30 high power omnidirectional Light Emitting Diodes (LEDs) which radiate exclusively in the infrared (940 nm) below the visible spectrum and therefore are not distracting to a human subject. A control unit serially excites each LED for 50 micro seconds at a frame repetition rate of 315 Hz. Two cameras with ordinary optics but with lateral photo-effect diode plates at their image planes observe the LEDs. Since the solid-state detectors are only sensitive to the wave length of the LEDs and the system includes a background illumination compensator, the data are unaffected by ambient illumination. As each LED is excited, its image appears optically at a discrete X-Y location in each image plane. The plate integrates over the area of incident light; thus the detected X-Y location result is at the centroid of the incident radiation. The resulting multiplexed analog voltages are digitized to a 10 bit level by the Selspot I controller and transmitted, together with timing pulses, thru a custom interface to computer memory at a rate of 40,000 words/sec. Figure 1 illustrates the Selspot-computer block diagram while Figure 2 shows the camera opto-electronics.

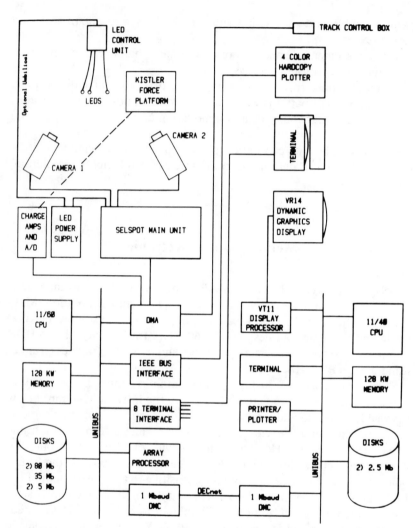

Figure 1—Block diagram of the TRACK system hardware.

Method: Kinematics

The software package for kinematic processing is called TRACK (*Te*lemetered *R*apid *AC*quisition of *K*inematics). The camera data combined with camera location described in absolute laboratory coordinates permits trigonometric calculation of the X-Y-Z position of each LED.

In contrast to the usual practice in gait and sports biomechanics of locating body markers at bony prominences near joint articulations to define segment endpoints, the LEDs are arranged so as to define a body-segment-centered Cartesian coordinate system for each segment. Three

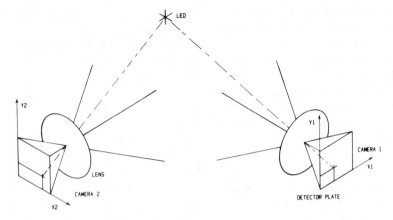

Figure 2 – Selspot camera optics and detectors.

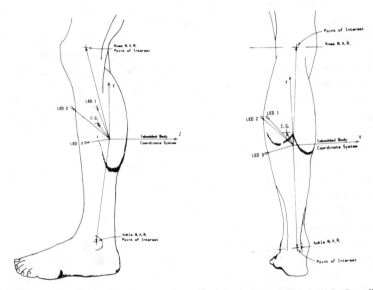

Figure 3 – Two views of a leg showing the vectors from the Imbedded Body Coordinate System to: the knee and ankle momentary axes of rotation (M.A.R.), the triplet of LEDs, and the Center of Mass (C.G.) of the Leg.

(or more) LEDs are accurately located in a non-colinear array on a lightweight, usually plexiglass, frame which is Velcro strapped at a convenient location on the body segment (see Figure 3). The TRACK software first accumulates a data set and outputs it to a disc in real-time to avoid the limitations of main memory size. It then looks up and applies corrections from a table to eliminate Selspot system errors; applies a forward and backward two-pass, sixth-order Butterworth, low-pass filter with selectable cutoff frequency and then calculates the X-Y-Z position of each LED. An error parameter arising from the over-specification of the camera data is constantly monitored to ensure data validity. This flags errors caused by reflections and anomalies. TRACK then applies a bad-point eliminator to check the measured vector length between all pairs of LEDs in an array and discards any LEDs deviating from the prescribed array by more than a predetermined amount. Any data gaps caused by either error assessment can be linearly interpolated provided they are sufficiently short. Again an option is provided to apply another digital sixth-order Butterworth, low-pass filter with a different selectable cutoff frequency. Now TRACK aggregates the LEDs by rigid array and applies an algorithm described by Schut (1960/61) to perform a least squares best fit of rotation if three or more LEDs have valid data. This algorithm has a surplus variable which is eliminated by choosing the solution that produces the most stable results, and thereby removes singularities. It also automatically adjusts to the number of valid LEDs remaining. Thus, the instantaneous position and rotation of each limb segment's body coordinate system is known in absolute laboratory coordinates.

If individual LEDs are temporarily obscured, TRACK treats this situation as bad data and calculates the segment position and rotation as long as data from three LEDs remain available. When fewer than 10 body segments are under study, more than three LEDs are assigned to each segment reducing the influence of noise and increasing the resolution. Six point Lagrangian interpolation for equally spaced abscissas followed by the same sixth-order Butterworth filter produces first and second derivatives of position and angle for each segment.

Since the translational and rotational time histories of adjoining linked segments are known in the absolute coordinate system, the Momentary Axis of Rotation (M.A.R.) can be calculated for short time segments (see Figure 3). Precise knowledge of this location is especially important to subsequent dynamic analyses.

Calibration of separate parts and the entire system is stressed. Computer-controlled stepper motor actuated rotation of a 30-LED array in the camera field of view produces independent position information which is compared with TRACK X-Y data for each LED to provide optical and image-plate calibration to within one discretization level out of the range of 1024. Four 51 by 51 computer lookup tables store the necessary corrections for X and Y for each camera. To date, system stability has been such that a new set of correction matrices need only be generated when optics or electronics are modified.

Overall system calibration employs a two-degree-of-freedom pendulum (see Figure 4) whose motion patterns are predicted from closed form, analytical Newtonian formulations, and precisely measured geometry and inertial parameters. Experimental data from LED arrays mounted on the pendulum are then processed through TRACK and compared to the analytical results.

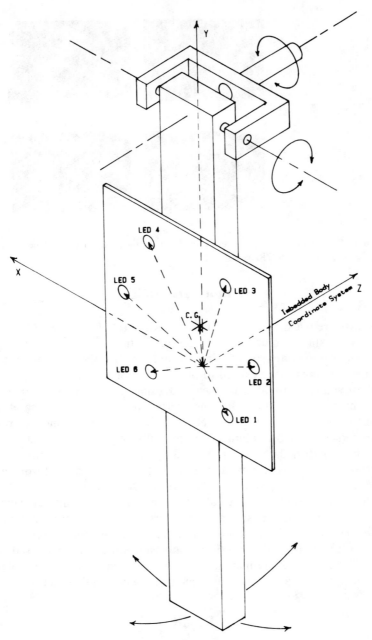

Figure 4 — The two-degree-of-freedom pendulum used for dynamic verification of the TRACK system.

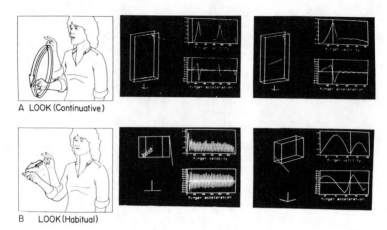

A LOOK (Continuative)

B   LOOK (Habitual)

Figure 5 — Position, velocity, and acceleration data from the fingertip during American sign language acquired by the TRACK system.

Method: Dynamics

The software package for dynamic processing is called NEWTON. The kinematic data combined with inertial properties for each segment permit calculation via a Newtonian formulation of the net forces and torques that must have existed to produce the observed motion.

The dynamic estimator was verified using the same two-degree-of-freedom pendulum. First, the geometry and inertias were carefully measured independently. Then Newtonian equations of motion were generated for that pendulum. This resulted in a perfect kinematic data set. These perfect data were processed by NEWTON and compared to the Newtonian analytical dynamic results. The two force and torque histories agreed exactly, verifying that NEWTON would produce correct results with perfect input kinematic data; thus, for a well calibrated data-acquisition system, the only source of error in dynamic estimation is noise. Next TRACK data sets were acquired for a carefully prescribed set of pendulum initial conditions and subsequently processed by NEWTON. These experimental results were compared with the analytical Newtonian solution with the same pendulum initial conditions.

Results

Thanks primarily to the averaging advantage of multiple LED arrays, plus careful attention to calibration, the overall kinematic accuracy of each segment of the system is within 1 mm and 1° in a roughly 2 m cube viewing volume. Figure 5 illustrates a TRACK study of finger, hand, and forearm movement in American Sign Language conducted in collaboration with Salk Institute investigators studying a possible genetic basis for sign language. Tachi et al. (1981) in these proceedings describes a real-time application of TRACK. For dynamic estimation, NEWTON force and torque calculations based on TRACK data are within 10% of analytical results in the global X and Y directions (those directions being closest to parallel to the focal planes of the cameras) and only slightly worse in Z, which has the noisiest kinematic data. TRACK and NEWTON have also been used in studies of optimal criteria in muscle recruitment using a 5-segment gait analysis (Hardt, 1978; Patriarco et al., 1981).

Conclusion

The TRACK kinematic acquisition system and the NEWTON dynamic estimator both have fulfilled the eight original criteria for mobility analysis defined at the outset. Attaining those goals has produced a facility of unparalleled speed, accuracy, objectivity, and automaticity.

Acknowledgments

This research was performed in the Eric P. and Evelyn E. Newman Laboratory for Biomechanics and Human Rehabilitation and supported by the Whitaker Professorship of Biomedical Engineering at M.I.T.

Footnote

1. The particular computer configuration supports not only this system, but also other computer-mediated, experimental apparatus, data reduction, and finite element analysis as well as serving as a multi-user general computation, word processing and graphics facility.

References

ANTONSSON, E.K. 1978. The Derivation and Implementation of a Dynamic Three-Dimensional Linkage Analysis Technique. M.S. Thesis, Dept. of Mechanical Engineering, M.I.T.

ANTONSSON, E.K., and Mann, R.W. 1978. Automatic 3-D Gait Analysis Using a Selspot Centered System. 1979 Advances in Bioengineering, N.Y.:ASME 51.

CONATI, F.C. 1977. Real-Time Measurement of Three-Dimensional Multiple Rigid Body Motion. M.S. Thesis, Dept. of Mechanical Engineering, M.I.T.

HARDT, D.E. 1978. Determining muscle forces in the leg during normal human gait – An application and evaluation of optimization methods. ASME Journal of Biomechanical Engineering 100:72-78.

HARDT, D.E. 1978. A Minimum Energy Solution for Muscle Force Control During Walking. Ph.D. Thesis, Dept. of Mechanical Engineering, M.I.T.

HARDT, D.E., and Mann, R.W. 1980. A five-body, three-dimensional dynamic analysis of walking. Journal of Biomechanics 13(5):455-457.

PATRIARCO, A.F., Mann, R.W., Simon, S.R., and Mansour, M.J. 1981. An evaluation of the approaches of optimization models in the prediction of muscle forces during human gait. Journal of Biomechanics, Vol. 14, No. 1, pp. 513-525.

SCHUT, G.H. 1960/61. On exact linear equations for the computation of the rotational elements of absolute orientation. Photogrammetria 17(1).

TACHI, S., Mann, R.W., and Rowell, D. 1981. A Quantitative Comparison Method of Display Scheme in Mobility Aids for the Blind. In: H. Matsui and K. Kobayashi (eds.), Biomechanics VIII-B, pp. 1181-1189. University Park Press, Baltimore, MD.

TETEWSKY, A.K. 1978. Implementing a Real Time Computation and Display Algorithm for the Selspot System. M.S. Thesis, Dept. of Mechanical Engineering, M.I.T.

B.
Methods

Alpha Diagram Using a Force Plate for the Study of Movement in Man and a Bipedal Model

Tomokazu Matake

Nagasaki University, Nagasaki, Japan

The movement of human walking is a dynamic motion in which one's own weight is carried forward. This motion may be quantified by measuring the dynamic force using a force plate equipped with load cells. Many researchers have studied the gait and/or the movement of man by using force plates as summarized by Asmussen (1976). Matake (1976) proposed also to draw diagrams combining the components of the step force by several methods. Among them, an α diagram which is the locus of the resultant force of combined compression and anterior-posterior shear components. It is an inverse heart shape and is the largest and the most important one in the movement (see Figure 1).

The step-force ratio is defined as the ratio of compression to the fore-and-aft shear component, and is a characteristic of an α diagram and can be used to predict its shape. If the ratio is small or large, the inverse heart will be inflated or deflated, respectively. The purpose of this study was to clarify the effect that cadence and step length have on the step-force ratio and what kind of an α diagram can be drawn for a simple bipedal model.

Methods

The forces for the steps of one stride were measured by two force plates, one for the right and one for the left foot. This method is useful in the case of an individual who has differences in the movement. These force plates are suitable for the direct measurements of step force components by load cells with Wheatstone bridge circuits since α diagrams can be drawn immediately.

These α diagrams were studied in the case of various cadences and step lengths performed by one subject and of a constant cadence and step length by 30 subjects. In the first case, cadences of 100, 124, and 148

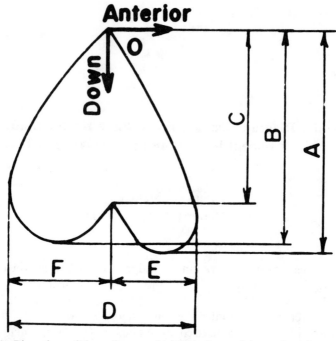

Figure 1—Dimensions of the α diagram which is the locus of the resultant force of combined compression and anterior-posterior components and yields an inverse heart shape.

steps/min were included; 124 steps/min was the standard cadence, under which a subject is able to walk comfortably, and 148 and 100 steps/min are 20% higher and lower than the standard value so as to create a noticeable difference. The trials were performed under five step lengths, 0.6, 0.7, 0.8, 0.9, and 1.0 m. There were 14 combinations of these cadences and step lengths used as experimental conditions. Every condition was repeated five times. In the second situation, the cadence condition was 120 steps/min and each subject's individual step length was utilized in order to make the landing condition of 30 subjects standard.

Although the typical bipedal model has been studied by Chow and Jacobson (1970), this model, shown in Figure 2, has massless links at the foot, lower leg and thigh, and spring joints at the ankle and knee, and a concentrated mass at the waist. A characteristic of this model is that there is no control system since it uses gravity and initial driving force. The movement of one step is divided into five different phases as shown in Figure 2. Then, each equation of motion is solved for the period and when its motion comes to some condition, it must be related to the next period. As compressive and anterior-posterior shear components are calculated in every period, it is possible to draw α diagrams from their results.

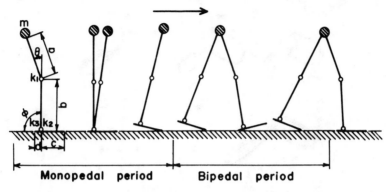

Figure 2—The five phases of walking and the dimensions of a bipedal model which is separated into a monopedal and bipedal period.

Results and Discussion

Figures 3a, b, and c are α diagrams of cadences 100, 124, and 148 steps/min, respectively, for various step lengths. In these figures, the five solid lines represent the results from the left foot and the broken ones from the right foot. They form two narrow bands which indicate that five trials are sufficient, except under the conditions of fast cadences and long strides.

Figure 4 shows the relation of step length to the mean value of 10 step-force ratios given in Figure 3. As shown in this figure, step-force ratio (R) is nearly constant, at about 2.6, within the range of normal step lengths in spite of a wide range of cadences from slow to fast. If the step-force ratio is constant, α diagrams are similar, indeed this is the situation as can be observed in Figure 3. In this case, the step-force ratio (R = 2.6) seems to be a personal constant as the result of the control required to maintain one's individual pace under a variety of cadences.

For the purpose of studying individual differences among the personal constant, 30 subjects were observed under the same condition. The frequency of these step-force ratios was almost normally distributed, with a mean value of 2.5. Therefore, general normal human walking is performed under a step-force ratio (R) of 2.5.

Since human walking motion consists of a period of retardation (E) and an acceleration period (F) (see Figure 1) or monopedal and bipedal period (see Figure 2), the movement is not necessarily performed with a uniform velocity. Actually, the mean velocity of movement under the conditions of a slow cadence and a long stride may be the same as performed under a fast cadence and a short stride, but these movements look quite different. It is interesting to note, however, that the movement using a uniform velocity can easily be calculated from the cadence and step length.

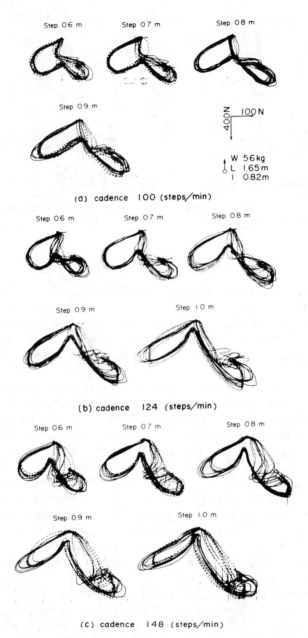

Step 0.6 m   Step 0.7 m   Step 0.8 m

Step 0.9 m

N 400N   100 N

W 56 kg
L 1.65 m
I 0.82 m

(a) cadence 100 (steps/min)

Step 0.6 m   Step 0.7 m   Step 0.8 m

Step 0.9 m   Step 1.0 m

(b) cadence 124 (steps/min)

Step 0.6 m   Step 0.7 m   Step 0.8 m

Step 0.9 m   Step 1.0 m

(c) cadence 148 (steps/min)

Figure 3 — α diagrams under various conditions are shown by solid lines from the left foot and broken lines from the right foot. The cadences are (a) 100, (b) 124, and (c) 148 steps/min.

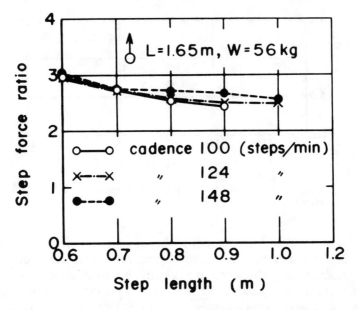

Figure 4 — Relation of the step-force ratio to step length from data in Figure 3.

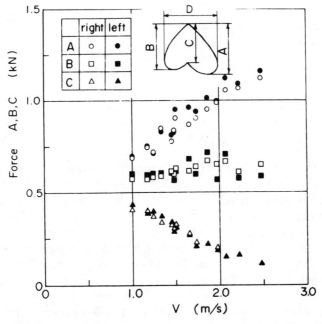

Figure 5 — Relation of uniform velocity to vertical forces A, B, and C in Figure 3; force A increases and force C decreases with increasing velocity, but force B remains constant.

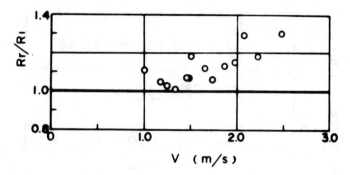

Figure 6 — Relation of uniform velocity to ratio of the right and left foot (R_r/R_l) calculated from data in Figure 3.

The relation of uniform velocity to the three extreme vertical forces in Figure 3 is shown in Figure 5. In this figure, force A increases and force C decreases linearly with increasing velocity, but force B remains almost constant. From these results, the relation of force ratio C/A (see Figure 1) to velocity ratio (v/v_o) is expressed quite well by Equation 1. Here v_o is the standard velocity of the subject in which the cadence is 124 steps/min and the step length is 0.8 m.

$$C/A = 0.271 \, (v_o)^{-2.08} = 0.27 \, (v/v_o)^{-2} \tag{1}$$

In the same manner, force ratio C/D (see Figure 1) can also be expressed by Equation (2).

$$C/D = 0.722 \, (v/v_o)^{-2.21} = 0.72 \, (v/v_o)^{-2} \tag{2}$$

Then, from Equations (1) and (2),

$$R \fallingdotseq A/D = 2.7 \tag{3}$$

That is, the step-force ratio becomes almost constant.

In order to distinguish the dominant foot, the ratio of right to left (R_r/R_l) is plotted in relation to the velocity in Figure 6. This figure shows that the step-force ratio of the right foot (R_r) is consistently larger than the left foot (R_l) and that the fore-and-aft component of the left foot increases with increasing velocity compared with the right. This coincides with the fact that the dominant foot of this subject is the left. Thus, step-force ratios in one stride can be used to distinguish the dominant foot. The differences will become clearer in abnormal walking, than in normal walking, since the sound leg might be used to compensate for the weak leg.

A result of calculation of the simple bipedal model is shown in Figure

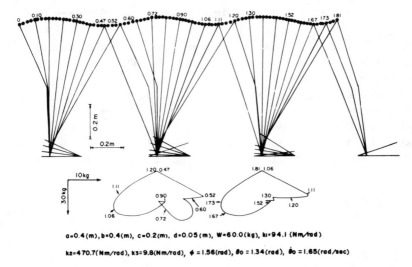

$a=0.4(m), b=0.4(m), c=0.2(m), d=0.05(m), W=60.0(kg), k_1=94.1(Nm/rad)$

$k_2=470.7(Nm/rad), k_3=9.8(Nm/rad), \phi=1.56(rad), \theta_0=1.34(rad), \dot{\theta}_0=1.65(rad/sec)$

Figure 7—Stick figures and α diagrams of a bipedal model (see Figure 2) from results of numerical calculations.

7, in which the link dimensions and initial condition are given, and the forward spring constant of the ankle is different from the backward. The condition in this movement is a cadence of 100 steps/min and a step length of 0.6 m. Numbers beside the mass and α diagrams are the time lapse (in sec) from the start. Indeed, stick figures and α diagrams are certainly similar in humans, but this step-force ratio is R = 1.57. Then, it is interesting to study how the step-force ratio varies with variation in the spring constants of the knee and ankle, and how this simulation can be applied to artificial limbs.

Conclusions

From the results of the study of α diagrams under varying movement conditions it is clear that the step-force ratio is a useful criterion to determine the state of walking; in normal walking the step-force ratio is nearly constant and seems to be an individual value.

If the step-force ratio is constant, α diagrams will be similar. Generally, the step-force ratio was found to be 2.5 from the data of a number of subjects in normal walking. Walking by humans is not necessarily performed at a uniform velocity, but the step-force ratio will be constant whether the velocity is uniform or not.

When the simple bipedal model is used, stick figures can be drawn and α diagrams are like those of humans. Therefore, this simulation method may be useful for the study of walking.

References

ASMUSSEN, E. 1976. A scanning review of the development of biomechanics. In: P.V. Komi (Ed.), Biomechanics V-A, pp. 23-40. University Park Press, Baltimore, MD.

CHOW, C.K., & Jacobson, D.H. 1970. Study of human locomotion via optimal programming. Technical Report No. 617, Division of Engineering and Applied Physics, Harvard University.

MATAKE, T. 1976. On the new force plate study. In: P.V. Komi (ed.), Biomechanics V-B, pp. 426-432. University Park Press, Baltimore, MD.

Horizontal Floor Reaction Forces and Heel Movements During the Initial Stance Phase

Håkan Lanshammar
Uppsala University, Uppsala, Sweden

Lennart Strandberg
National Board of Occupational Safety and Health
Solna, Sweden

The ground reaction forces during normal level walking have been investigated by several authors and are described, for example, in Cunningham (1958) and Harper et al. (1961). The most prominent peaks in curves of force components plotted versus time are well understood and correspond to the deceleration and acceleration phases of the gait cycle. There is also an initial spike in the ground reaction force. This spike is of shorter duration and has a smaller magnitude than the two peaks mentioned previously.

The initial spike has been noted by several authors (Harper et al., 1961; Andriacchi et al., 1977; Perkins, 1978; Cavanagh et al., 1980), but to cite Perkins (1978): "no reason for its existence has been suggested."

The spike in the horizontal fore-aft component of the ground reaction force is especially remarkable, because its direction is usually opposite to the backward directed external force (acting from the ground on the subject) which dominates the decelerating part of stance phase.

The purpose of this paper is to explain the occurrence of this initial spike in the fore-aft component of the horizontal ground reaction force. This component will be referred to as the horizontal GRF, and the sign convention is such that a positive force means an external force acting in the direction of progression.

Results from recent gait experiments will be discussed below. It will be demonstrated that in most cases the spike in the horizontal GRF occurred simultaneously with a backward motion of the rear edge of the shoe, which was the part of the shoe that was in contact with the ground during the early stance phase. This backward motion, which has not been

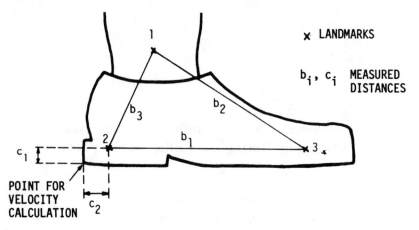

Figure 1 — Illustration of landmark location and manually measured distances (b_i, c_i) used for calculation of the heel rear end velocity.

noted by other authors, provides an explanation to the positive spike in the horizontal GRF, since such motions create forward directed frictional forces.

Method

The test subjects walked in a straight line, thereby passing a zone where the location of selected landmarks on the body were measured, as well as the components of the ground reaction force.

In the investigation the ENOCH system for measurement and analysis of human gait was used. This system, described in Gustafsson and Lanshammar (1977) consists of a minicomputer (HP 21 MX), with a piezoelectric force plate (Kistler) for measurement of floor reaction forces and an optoelectronic device (Selspot) for measurement of the displacement of body markers. The force plate was sampled at 105 Hz and the optoelectronic device at 315 Hz. The resolution of Selspot was 0.003 m.

The present study is based on 126 experiments with four different subjects, three male and one female. Two different shoe designs, clogs and soft leather shoes, were used. Since the experiments were part of a slipping accident research project (Strandberg & Lanshammar, 1983), the surface of the force plate was covered with soap solution during some of the experiments. Landmarks (light emitting diodes) were located on the ankle joint, close to the back of the shoe heel, and at the toe base of the shoe. In Figure 1 the positions of these diodes are illustrated. One major variable in the investigation was the velocity at the contact area between the shoe and the ground during the initial stance phase. It was concluded that the velocity of the heel landmark (No. 2 in Figure 1) could not be

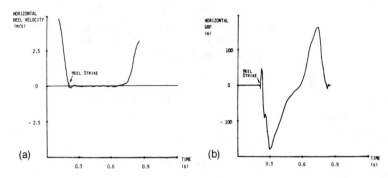

Figure 2 — Examples of the investigated variables. Both curves originates from the same experiment. a = horizontal velocity of the heel rear end, b = horizontal fore-aft ground reaction force.

used to represent the desired velocity since the foot is rotating rapidly in early stance phase. Therefore the distances b_i between the diodes and the distance c_i were measured with a stick before the experiments.

From the coordinates of the landmarks measured during locomotion the velocities of these landmarks were estimated by numerical differentiation of the displacement data. A digital filter described in Gustafsson and Lanshammar (1977) with a bandwidth of 13 Hz was used for the differentiation. By applying geometrical relations the coordinates and velocity of the rear edge of the heel were then calculated.

All experiments were examined with respect to the existence of negative (i.e., backwards) velocities at the rear edge of the heel within 50 msec after heel strike. To determine whether the velocity was significantly negative a threshold value of -0.05 m/sec was used. This specific value was chosen because the error in calculated velocities was approximately 0.05 m/sec. A more detailed error analysis is presented by Lanshammar and Strandberg (1980).

When the minimal value of the velocity was below -0.05 m/sec the backward velocity was classified as existing. If this value was between zero and -0.05 m/sec the existence of a backward velocity was classified as doubtful. Finally, when the minimal velocity was above zero, the backward heel movement was classified as nonexisting.

The ground reaction forces were measured with the force plate located in the center of the 3 m measurement zone of the walkway. The measured force data were more accurate than the calculated velocities, and therefore the existence of a positive spike in the horizontal GRF was determined directly by the sign of the data. If there was a positive spike, however, it was classified as doubtful if its time duration was less than 0.01 sec.

Table 1

**Frequency Table for the Number of Experiments in
Each of the Four Categories. The Total Number of
Analyzed Experiments was 111.**

Existence of an initial backward heel movement	Existence of an initial positive spike in the horizontal GRF	
	Yes	No
Yes	69	11
No	13	18

Results

An example of the horizontal GRF and of the heel velocity for the same step is given in Figure 2. The initial spike in the force diagram is very prominent in this case. From Figure 2a it is clear that the heel velocity was negative shortly after heel strike, meaning that the rear edge of the heel was moving backwards.

In Table 1 the detections of backward heel movement and initial spikes in the horizontal GRF are presented. Fifteen experiments were classified as doubtful according to the definitions given above. These experiments were therefore excluded from the analysis below. For the remaining 111 experiments there seem to be a strong correlation between the existence of a backward heel movement and a forward spike in the horizontal GRF.

To test the data in Table 1, the following null hypothesis was statistically tested: "The initial spike in the horizontal GRF data is occurring in the same proportion of experiments, whether there is a backward movement at the shoe heel close to heel strike or not."

A test variable given in Box et al. (1978) was used: If the four frequencies in a 2 × 2 table are labeled a, b, c, d then a test variable is given by

$$\chi^2 = \frac{(ad-bc)^2(a+b+c+d)}{(a+b)(c+d)(a+c)(b+d)} \tag{1}$$

If the null hypothesis is true, χ^2 is approximately chi-square distributed with 1° of freedom. In our case we get

$$\chi^2 = 22.7 \tag{2}$$

The high value of χ^2 means that the null hypothesis must be rejected at the significance level 0.00001.

Thus, if the data represent random samples there is statistically little doubt that the spike in the horizontal GRF is occurring more frequently when there is a backward heel motion. One must of course be very careful in interpreting these results since the low number of subjects and shoe types might bias the results.

Discussion

It has here been verified for the investigated material that the existence of an initial spike in the fore-aft component of the ground reaction force immediately after heel strike can be explained by a simultaneous backward motion of the rear edge of the shoe heel. The horizontal velocity of the rear heel edge has been assumed to be zero at heel strike by other authors. The present study, however, exhibits small but non-zero values in most of the experiments.

As demonstrated in Figure 2a the heel velocity is several msec shortly before heel strike, while it is desirable that the velocity is close to zero at heel strike. Thus the foot must be strongly decelerated shortly before heel strike. According to control systems theory an overshoot, such as the one depicted in Figure 2a, is not at all unexpected. Actually this is a typical behavior of a high order dynamical system such as the human locomotor apparatus.

References

ANDRIACCHI, T.P., Ogle, J.A. & Galante, J.O. 1977. Walking speed as a basis for normal and abnormal gait measurements. J. Biomechanics 10:261-268.

BOX, G.E.P., Hunter, W.G., & Hunter, J.S. 1978. Statistics for Experimenters. John Wiley & Sons, New York.

CAVANAGH, P.R., Williams, K.R., & Clarke, T.E. 1980. A comparison of ground reaction forces during walking barefoot and in shoes. In: K. Fidelus and A. Morecki (eds.), Biomechanics VII-B, pp. 151-156. University Park Press, Baltimore, MD.

CUNNINGHAM, D.M. 1958. Components of floor reaction forces during walking. Inst. of Eng. Research, University of California, Berkeley, CA.

GUSTAFSSON, L., & Lanshammar, H. 1977. ENOCH—An integrated system for measurement and analysis of human gait. Ph.D. dissertation. UPTEC 77 23R, Institute of Technology, Uppsala University, Uppsala, Sweden.

HARPER, F.C., Warlow, W.J., & Clarke, B.L. 1961. The forces applied to the floor by the foot in walking. Walking on a level surface. National Building Study Research Paper 32, London.

LANSHAMMAR, H., & Strandberg, L. 1980. Halkmekanik — Etapp 1 (Slipping Accident Mechanics — Stage 1). Investigation report 1980:30. National Board of Occupational Safety and Health, S-171 84 Solna, Sweden.

PERKINS, P.J. 1978. Measurement of slip between the shoe and ground during walking. In: C. Anderson and J. Senne (eds.), Measurement of slip resistance, pp. 71-87. ASTM, STP 649, Philadelphia, PA.

STRANDBERG, L., & Lanshammar, H. 1983. On the biomechanics of slipping accidents. In: H. Matsui and K. Kobayashi (eds.), Biomechanics VIII. University Park Press, Baltimore, MD.

Analysis of Constrained Dynamic Systems via Penalty Function Method with Application to Biped Locomotion

Tatsuo Narikiyo and Masami Ito

Nagoya University, Nagoya, Japan

In this research, the general problem of constrained dynamic systems was considered. Such constrained systems occur frequently in biped locomotion systems (BLS). For example, in a locomotion cycle, a double support phase (DSP) and a single support phase (SSP) alternately appear. Thus in analysis of BLS the dynamic systems where the constraints are either maintained or deliberately violated must be considered. In such cases, since these systems frequently switch back and forth between systems of different dimensionality, the formulation of reduced state space is very difficult. Therefore, an analysis of a constrained system of this type is forced to go back to Lagrange (Hemami & Wyman, 1979), since constrained systems are described by mixed algebraic and differential equations. These theoretical analyses or numerical analyses are very difficult (Brayton et al., 1972).

In this research the constraints which were imposed on the mechanical systems were defined as a smooth manifold in a configuration space. The imposition of constraints was defined as the existence of an infinite force field toward the manifold. By this definition a clear and intuitive theorem was derived (Arnold, 1978). Under the basis of this theorem, a penalty function containing the constraints was introduced and BLS was described in a non-reduced state-space by identifying the penalty function with Lagrangian. By several simulation results the usefulness of this method is shown and the motions of BLS and DSP are discussed.

Description of the Constrained System

Generally, in mechanical systems two classes are defined according to the

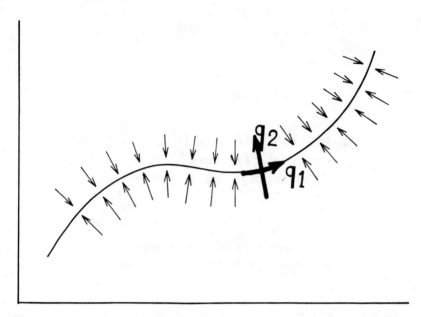

Figure 1 — Constraints as an infinitely strong force field and curvilinear coordinates.

characteristics of the constraints. The first class is a holonomic system whose constraints are integrable, the second class is a nonholonomic system whose constraints are not integrable. In this research only the holonomic system is discussed for the sake of simplicity.

Let γ be a smooth manifold in the configuration space (see Figure 1). If there is a very strong force field in the neighborhood of γ, in the limit case of an infinite force field, a constraint is put on the system.

To formulate this precisely, curvilinear coordinates (q_1 and q_2) are introduced on a neighborhood of γ. q_1 is in the direction of γ, and q_2 is a distance from the manifold (see Figure 1).

Potential energy is defined by

$$U_N = Nq_2^2 + U_0(q_1, q_2) \tag{1}$$

$$(q_2^2 \stackrel{\triangle}{=} q_2^T q_2)$$

depending on the parameter N (tending to infinity), where $U_0(q_1, q_2)$ is the potential energy in the unconstrained case. For this system the following theorem is given by Arnold (1978).

Theorem (Arnold)

Denote by $q_1 = \varphi(t, N)$ the evolution of the coordinate q_1 under a motion with the following initial condition in the field U_N.

$$q_1(0) = q_1^0, \; \dot{q}_1(0) = \dot{q}_1^0, \; q_2(0) = 0, \; \dot{q}_2(0) = 0 \tag{2}$$

Then the following limit exists, as $N \to \infty$

$$\lim_{N \to \infty} \varphi(t, N) = \psi(t) \tag{3}$$

The limit $q_1 = \psi(t)$ satisfies Lagrange's equation

$$\frac{d}{dt} \left(\frac{\partial L^*}{\partial \dot{q}_1} \right) = \frac{\partial L^*}{\partial q_1} \tag{4}$$

where $L^*(q_1, \dot{q}_1) = T|_{q_2 = \dot{q}_2 = 0} - U_0|_{q_2 = 0}$ (T is the kinetic energy of motion along γ). Using this theorem a corollary is easily derived.

Corollary

$q = [q_1, \ldots, q_n]^T$: generalized coordinate

$f = [f_1, \ldots, f_r]^T$: constraints $(f = 0)$

$T = T(q, \dot{q})$: kinetic energy, $U = U(q)$: potential energy

$U_N = U + Nf^2$: penalty function,   N: weight of penalty

If $\dfrac{d}{dt} \left(\dfrac{\partial L}{\partial \dot{q}} \right) = \dfrac{\partial L}{\partial q}$ $(L = T - U_N)$ $\tag{5}$

have a solution for a sufficiently large N, then as $N \to \infty$, there exists limit function $\psi(t)$ and this satisfies following equations.

$$\frac{d}{dt} \left(\frac{\partial L^*}{\partial q} \right) - \frac{\partial L^*}{\partial q} = \left(\frac{\partial f}{\partial q} \right)^T \lambda \tag{6}$$

$$f(q) = 0$$

where

$$L^* = T - U, \quad \lambda : \text{Lagrange's multiplier.}$$

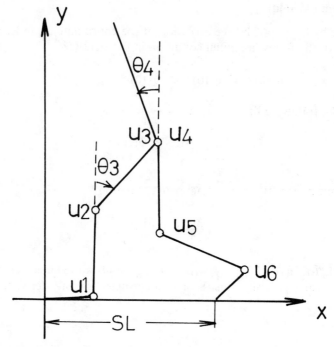

Figure 2—Seven-links biped model.

Table 1

Parameters

No.	m (Kg)	ℓ (m)	r (m)	I (Kgm²
1	2.0	0.25	0.125	0.0026
2	6.0	0.4	0.2	0.0200
3	8.0	0.4	0.2	0.0270
4	30.0	0.6	0.3	0.2250
5	8.0	0.4	0.2	0.0270
6	6.0	0.4	0.2	0.0200
7	2.0	0.25	0.125	0.0026

Simulation Results

Figure 2 shows a 7-link biped model and Table 1 shows its parameters. The i-th element is described by its mass (m_i), moment of inertia about its center of mass (I_i), its length (ℓ_i), length from the lower joint to its center of mass (r_i), and its angle from vertical (θ_i). The actuators acting on the i-th joint produce torque (u_i).

Suppose that the following five constraints are imposed on this system:

$f_1 : \theta_4 = 0°$ (degree), $f_2 : \theta_1 - \theta_2 = 90°$ (degree)

$$f_3 : \ell_1 \sin\theta_1 + \ell_2 \sin\theta_2 + \ell_3 \sin\theta_3$$
$$- \ell_5 \sin\theta_5 - \ell_6 \sin\theta_6 - \ell_7 \sin\theta_7 = SL$$

$$f_4 : \ell_1 \cos\theta_1 + \ell_2 \cos\theta_2 + \ell_3 \cos\theta_3$$
$$- \ell_5 \cos\theta_5 - \ell_6 \cos\theta_6 - \ell_7 \cos\theta_7 = 0$$

$$f_5 : \ell_1 \cos\theta_1 \geq 0.$$

These constraints mean that the torso is maintained vertically, the relative angle between foot and shank is maintained at 90°, and DSP is maintained.

By the corollary in the previous section, the equation of this system is described by

$$A(\oplus)\ddot{\oplus} + B(\oplus)\dot{\oplus}^2 + C(\oplus) + \left(\frac{\partial f}{\partial \theta}\right)^T Nf = Du$$

where $\oplus = [\theta_1,...,\theta_7]^T, \dot{\oplus}^2 = [\dot\theta_1^2,...,\dot\theta_7^2]^T, u = [u_1,...,u_6]^T$

$$A(\oplus) = [k_{ij} \cos(\theta_i - \theta_j)], B(\oplus) = [k_{ij} \sin(\theta_i - \theta_j)], C(\oplus) = [g_{ij} \sin\theta_i]$$

$$k_{ij} = I_i + m_i r_i^2 + \sum_{p=i+1}^{7} m_p \ell_i^2$$

$$\begin{cases} k_{ij} = (m_j r_j + \sum_{p=j+1}^{7} m_p \ell_j)\ell_i & (i<j) \\ k_{ij} = k_{ji} & (i>j) \end{cases}$$

$$g_{ij} = \begin{cases} -(m_i r_i + \sum_{p=i+1}^{7} m_p \ell_i)g & (i=j) \\ 0 & (i \neq j) \end{cases}$$

and

$$D = \begin{bmatrix} -1 & & & & & \\ 1 & -1 & & & & \\ & 1 & -1 & & & \\ & & 1 & -1 & & \\ & & & 1 & -1 & \\ & & & & 1 & -1 \\ & & & & & 1 \end{bmatrix} \quad , N = \mathrm{diag}[n_1, n_2, n_3, n_4].$$

Figure 3 shows the motion of the constrained biped system in the case of all torques being zero. It is clear that the center of gravity of this system

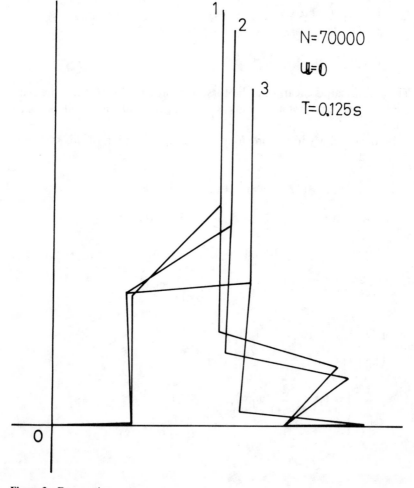

Figure 3—Free motion.

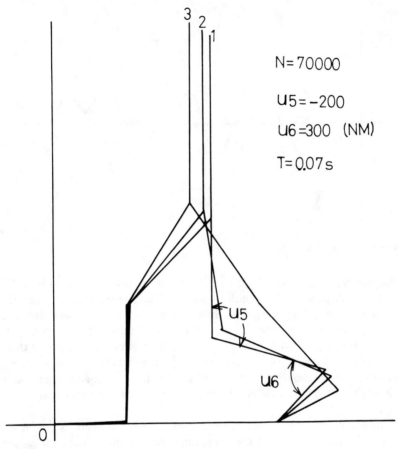

Figure 4—Forced motion.

is falling without violating the constraints. Figure 4 shows the motion of the system with torques at the knee joint and the ankle joint of the right leg. Also in this case the constraints are maintained.

More careful inspection of the numerical solutions gives serious results. That is, numerical solutions contain high frequency components. Figure 5 shows trajectories of θ_4 with respect to a sufficiently large N. Since θ_4 is constrained to zero, these curves have errors. Also other high frequency components have some errors depending on N. These values, however, are sufficiently small when compared to the size of motion.

Discussion and Conclusion

Dynamic systems with certain holonomic constraints (including nonlinear constraints) are precisely modeled by the algorithm in the cor-

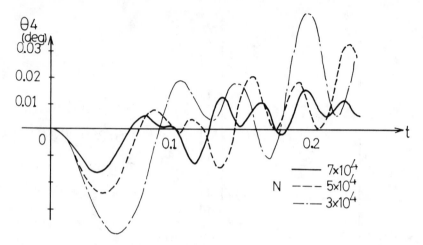

Figure 5 — Trajectories of θ_4.

ollary of Section 2. Simulation results show that the algorithm is sufficiently applicable to these multi-freedom dynamic systems with certain nonlinear constraints imposed. Still further, it can be shown that the solutions obtained by the simulation contain high frequency components. Therefore, stability considerations of numerical solutions and their error analysis become very important. But since convergence of solutions is assured, if the step size of the integration in the Runge-Kutta method is suitably selected for some N, errors of solutions can be reduced to small ones.

More precise analysis of the linearized system obtained by perturbation in the vicinity of solutions gives important results. That is, it can be shown that generalized coordinates of constrained systems are separated into two arrays. The first includes unconstrained variables. The second has constrained variables whose frequencies increase in proportion to Order ($N^{1/2}$), and whose amplitudes decrease in proportion to Order ($N^{-1/2}$) or more rapidly, as $N \to \infty$. This system provides some contributions to the design of locomotion robots and understanding human locomotion.

References

ARNOLD, V.I. 1978. Mathematical method of classical mechanics. Springer-Verlag pp. 75-76.

BRAYTON, R., Gustavson, F.G., & Hachtel, G. 1972. A new efficient algorithm for solving differential-algebraic systems using implicit backward differentiation formulas. Proc. IEEE 60(1):98-108.

HEMAMI, H., & Wyman, B.F. 1979. Modeling and control of constrained dynamic systems with application to biped locomotion in the frontal plane. IEEE Trans. Auto. Cont., AC-24(4):526-535.

A Modified Release Method for Measuring the Moment of Inertia of the Limbs

V.V. Stijnen, E.J. Willems, A.J. Spaepen,
L. Peeraer, and M. Van Leemputte
Katholieke Universiteit, Leuven, Belgium

Several techniques have been proposed for determining the moment of inertia of the extremities of living subjects. It is the purpose of this paper to present a modified release method that enables the measurement of the moment of inertia for the whole leg, the whole arm, the forearm plus hand, and the shank plus foot. In contrast with the quick release method, only the gravitational force is allowed to produce a moment relative to the joint axis. The results of this method are compared to values obtained from a pendulum test (Hatze, 1975) and those from two mathematical models of the human body (Dempster, 1955; Hanavan, 1964).

Procedure

Thirty-four subjects, all students in physical education, 18 to 23 years of age, submitted to two series of tests (pendulum and release). The anthropometric measurements (Hanavan, 1964) were taken in order to dimension the model of the human body.

In the first test (release) a relaxed body-segment was splinted and supported in an horizontal position. The static force (F) was measured by a force transducer (Hottinger & Baldwin type U-1) together with the distance (1) between the joint axis and the point of application of that force. The moment of the segmental weight relative to the joint axis (M) was then given by: $M = - F.1$

The release method was applied to the right extremities: the whole leg, the whole arm, forearm plus hand, and shank plus foot (see Figure 1).

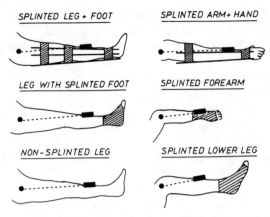

Figure 1—Positions of the body segments for the modified release method.

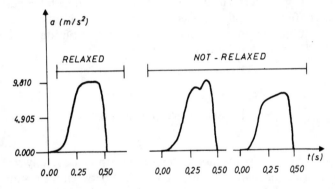

Figure 2—Typical accelerometer record of a release, showing accelerometer outputs of a relaxed and a not-relaxed body segment.

An accelerometer (Sundstrand model 2180 mini-pal) was fixed near the estimated (by mathematical model) center of percussion of the segment. The distance (r) between the joint axis and the accelerometer was determined. To enable the relaxation of the segment during the downward movement, a suitable position was determined. Then the relaxed body-segment, supported with a girth, was elevated approximately 10° above its horizontal position. After the sudden interruption of the support the segment immediately began its rotational acceleration around the joint axis concerned. The tangential acceleration (a) was measured during the first 0.5 sec of the movement (sampling frequency 1 KHz, 12 bits).

The resulting acceleration-time curve was immediately evaluated visually on an oscilloscope. Since the acceleration is constant when no forces except the gravitational force influence the downward movement of the segment, a good acceleration-time curve must show a constant level of acceleration (see Figure 2). The test was repeated when irregularities or an obvious insufficient level of acceleration occurred. The moment of inertia of the segment (I) relative to the joint axis was then calculated from: I = M.r/a.

A second test was based on a pendulum technique (Hatze, 1975). The influence of two variables on the resulting moment of inertia was examined (distance between joint axis and suspension, spring constant). The period of oscillation was derived from the acceleration-time curve, and the force-time registration. This test was only executed for the whole leg and the whole arm. In this test too, immediate evaluation of the resulting acceleration-time curves was made on oscilloscope.

Prior to our experiments on human body segments, the pendulum technique and the modified release technique were used to determine the moment of inertia of a rigid object (iron rod: comparable length and mass with respect to the human leg).

These tests were performed under the same experimental conditions as those used for the body segments. The results obtained from both methods, subjected to the same experimental variations, all ranged within 0.5% of the known moment of inertia.

Results and Discussion

The results obtained by the modified release method are highly reproducible considering the high correlation coefficient between the results of the test and those of the retest (whole leg: .96 − .97; whole arm: .93; the forearm plus hand: .87; the shank plus foot: .91.) No statistically significant differences between the means of the test and the retest (see Tables 1 and 2) were found except for those of the non-splinted leg (significant at the 5% level).

The mean values of the moment of inertia for the whole leg, as calculated from the release method under various conditions, are of comparable magnitude (see Table 1). The results of the non-splinted leg, however, show a statistical difference at the 1% level with the two previous test situations of the leg. This could be explained by the fact that during the downward movement of the leg the acceleration is influenced by modifications in the multi-link body segments. For this reason splints were also used for the test of the forearm and the lower leg.

As shown by Tables 1 and 2 the estimated moments of inertia given by Dempster, and those calculated from the mathematical model of Hanavan, differ significantly from those obtained by the release method (5% or 1% level). The same conclusion can be drawn concerning the pendulum test (see Tables 1, 2, and 3).

Table 1

**Mean Values (kg.m²) and Standard Deviations of the Moment of Inertia
of the Leg Relative to the Proximal Joint Axis (N = 34)**

REFERENCE	RELEASE METHOD						MATHEMATICAL MODELS	
	SPLINTED LEG + FOOT		LEG WITH SPLINTED FOOT		NON-SPLINTED LEG		Dempster	Hanavan
	test	retest	test	retest	test	retest		
MEAN	2.445	2.442	2.448	2.436	2.645	2.607	2.792	2.827
S.D.	0.342	0.330	0.298	0.316	0.318	0.338	0.364	0.380

The pendulum test also resulted in significantly different moments of inertia (1% level) when various conditions (suspension points and spring constants) are used (see Table 3).

Increasing the distance between joint axis and suspension point, or increasing the spring constant, enlarges the estimated moment of inertia. The latter seems to be inversely proportional to the period of the oscillation ($r = -.97$). This disparity was unexpected considering the previously mentioned results obtained from a rigid object. Application of the pendulum method on the body segments of human living subjects is therefore possibly influenced by the moving masses of the muscles during the oscillation, the specific shape of the joint, small rotations around the longitudinal axis of the segment, etc. Similar remarks can be made concerning the results from the pendulum tests of the whole arm.

Among other experimental methods we may consider the release method as a useful method in determining the moment of inertia of the human body segments in situ, taking into account the previous results obtained with the rigid object, the reproducibility on human limbs, the conformity of the results obtained in different test situations (the leg), the immediate evaluation of experimental results, and the simplicity of calculation.

Factors influencing experimental results should be kept in mind during the test. These factors can be 1) sufficient relaxation of the measured limb and the muscles across the joint, 2) accurate location of the joint axis, 3) alignment of the accelerometer axis with respect to the body segment, and 4) the fixation of the multi-link segments.

Table 2

Mean Values (kg·m²) and Standard Deviations of the Moment of Inertia of the Arm,
the Forearm plus Hand and the Shank plus Foot Relative to the Proximal Joint Axis (N = 34)

REFERENCE	THE WHOLE ARM				THE FOREARM PLUS HAND				THE SHANK PLUS FOOT			
	RELEASE (SPLINTED ARM + HAND)		MATHEM. MODELS		RELEASE (SPLINTED FOREARM)		MATHEM. MODELS		RELEASE (SPLINTED LOWER LEG)		MATHEM. MODELS	
	test	retest	Dempster	Hanavan	test	retest	Dempster	Hanavan	test	retest	Dempster	Hanavan
MEAN	.428	.422	.521	.464	.081	.080	.092	.072	.378	.380	.435	.349
S.D.	.065	.070	.075	.063	.015	.015	.015	.011	.051	.055	.064	.050

Table 3

The Periods of Oscillation (Means in s.) and the Moments of Inertia (Means and Standard Deviations in kg.m^2) of the Whole Leg and the Whole Arm from the Pendulum Method in Various Test Situations (N = 34)

TEST SITUATION	THE WHOLE LEG							THE WHOLE ARM	
	SUSPENSION POINTS			SPRING CONSTANTS (c) (N/m)				SPRING CONSTANTS (N/m)	
	near	mid	far	c=184	c=1080	c=1264	c=2170	c=184	c=387
PERIOD OF OSCILLATION	0.528	0.415	0.350	0.775	0.363	0.350	0.277	0.496	0.389
MOMENT OF INERTIA mean	2.296	2.455	2.646	1.809	2.483	2.646	2.921	0.322	0.375
standard deviation	0.326	0.314	0.367	0.230	0.319	0.367	0.398	0.075	0.070

References

DEMPSTER, W. 1955. Space requirements of the seated operator. WADC Technical Report 55-159. Wright-Patterson Air Force Base, OH.

HANAVAN, E.P. 1964. A mathematical model of the human body. AMRL Technical Report 64-102. Wright-Patterson Air Force Base, OH.

HATZE, H. 1975. A new method for the simultaneous measurement of the moment of inertia, the location of the center of mass, and the damping coefficient of a body segment in situ. Eur. J. Appl. Physiol. 34:217-226.

Synchronization of Partial Impulses as a Biomechanical Principle

Kornelia Kulig, Zbigniew Nowacki, and Tadeusz Bober
Biomechanics Laboratory, Wroclaw, Poland

The utilization of motion transfer phenomena has interesting effects. A whip is an example. A motion impulse, adequate in magnitude and character, acting on one whip results in a hypersonic motion in the other. The human upper extremity, as a kinematic chain constructed of clearly separable links coupled together by movable joints, can also be considered, though only as to the character of the motion at its end. The human extremity is a complex formation with autonomous actuators (muscles) that may both sustain and impair the process of movement transfer. Some understanding of that yields the impulses transfer phenomenon. Its description, however, is usually limited to contributions by the subsequent segments of the extremity to the resultant velocity of the last segment (Toyoshima et al., 1974; Hoshikawa and Toyoshima, 1976; Zernicke and Roberts, 1976). Miller (1981) in her extensive monograph, has pointed out some deficiencies of such an approach, especially when the results are obtained by blocking subsequent parts of the body. It has also been stressed that the existing works do not discuss moments of muscular forces in relation to time. Thus, the aim of this research was to create a kinematic and dynamic model based on experimental studies of the nature of the velocity transfer process arising between the upper extremity segments (biokinematic chain) in a throwing task.

Material and Method

A medicine ball throw was used for the description of a motor impulse transfer through the links of a biokinematic chain. In this test the biokinematic chain involved the feet, lower extremities, trunk, and upper extremities with the hands and a ball. An adequate ball diameter insured

1144

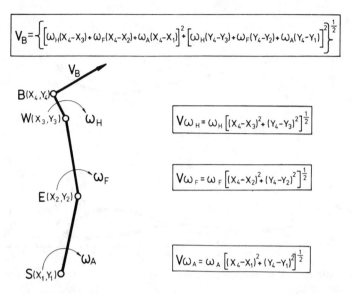

$$V_B = \left\{ \left[\omega_H(X_4-X_3) + \omega_F(X_4-X_2) + \omega_A(X_4-X_1) \right]^2 + \left[\omega_H(Y_4-Y_3) + \omega_F(Y_4-Y_2) + \omega_A(Y_4-Y_1) \right]^2 \right\}^{\frac{1}{2}}$$

$$V\omega_H = \omega_H \left[(X_4-X_3)^2 + (Y_4-Y_3)^2 \right]^{\frac{1}{2}}$$

$$V\omega_F = \omega_F \left[(X_4-X_2)^2 + (Y_4-Y_2)^2 \right]^{\frac{1}{2}}$$

$$V\omega_A = \omega_A \left[(X_4-X_1)^2 + (Y_4-Y_1)^2 \right]^{\frac{1}{2}}$$

Figure 1—Kinematic model of the segments of the kinematic chain (upper extremity). Reference points: x_1, y_1 = shoulder; x_2, y_2 = elbow; x_3, y_3 = wrist; x_4, y_4 = center of gravity of the hand and ball; ω_H, ω_F, and ω_A = angular velocities of the hand, forearm, and arm, respectively.

a symmetrical motion of both extremities in the sagittal plane. The kinematic characteristics were obtained by filming the given motion in the sagittal plane and determining the coordinates of the frontal axes of the kinematic chain. A Bolex camera with f 50 mm lens was used with a filming rate of 48 frames/sec. The test subjects included 25 students of physical education.

Results and Discussion

The first stage in the analysis of the films involved the determination of angular changes ($\alpha[t]$) in the joints as well as angular velocities ($\omega[t]$), and the angular acceleration ($\epsilon[t]$) of the subsequent segments of the kinematic chain. As a result the following was decided: 1) limit the subject of study to the upper extremity kinematic chain with the shoulder girdle (acromion) as a reference point; the segments were now as follows: arm (A), forearm (F), and hand and ball (H); and 2) analyze only the phase of throw, lasting $\approx .2$ sec, with the omission of the pre-stretch. Note that the total positive movements (extension) were included in the analysis.

The second stage of analysis consisted of constructing a kinematic model in which the velocity of the ball was the total of the component velocities originating from segments A, F, and H (see Figure 1). This

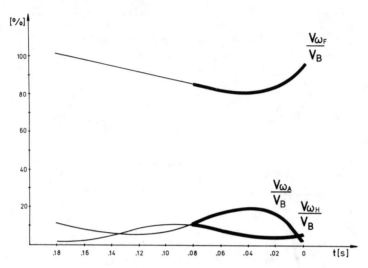

Figure 2 – Percentage contribution of body segments' velocities to the velocity of ball as a function of time.

model allows one to determine a percentage contribution of the A, F, and H velocities to the velocity of the ball as a function of time. Typical contributions obtained for the majority of the subjects are presented in Figure 2. The figures change with time, and at ball release, they amount to A = 3%, F = 90%, H = 7%. It is worth noting that the forearm's contribution increases at the cost of the arm's contribution. Toyoshima and Hoshikawa (1974), while studying a throwing task with rubber balls of diverse weights by means of the segment blocking method, found the A and F + H values to be 20% and 80%, respectively.

The data from Figure 2, however, may serve as additional information since any quantitative comparison of different velocity vectors with the resultant vector is impossible. On the other hand, a formula including values, angles, and directions of the component velocities could not be analyzed. Hence, the percentage contribution analysis was concerned exclusively with the last four frames of film (.08 sec), as in this phase of the throw the angular changes in component velocities (A, F, and H) contribute least to the resultant velocity, thus producing a minimum error. Thanks to the analysis, it has been found that the contribution of the velocity of the hand (H) to the velocity of ball is negligible, meaning that the upper extremity kinematic chain can be reduced to a two-segment system.

The dynamic model was created based on the results of two-stage kinematic analysis, as well as on the following four assumptions:

1. The studied kinematic chain of the upper extremity was reduced to two segments: proximal segment (P = arm), and distal segment (D =

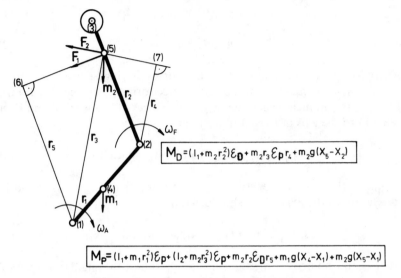

$$M_D = (I_1 + m_2 r_2^2)\mathcal{E}_D + m_2 r_3 \mathcal{E}_p r_4 + m_2 g (X_5 - X_2)$$

$$M_p = (I_1 + m_1 r_1^2)\mathcal{E}_p + (I_2 + m_2 r_3^2)\mathcal{E}_p + m_2 r_2 \mathcal{E}_D r_5 + m_1 g (X_4 - X_1) + m_2 g (X_5 - X_1)$$

Figure 3—Two-segment dynamic model of throw with upper extremity. 1 = shoulder; 2 = elbow; 3 = center of gravity of the hand and ball; 4 = center of gravity of the arm; 5 = center of gravity of the forearm, hand, and ball; F_1 = inertial force ($F_1 = m_2\omega_D r_2$) of mass m_2 as a result of the angular motion (ω_D) of segment [2;3]; F_2 = inertial force ($F_2 = m_2\omega_p r_3$) of mass m_2 as a result of the angular motion (ω_p) of segment [2;3]; r_4 = radius of gyration of F_2 in relation to the axis at the elbow joint; r_5 = radius of gyration of F_1 in relation to the axis at the shoulder joint.

forearm, hand, and ball) extending from elbow joint to the ball's center of gravity.

2. The subject of study was limited to an 11-frame film presenting the last phase of ejection (last 0.2 sec).

3. There exist two moments of propulsion in the system: P segment muscles and D segment muscles.

4. There are mutual interactions between the segments.

First, based on the data published by Dempster and then adopted by Hay (1973), the moment of inertia, I_1, and its radius, r_1 (Segment P), as well as the moment of inertia, I_2, and its radius, r_2, measured from the center of gravity to the elbow joint (Segment D), was calculated. The moments of external forces were replaced in calculations by the moments of inertia of the segments and the moments of inertial forces. A schematic diagram of a two-link biokinematic chain 'upper extremity' with marked radii of the moment of inertia, the parameters of joints' frontal axes, the medicine ball center of gravity, and the inertial forces' directions and their radii is presented in Figure 3. This model, describing a state of equilibrium between muscular moments and the moments generated by resistance to motion, can be generalized in the following way:

$$M_P = M_{IPD} + M_{GPD} + M_{FD} \tag{1}$$

$$M_D = M_{ID} + M_{GD} + M_{FP} \tag{2}$$

where M_P = the moment of muscular forces driving Segment P (proximal); M_D = the moment of muscular forces driving Segment D (distal); M_{IPD} = the moment generated by gravity forces of Segments P and D in relation to the axis at the shoulder joint:

$$M_{IPD} = (I_1 + m_1 r_1^2)\, \epsilon_P + (I_2 + m_2 r_3^2)\, \epsilon_P \tag{3}$$

M_{GPD} = the moment generated by gravity forces of Segments P and D in relation to the axis at the shoulder joint:

$$M_{GPD} = m_1 g\, (x_4 - x_1) + m_2 g\, (x_5 - x_1) \tag{4}$$

M_{FD} = the moment of inertial force as a result of the angular motion of Segment D at the elbow joint in relation to the axis at the shoulder joint:

$$M_{FD} = m_2 \epsilon_D r_2 r_s \tag{5}$$

M_{ID} = the radius of gyration of Segment D in relation to the axis at the elbow joint:

$$M_{ID} = (I_2 + m_2 r_2^2)\, \epsilon_D \tag{6}$$

M_{GD} = the moment of gravity force of Segment D in relation to the axis at the elbow joint:

$$M_{GD} = m_2 g\, (x_5 - x_2) \tag{7}$$

M_{FP} = the moment of inertial force as a result of the angular motion of Segment D at the shoulder joint in relation to the axis at the elbow joint:

$$M_{FP} = m_2 \epsilon_P r_3 r_4 \tag{8}$$

The mean characteristics computed based on kinematic and dynamic models are shown in Figure 4.

The study of the characteristics and the results obtained by way of the kinematic and dynamic investigations proved that simplification of the human upper extremity model as well as restriction of the throwing task analysis to the phase of throw (~ 0.2 sec) did not prevent inferences of some interesting observations:

1. A ball throw involves two stages lasting ~ 0.08 sec and ~ 0.12 sec, respectively.

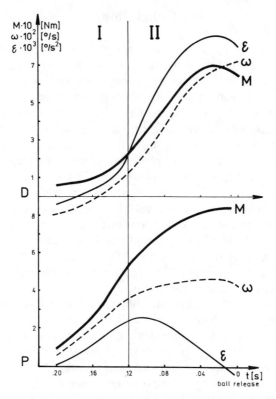

Figure 4—Kinetmatic and dynamic characteristics of two-segment model upper extremity in throwing task. P = proximal segment (arm); D = distal segment (forearm, hand, and ball); ω = angular velocity $d\alpha/dt$; ϵ = angular acceleration $d^2\alpha/dt^2$; M = muscular moment of force equal to a total moment impeding the motion.

2. The first stage is characterized by large activity in Segment P driving forces as opposed to those of Segment D. Moreover, rapid acceleration of Segment P causes some 'recession' of Segment D which may increase tension in the given muscles and produce the effect similar to pre-stretch.

3. The second stage is characterized by large activity of the muscles acting on Segment D. This causes retardation of the Segment P motion though its muscles remain active. The acceleration of Segment D that retards the motion of Segment P is caused by the component force (M_{F_D}) of the generalized dynamic model. As $\epsilon_D r_5$ starts to increase to the 2nd or even higher power, it makes M_P increase, too (see Figure 4), so that diminishing ϵ_P (see Figures 2 and 4) does not stop M_P from supporting the motion of Segment D. From this analysis, it follows that the influence of Segment P on Segment D causes M_{F_P} to remain stable. This is because the increases in radii, r_3 and r_4, are being compensated for by a decrease in acceleration ϵ_P.

Conclusions

It has been shown that during an upper extremity throw, with the trunk as a reference point, the movement starts from the arm, i.e., the link closest to the system's center of gravity, and next, with a delay, it is passed to the succeeding link. At this point, the results are inconsistent with the theory testifying that a greater effect of the kinematic chain action is obtained while diminishing the time interval between the maximal acceleration of the adjacent links. But one also must consider the criterion of motion that takes place during the experiment. For a ball throw, the effect depends on two variables, the angle and the velocity of throw. These two criteria arouse contradictory tendencies; while from the viewpoint of the first one the distal link (D) should be activated in the shortest time possible, the other demands that the time be long enough for the proximal link (P) to attain its maximum speed. For this reason, the activation of the second link must take place at the optimum moment under particular conditions, thus causing the acceleration peaks to be displaced, by .08 sec in this experiment.

It should also be stressed that the objective interpretation of the activation and transfer of motor impulses in a kinematic chain requires simultaneous kinematic and dynamic analyses submitting causes and effects. Mutual, as they are in most cases, interactions between the kinematic chain links are difficult to interpret. Such a situation is presented in Figure 4, where the time necessary for the P link to attain the maximum acceleration is affected by the D link rather than still increasing the muscular impulse.

While analyzing the process of partial motor impulse transfer, the duration and character of the activating muscular impulses (moments of muscular forces) are mainly considered, the accelerations and velocities being additional values. It can also be assumed that the extent of influence of one link upon another is reciprocal in the sequence of their motor impulses. This means that the links distal to the main body mass affect the proximal links more than the other way around.

References

HAY, J.G. 1973. The Biomechanics of Sports Techniques. Prentice-Hall, Inc., Englewood Cliffs, N.J.

HOCHMUTH, G. 1971. Biomechanik Sportlicher Bewegungen (Biomechanics of Athletic Movements). 2nd ed. Sportverlag Berlin.

HOSHIKAWA, T., and Toyoshima, S. 1976. Contribution of body segments to ball velocity during throwing with nonpreferred hand. In: P.V. Komi (ed.), Biomechanics V-B, pp. 109-117. University Park Press, Baltimore.

MILLER, D.I. 1980. Body segment contributions to sport skills performance: Two contrasting approaches. Research Quarterly for Exercise and Sport 51(1): 219-233.

TOYOSHIMA, S., Hoshikawa, T., Miyashita, M., and Oguri, T. 1974. Contribution of the body parts to throwing performance. In: R.C. Nelson and C.A. Morehouse (eds.), Biomechanics IV, pp. 169-174. University Park Press, Baltimore.

ZERNICKE, R.F., and Roberts, E.M. 1976. Human lower extremity kinetic relationships during systematic variations in resultant limb velocity. In: P.V. Komi (ed.), Biomechanics VB, pp. 20-25. University Park Press, Baltimore.

The Mass and Inertia Characteristics
of the Main Segments of the Human Body

V. Zatsiorsky and V. Seluyanov
Central Institute of Physical Culture, Moscow, U.S.S.R.

In the early 1970s suggestions were made regarding some of the variants of the radioisotope (gamma-scanner) method for determining mass inertia (MI) characteristics based on the recording of the weakening of the mono-power gamma-quantum beam when it passes through the patient's body (Baster, 1971; Casper, 1971; Zatsiorsky, Sereda and Sarsaniya, 1972). This method achieves a high degree of accuracy (Casper, 1971; Brooks and Jacobs, 1975) but it also presents many difficulties. As far as we know no experimental paper has been published until now on the application of the radioisotope method for a lifetime determination of the MI characteristics.

The aim of this investigation was to determine the mass and inertial characteristics of male body segments through the use of the gamma-scanner method. Special modifications of this method have been devised to measure body segment parameters without exposing subjects being tested to danger from radiation.

Methods

Basic Principle

When the gamma-radiation beam passes through a substance, it becomes less strong. If the intensity before and after the passage of the gamma-radiation beam through a material layer is measured one can calculate its surface density, i.e., the mass per unit of surface area.

The methods used are the following: a person is exposed to gamma-ray radiation, while the intensity of absorption of the beam and the distance from previously arranged reference points are measured. The mass of the

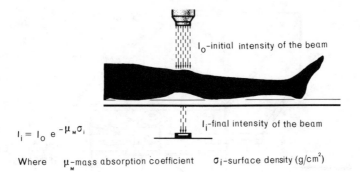

$$I_i = I_0 \, e^{-\mu_M \sigma_i}$$

I_0-initial intensity of the beam

I_i-final intensity of the beam

Where μ_M-mass absorption coefficient σ_i-surface density (g/cm^2)

Figure 1 — The gamma scanner method.

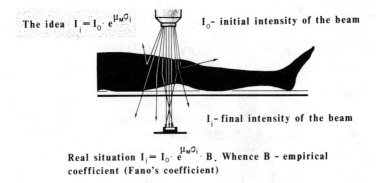

The idea $I_i = I_0 \cdot e^{\mu_M \sigma_i}$

I_0- initial intensity of the beam

I_i- final intensity of the beam

Real situation $I_i = I_0 \cdot e^{\mu_M \sigma_i} \cdot B$. Whence B - empirical coefficient (Fano's coefficient)

Figure 2 — The gamma scanner method is actually more complicated than shown in Figure 1.

underlying tissues is evaluated by the intensity of absorption (see Figure 1).

The mass absorption coefficient (μ) depends on the energy of gamma-quantums and on the elementary (non-molecular) material composition. With the energy of gamma-quantums equal to 0.5/1.3 MeV, the absorption quantity in the human body, in fact, coincides.

It should be emphasized that the practical application of the radioisotope method is associated with a number of difficulties. First, if the gamma-radiation beam does not appear to be narrow, its interaction with absorbers produces secondary gamma-quantums (see Figure 2). Second, if the spectrum differs from the actual energy spectrum of the radioactive substance, the photo peak remains unsteady due to the unstable equipment operation. In view of these facts the direct application of the formula which is indicated in Figure 1 is hardly possible. Therefore, some experiments were conducted which were designed to determine the optimal working conditions (for instance, the most acceptable zone of the energy spectrum, the position of the radiation source over the collimation hole, etc.) and accumulation factor. The results of these in-

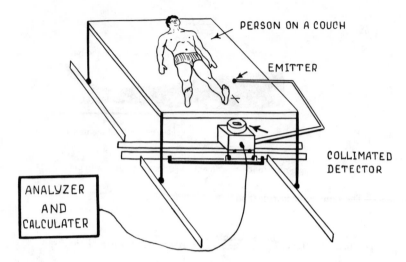

Figure 3—Subject being scanned by emitter and collimated detector.

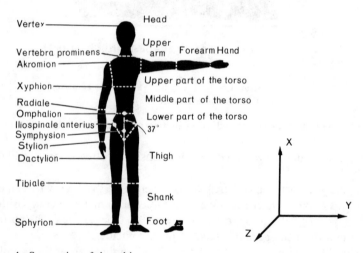

Figure 4—Segmenting of the subjects.

vestigations and the detailed description of the procedure and methods of the experiments are given elsewhere (Zatsiorsky, Sereda and Sarsaniya, 1972; Seluyanov, 1978; Zatsiorsky and Seluyanov, 1979).

The Radioisotope Plant

The principle of operation of this plant involves scanning a patient's body and obtaining a simultaneous recording of the surface density and coordinates of the body section affected by radiation.

Table 1

Characteristics of Subjects (n = 100)

Anthropometric features	Age (yrs)	Height (cm)	Weight (kg)	Circumference of the chest (cm)	Relative masses of tissues, %		
					fat	muscles	bones
$\overline{X}$	23.8	174.1	73.0	91.9	11.4	48.4	16.7
SD	6.2	6.2	9.1	5.3	3.4	2.9	1.3

Table 2

The Mean Values of Mass and Inertial Characteristics of the Human Body Segments (n = 100) X (± SD)

Segment	M (kg)	C.G. (%)	M (%)	K_1 (%)	K_2 (%)	K_3 (%)
Foot	0.997	55.85	1.370	25.70	24.500	12.40
	(0.141)	(3.65)	(0.155)	(0.95)	(0.800)	(1.21)
Shank	3.160	40.47	4.330	28.10	27.500	11.40
	(0.439)	(2.81)	(0.305)	(0.63)	(0.61)	(1.96)
Hip	10.360	45.49	14.165	26.70	26.700	12.10
	(1.568)	(1.94)	(0.998)	(0.99)	(0.990)	(0.93)
Hand	0.447	63.09	0.614	28.50	23.300	18.20
	(0.072)	(4.85)	(0.083)	(2.16)	(1.710)	(2.30)
Forearm	1.177	57.26	1.625	29.50	28.400	13.00
	(0.161)	(3.26)	(0.140)	(0.86)	(0.650)	(1.51)
Upper arm	1.980	55.02	2.707	32.80	31.000	18.20
	(0.319)	(4.19)	(0.243)	(1.61)	(1.245)	(3.27)
Head	5.018	50.02	6.940	30.30	31.500	26.10
	(0.393)	(2.23)	(0.707)	(1.29)	(1.340)	(2.10)
Upper part of the torso	11.654	50.66	15.955	50.50	31.990	46.50
	(1.873)	(2.24)	(1.529)	(3.74)	(1.710)	(4.64)
Middle part of the torso	11.953	45.02	16.327	48.20	38.300	46.80
	(2.176)	(2.12)	(1.725)	(4.24)	(2.430)	(5.60)
Lower part of the torso	8.164	35.41	11.174	35.60	31.900	34.00
	(1.492)	(3.01)	(1.428)	(2.05)	(2.370)	(3.00)

M (kg) = mass of the segment; C.G. (%) = center of gravity along the longitudinal axis; M (%) = ratio of mass of the segment to the weight of the body (relative weight); K_1 (%) = the ratio of the radius of gyration about the anteroposterior axis of the segment to the length of the segment; K_2 (%) = the ratio of the radius of gyration about the transverse axis to the length of the segment; K_3 (%) = the ratio of the radius of gyration about the longitudinal axis to the length of the segment.

Table 3

Coefficients of Multiple Regression Equations for Estimating the Mass of Segments of the Body[a]

Segment	B_0	B_1	B_2	R	SD
Foot	−0.829	0.0077	0.0073	0.702	0.101
Shank	−1.592	0.0362	0.0121	0.872	0.219
Hip	−2.649	0.1463	0.0137	0.891	0.721
Hand	−0.1165	0.0036	0.00175	0.516	0.063
Forearm	0.3185	0.01445	−0.00114	0.786	0.101
Upper arm	0.250	0.03012	−0.0027	0.837	0.178
Head	1.296	0.0171	0.0143	0.591	0.322
Upper part of the torso	8.2144	0.1862	−0.0584	0.798	1.142
Middle part of the torso	7.181	0.2234	−0.0663	0.828	1.238
Lower part of the torso	−7.498	0.0976	0.04896	0.743	1.020

[a]Multiple regression equations are in the form, $y = B_0 + B_1 X_1 + B_2 X_2$, where X_1 = body weight in kg, X_2 = body height in cm., and y = predicted mass of segment in kg.

Table 4

Coefficients of Multiple Regression Equations for Estimating the Center of Gravity of Segments along the Longitudinal Axis[a]

Segment	B_0	B_1	B_2	R	SD
Foot	3.767	0.065	0.033	0.530	1.1
Shank	−6.05	−0.039	0.142	0.510	1.25
Hip	−2.42	0.038	0.135	0.600	1.31
Hand	4.11	0.026	0.033	0.383	1.12
Forearm	0.192	−0.028	0.093	0.371	1.14
Upper arm	1.67	0.03	0.054	0.368	1.4
Head	8.357	−0.0025	0.023	0.288	0.69
Upper part of the torso	3.32	0.0076	0.047	0.258	1.19
Middle part of the torso	1.398	0.0058	0.045	0.437	1.18
Lower part of the torso	1.182	0.0018	0.0434	0.320	1.0

[a]The form of the multiple regression equation is the same as in Table 3.

In this experiment the object under investigation was subjected to a scan (see Figure 3). The person being tested reclined on a couch. Above the couch were a moving emitter and, below it, a collimated detector. The entire body was scanned in this way. The information recorded during a test was put onto punched tape and processed in a computer.

In this case, the surface density of a specific section and its coordinates

Table 5

Coefficients of the Multiple Regression Equations for Estimating the Principal
Moment of Inertia about the Anteroposterior Axis (kg · cm^2)[a]

Segment	B_0	B_1	B_2	R	SD
Foot	−100.	0.480	0.626	0.75	6.8
Shank	−1105.	4.59	6.63	0.85	48.6
Hip	−3557.	31.7	18.61	0.84	248.
Hand	−19.5	0.17	0.116	0.50	3.7
Forearm	−64.	0.95	0.34	0.71	10.2
Upper arm	−250.7	1.56	1.512	0.62	27.6
Head	−78.	1.171	1.519	0.40	42.5
Upper part of the torso	81.2	36.73	−5.97	0.73	297.
Middle part of the torso	618.5	39.8	−12.87	0.81	237.
Lower part of the torso	−1568.	12.	7.741	0.69	156.

[a]The form of the multiple regression equation is the same as in Table 3.

Table 6

Coefficients of Multiple Regression Equations for Estimating the Principal
Moment of Inertia about the Transverse Axis (kg · cm^2)[a]

Segment	B_0	B_1	B_2	R	SD
Foot	−97.09	0.414	0.614	0.77	5.77
Shank	−1152.	4.594	6.815	0.85	49.
Hip	−3690.	32.02	19.24	0.85	244.
Hand	−13.68	0.088	0.092	0.43	2.7
Forearm	−67.9	0.855	0.376	0.71	9.6
Upper arm	−232.	1.525	1.343	0.62	26.6
Head	−112.	1.43	1.73	0.49	40.
Upper part of the torso	367.	18.3	−5.73	0.66	171.
Middle part of the torso	263.	26.7	−8.0	0.78	175.
Lower part of the torso	−934.	11.8	3.44	0.73	117.

[a]The form of the multiple regression equation is the same as in Table 3.

vis-a-vis segment boundaries were recorded simultaneously. These data
were put into a computer which calculated all the necessary MI
characteristics.

Radiation Protection

The sources of gamma-radiation used in these experiments included

Table 7

Coefficients of Multiple Regression Equations for Estimating the Principal Moment of Inertia about Longitudinal Axis (kg · cm^2)[a]

Segment	B_0	B_1	B_2	R	SD
Foot	−15.48	0.144	0.088	0.55	2.7
Shank	−70.5	1.134	0.3	0.47	22.
Hip	−13.5	11.3	−2.28	0.89	49.
Hand	−6.26	0.0762	0.0347	0.43	1.8
Forearm	5.66	0.306	−0.088	0.66	2.9
Upper arm	−16.9	0.662	0.0435	0.44	12.5
Head	61.6	1.72	0.0814	0.42	35.6
Upper part of the torso	561.	36.03	−9.98	0.81	212.
Middle part of the torso	1501.	43.14	−19.8	0.87	188.
Lower part of the torso	−775.	14.7	1.685	0.78	116.

[a]The form of the multiple regression equation is the same as in Table 3.

zinc-65 and cesium-137. The activity of these two elements does not exceed 250 microcurie. In this case, the radiation dose during the experiment does not exceed 10 millirads (50 times lower than maximum permissible dose). The researcher is subjected to a dose of 10 millirad/year.

The Patient's Segments

The mass and inertial parameters of 10 segments of the human body including the upper, middle, and lower part of the trunk were determined. These segments are shown in Figure 4.

The method of dissecting the human body into segments was comparable to the historically-evolved method of dissecting cadavers, thus making it possible to compare the obtained results. There were, however, two distinctions:

1. In this investigation the shank was separated from the thigh along the joint line of the knee joint (while dissecting cadavers some authors have included a portion of the femur in the shank mass, thus distorting the actual mass of the thigh and shank).

2. The thigh was dissected from the trunk along a plane passing through the spina iliaca anterior superior at an angle of 37° to the saggital plane of the body. The analysis of the X-ray pictures showed that this plane lies near the caput femoris.

Subjects

The subjects who underwent testing totalled 100 men, including 56 students of physical education, 26 students from a technical college, and 18 researchers.

The information on these subjects is given in Table 1.

Results

The main results of this study included the biomechanical body segment parameter data, specifically the weights of the segments, locations of the centers of mass, moments of inertia, radii of gyration and more than 150 regression equations.

The mass and inertial characteristics are given in Table 2.

In practice, the researcher who is in need of the body mass-inertial characteristics should reach a reasonable compromise between the requirements for accuracy of the estimates and the time required to complete the measurements. To satisfy the demands of various researchers, regression equations have been generated using the following independent variables: 1) the weight and length of the human body (Tables 3 to 7), and 2) the most efficient anthropometric predictors (different for each of the body segments).[1]

Footnote

1. These results are to be published in full in English in Journal of Biomechanics (1983). They have already been published in Russian (Zatsiorsky and Seluyanov, 1979).

References

BASTER, C.M. 1971. The use of the gamma ray scanner in determining body mass parameters. In: R.C. Nelson (ed.), Proceedings of the Pennsylvania State University Biomechanics Conference. University Park, Pennsylvania.

BROOKS, C., and Jacobs, A. 1975. The gamma mass scanning technique for inertial anthropometric measurement. Med. Sci. in Sports 7:290-294.

CASPER, R.M. 1971. On the use of gamma ray images for determination of human body segment parameters. Unpublished Master's Thesis, Pennsylvania State University. Cited by D.J. Miller and R.C. Nelson, 1973. Biomechanics of Sport. Lea and Febiger, Philadelphia.

SELUYANOV, V.N. 1978. (Mass-inertional characteristics of the human body segments and their relationship with anthropometric measurements). Ph.D. dissertation, Central Institute of Physical Culture, Moscow.

ZATSIORSKY, V.M., Sereda, M.G., and Sarsaniya, S.K. 1972. (Method of measuring moments of inertia of unhomogeneous unfree bodies). Author's certificate N.427698.

ZATSIORSKY, V.M., and Seluyanov, V.M. 1979. (Mass-inertional parameters of human body). Voprosi anthropologii (Problems of anthropology). N 62, pp. 91-103. (in Russian)

Computer-Assisted Three-Dimensional Motion Analysis and Dynamic Body Force Evaluation in Simple Loading Operations

Kazuaki Iwata, Toshimichi Moriwaki, Tsuneo Kawano, Norimasa Misaki, Haruo Nomura, and Yasuyoshi Yanagida
Kobe University, Kobe, Japan

Simple loading and unloading of a weight are often encountered in daily activities and constitute the basis of many other operational motions. There is much unknown, however, about the basic motion patterns and the kinematic characteristics of the body during movement. Most of the previous studies concerning dynamic motion analysis of the human body were restricted to: 1) studies of partial body motions (Smidt, 1973; Winter and Kuryliak, 1978), such as those of upper and lower limbs only, and 2) two-dimensional analysis for the sake of simplicity (Danis, 1975; Ghosh and Boykin, 1976; Iwata et al., 1980). Many of the two-dimensional motion analyses deal with the motions projected onto the sagittal plane.

The present study aims at establishing a three-dimensional mathematical model of the human body to carry out detailed analysis of gross body motion under normal working conditions and to estimate forces and torques acting at the joints of the body during the motions.

Modeling, Methods, and Procedure

In order to estimate the forces and the torques acting at the individual joints of the body during motion, a three-dimensional mathematical model of the human body was established. The model basically consisted of 17 equivalent cylindrical elements each of which represented individual body segments, such as the head, body, upper arm, forearm, hand, hip, thigh, leg, foot, etc. Newton's equations of motion were derived for all of the body segments. For example, the equations of motion for an arbitrary segment i are given as follows:

$$F_{i-1,i} - F_{i,i+1} - W_i \cdot j = W_i/g \cdot \alpha_{Gi} \tag{1}$$

$$T_{i-1,i} - l_i \cdot r_i \times F_{i-1,i} - T_{i,i+1} - l_i \cdot r_i \times F_{i,i+1} = M_i \tag{2}$$

where $F_{i,i+1}$ = force acting from segment i to segment $i+1$; $T_{i,i+1}$ = torque acting from segment i to segment $i+1$; W_i = weight of segment i; l_i = length of segment i; r_i = unit vector which gives direction of segment i; g = acceleration due to gravity; α_{Gi} = acceleration of the center of mass (G_i) against ground; and M_i = externally applied moment. The equations of motion are solved based on the measured motion data in order to obtain the forces and torques acting at the individual body joints.

Sequences of loading and unloading motions of seven subjects were examined. The subjects were asked to lift a weight of 5.5 kg from the floor, put it on a table in front of them, and then return it to the floor in the reverse order of loading. The height of the table was 0.75 m from the floor. Figure 1 shows a schematic illustration of the experimental arrangements. A 16 mm high-speed movie camera was employed to take pictures of the motion from the side of the subject at a rate of 45.8 frames/sec. A life-size mirror was placed in front of the subject, and the image in the mirror was also taken simultaneously so that the three-dimensional information of the motion in both the frontal and sagittal planes was obtained. The height of camera and the distance to the subject were 0.97 m and 7.72 m, respectively. The three components of forces and torques acting between the feet and the floor were measured employing two force plates and recorded onto magnetic tapes, simultaneously.

The shape of each body segment was assumed to be cylindrical. The equivalent diameter and the length of each segment were estimated on the basis of photographs and direct measurements of the body. The volume of each segment was also confirmed by the displacement of water. The weights of the individual segments were calculated from the estimated volumes of the equivalent cylinders assuming that the specific gravity was equal to 1.

The motion pictures taken were projected onto a graphic tablet and the coordinate information of each joint in both the sagittal and the frontal planes was input into a minicomputer frame by frame by pressing the graphic pen in a given sequence. The time series data of the joint coordinates were processed by employing the digital low-pass filtering algorithm and fitted to spline functions to obtain the velocities and the accelerations.

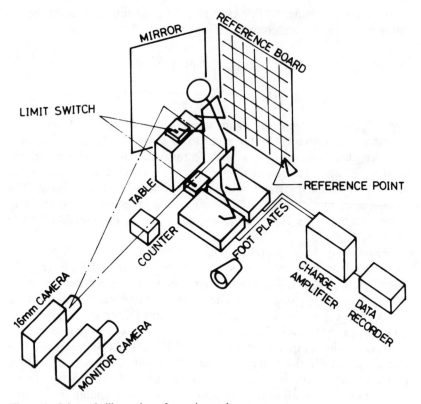

Figure 1—Schematic illustration of experimental arrangements.

Results and Discussion

The three-dimensional line drawing of the subject was reconstructed within the computer based on the digitized motion data. Figure 2 shows a typical example of a sequence of loading and unloading motions reproduced on the CRT frame by frame. The time interval of the succeeding skeletal diagrams is 0.14 sec and the location of viewpoint to the subject in this case is a slant front. The result in Figure 2 shows that the unloading motion took a little longer time in comparison with the loading motion, even though instructions on the timing of each motion were given to the subject during the experiment. This is attributable to a longer half bending posture in the unloading sequence. The upper half body is rather bent forward with much inclination of the head during the unloading process, while the upper half body is rather bent backward during the loading process. The animated three-dimensional display of body motions on the CRT is thus quite effective for detailed analyses of the motion patterns.

The loci of the body center of gravity during the loading and the

Time

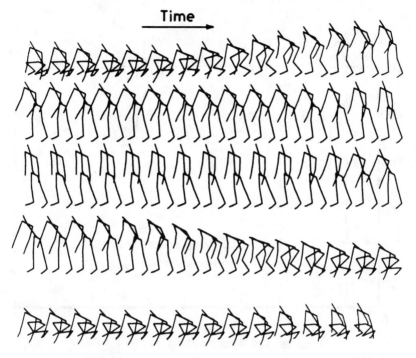

Figure 2 — Plots of loading and unloading motions on CRT generated by computer. The location of viewpoint to the subject is slant front. The subject is 23 years old. Motion time is 8.1 sec.

unloading motions were calculated and they were related to the motion patterns. The displayed loci showed almost no side movement of the body center of gravity in the right and left directions. The loci in the sagittal plane were almost linear and quite similar for both loading and unloading motions. These indicate that the motion analyzed here is quite stable for an operator and safe for the subject. The locus of the body center of gravity gives a good measure of stability of the motion.

The forces and torques acting at individual joints were computed by solving the equations of motion based on the motion data measured and mathematically treated. The forces and torques acting at the lower limbs cannot be computed purely theoretically, since the assignments of the forces and torques at the hip joint to the right and the left thighs cannot be determined. In order to solve this problem, the forces and the torques acting at the thighs were computed individually. First, taking into account the forces and the torques measured by the foot plates, the ratios of the forces and the torques between the right and the left thighs were determined. The forces and torques acting at the lower limbs were then finally calculated taking into account the ratios; however, the absolute

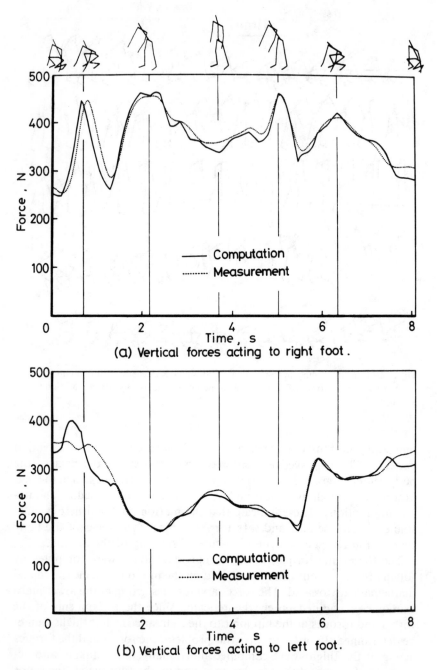

(a) Vertical forces acting to right foot.

(b) Vertical forces acting to left foot.

Figure 3—Comparison of measured and computed vertical forces acting on the feet.

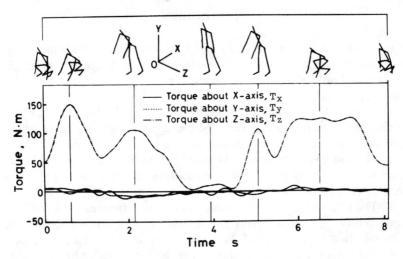

Figure 4—Typical examples of computed three components of torques at hip joint.

values of the measured forces and torques were not taken into consideration.

Figure 3 shows the comparison of the vertical forces at the feet computed thus and those measured by the foot plates. The computed forces acting on the floor coincide well with the measurements. This provides evidence that the analytical estimations are reliable. The vertical force acting on the right foot, which was positioned forward, shows four characteristic peaks observed when the subject: 1) lifts the weight from the floor, 2) places the weight on the table, 3) lifts the weight from the table, and 4) puts the weight on the floor (see the motion patterns shown in Figure 2). Relatively larger forces act on the left foot when the subject lifts the weight from the floor and also when bending down with the weight.

Figure 4 shows typical examples of the computed three components of torques at the hip joint. It is clear that the torque component about the Z axis, which is perpendicular to the sagittal plane, is predominant. The positive sign of T_z in the figure indicates that the torque acts to raise the upper body in this case. It is recognized that the pattern of the torque (T_z) is similar to that of the vertical force acting to the right foot shown in Figure 3.

The forces and the torques acting at various joints were computed and reviewed. One notable observation is that the magnitude of the torque acting at the right knee joint decreases with an increase in the number of cycles of motion, while that at the left knee joint decreases. This is presumably attributable to the effect of fatigue.

Conclusions

A system was established to analyze and evaluate human body motion in three dimensions under normal working conditions. Sequences of simple loading and unloading motions were analyzed and the following conclusions are drawn:

1. A three-dimensional mathematical model of the human body was established, which was proved to be useful for kinematic analysis and evaluation of the gross body motions.

2. Methodology was developed to obtain the three-dimensional information of the body motion and to estimate the dynamic forces and torques acting at the body joints.

3. The characteristics of the basic motion patterns and the loci of the body center of gravity during simple loading and unloading motions were clarified with the aid of the three-dimensional computer graphics.

4. The dynamic forces and the torques acting at the individual body joints during motion were computed and evaluated in relation to the motion patterns and fatigue.

References

DANIS, A. 1975. Analysis and synthesis of body movements utilizing the simple n-link system. In: R.C. Nelson and C.A. Morehouse (eds.), Biomechanics IV, pp. 513-518. University Park Press, Baltimore.

GHOSH, T.K., and Boykin, W.H., Jr. 1976. Analytic determination of an optimal human motion. J. Optimization Theory and Applications 19(2):327-346.

IWATA, K., Moriwaki, T., & Kawano, T. 1980. Computer graphics applied to human motion analysis and body force evaluation. Proc. of EUROGRAPHICS 80:167-178.

SMIDT, C.L. 1973. Biomechanical analysis of knee flexion and extension. J. Biomechanics 6:79-92.

WINTER, D.A., and Kuryliak, W.M. 1978. Dynamic stabilization in human gait: The biomechanical relationships between the triceps surae and the metatarsophalangeal joint. In: E. Asmusson and K. Jurgenson (eds.), Biomechanics VI-A, pp. 280-286. University Park Press, Baltimore.

A New Method for Obtaining the Position, Velocity, and Acceleration of a Moving Object

Hisao Kato and Masahiko Kojima
Nagoya Municipal Industrial Research Institute
Nagoya, Japan

Mikio Noda
Aichi University of Education
Aichi, Japan

Kazuo Tsuchiya
Labour Accident Prosthetic and Orthotic Center
Nagoya, Japan

There have been two methods for obtaining the position, velocity, and acceleration of moving objects (Tsuchiya, 1977), the classical method (the cyclograph), and the popular method of using a movie camera, or a camera and a stroboscope, in combination with reflective tapes. The accuracy of the former method is good but the results are often too complicated to distinguish the behaviors of many points on a moving object. The latter method is simple but if the velocities of the two points on a moving object are very different, simultaneous measurement of these points becomes inaccurate.

Recently two other methods have been developed (Tsuchiya, 1977). One, which uses a combination of a semiconductor position sensor in a camera and infrared light-emitting diodes (LEDs), is called the SELSPOT. The other uses a TV camera in conjunction with small reflective marks to determine the movement of an object. The combination of a semiconductor position sensor and infrared LEDs seems promising, but still there are many problems. The cost of the instrument is high and the measurable area of an object is limited. The distance between the camera and object is small, and position linearity is not sufficient. The TV camera method with reflective marks has the problem at present of the accuracy of positioning the points on the trajectory.

In this article, a new method using only a camera and visible light-emitting diodes to obtain the position, velocity, and acceleration of a moving object is introduced. This is a simple and accurate method. The operating frequencies of the points on the moving object can be high, and the cost of the instrument required is modest. With this method the measurable area and distance between the camera and object can be large. The electronic circuits including the battery are small enough to attach them to the moving object, for example, the human body.

Methods

Green light-emitting diodes (LEDs) were positioned on the object under investigation. The LED's were individually fed by an oscillating rectangular waveform voltage. The frequency of each flashing LED was adjusted depending upon the velocity of the point located on the object. To obtain the best result, the higher the velocity of the point positioning the LED, the higher the frequency fed to the LED, so that when different points on the object needed to be monitored consecutively at their respective velocities, this could be done quite accurately.

An astable multivibrator was used as the master oscillator and the generated frequency was divided into Y_2, Y_4, Y_6, Y_8, - - - Y_{2m} frequencies by a shift register. The fundamental frequency and the outputs from the register were fed to power transistors via one-shot monostable vibrators to drive the LEDs. This set-up permitted the adjustment of the signal pulse width and height to the LEDs. Marker LEDs were positioned just above the diodes already mentioned and pulsed at a lower frequency so as to indicate corresponding positions in time of the trajectories. Reference LEDs were positioned on a cross-type framework throughout the experiment to indicate the vertical and horizontal axes. The picture of the moving object with flashing LEDs was taken in a darkened room by a camera. While the object was in motion the shutter was open. A push button switch was used to start and stop the flashing diodes.

Application

The movement of a human arm was reproduced as shown in Figure 1, where the trajectory of the diodes on the wrist, upper arm, forearm, and shoulder of the arm are shown. In Figure 2, the positions of the diodes on the arm are indicated. The frequencies of the diodes were 96 Hz, 48 Hz, 48 Hz, and 24 Hz, respectively, and that of the marker diodes was 12 Hz.

The trajectory of the wrist was nearly a circle, but those of the upper

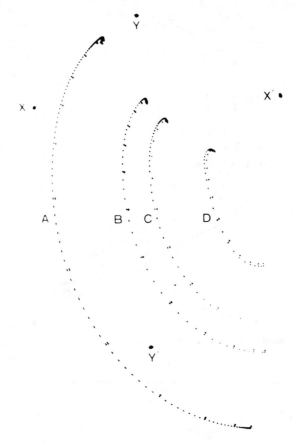

Figure 1—Trajectories of shoulder flexion. The arm was swung up at normal speed. A = wrist; B = forearm; C = upper arm; D = shoulder; XX' = vertical (distance: 80 cm); YY' = horizontal (distance: 60 cm).

and a forearm formed a part of an ellipse. The trajectory of the shoulder appeared as a part of a cardioid. An approximate estimation of torque and power of the shoulder was carried out by the following equations:

$$T = I_O \ddot{\theta} \pm mgr \sin \theta \tag{1}$$

$$P = T \dot{\theta} \tag{2}$$

where T = the torque; I_O = the moment of inertia about the pivot; $\ddot{\theta}$ = the angular acceleration; g = the gravitational constant; r = the distance between the center of gravity and the pivot along the axis; θ = the angle of the object to the vertical axis (YY'); the plus sign in Equation (1) indicates the arm is swinging up and the minus sign means it is swinging

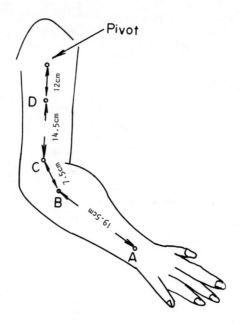

Figure 2—The positions of the diodes on the arm.

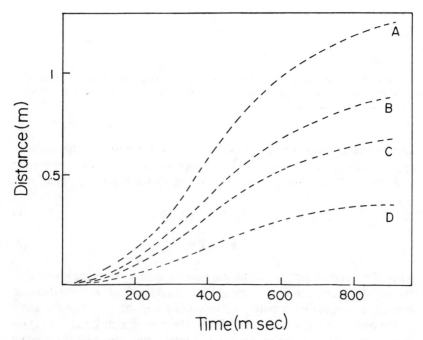

Figure 3—Distance vs time.

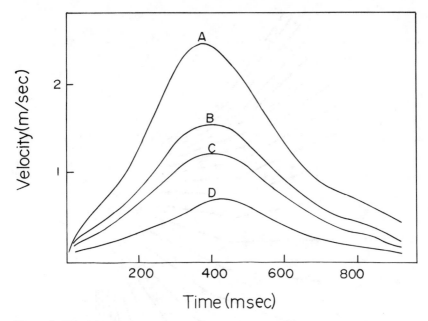

Figure 4—Velocity vs time.

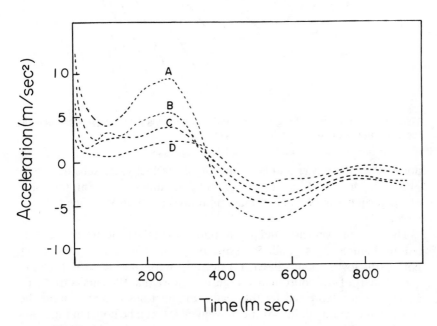

Figure 5—Acceleration vs time.

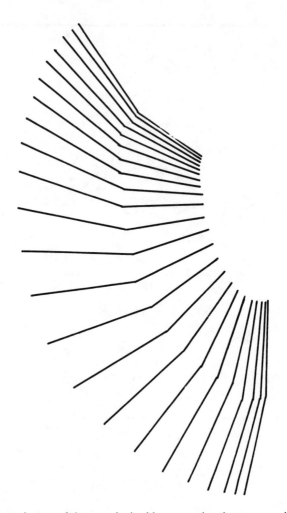

Figure 6 — The trajectory of the arm obtained by connecting the corresponding diodes at successive instants in time.

down; P = the power and $\dot{\Theta}$ = the angular velocity. By substituting m = 3.58 kg, g = 9.81 m/sec², I_O = 0.5 kg m², and r = 0.32 m, the value of the torque T is 19 Nm and maximum power P is 56 W, when t = 277 msec.

The position, velocity, and acceleration along the trajectories are plotted in Figures 3, 4, and 5, respectively. In Figure 3, the starting points (the lowest points in each trajectory) were considered zero or the vertical axis of the coordinates. Figure 4 shows that the curves of A, B, C, and D are expressed by the Gaussian functions except around the starting and ending points of the movements. It can be noted in Figures 4 and 5 that the maximum velocity and the maximum and minimum ac-

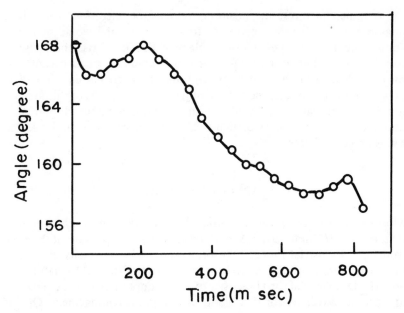

Figure 7 — The angle of the elbow vs time.

celeration appear later for the measuring points on the arm located nearer to the shoulder. In Figure 5, the curves of A, B, C, and D are almost symmetrical. By connecting the corresponding points in the trajectories of A, B, C, and D, Figure 6 is obtained. From this figure, the angle of the elbow as a function of time is obtained as shown in Figure 7. The angle of elbow at the start of the movement is about 168°. When movement begins, first, the elbow flexes to start the movement and extends again to the starting value. After the extension, flexion begins until the movement is almost completed. Just before the finish of the movement the elbow extends a little bit again. This characteristic is nearly symmetric but if there is trouble with the shoulder, the angle of elbow at the finishing point becomes much smaller. The detailed results will be reported elsewhere.

Summary

A new method for measuring the position, velocity, and acceleration of a moving object by using only a camera and visible light-emitting diodes has been developed. This method is very simple and accurate because the frequencies of the LEDs on the moving object are variable depending on their velocities. The cost of the instrument is modest. The movement of the human arm was measured by this method. From the trajectories in the obtained pictures, the position, velocity, acceleration, and elbow angle were plotted as a function of time. This method would be useful for a clinic in order to diagnose problems with joints. Also this is useful to measure the human movements in many kinds of sports because the measurable area can be made very large and the electronic circuit with its power supply can be attached to the object. The problems of this method are that the measurement should be carried out in a darkened room, and automatic measurement is difficult. The former problem can be solved by using color filters.

Acknowledgment

Most of this work was carried out at the University of Southampton in Britain in 1977 and 1978. A part of this work was done at Nagoya Municipal Industrial Research Institute and Aichi University of Education. The authors would like to express thanks to Prof. J.M. Nightingale and Mr. D. Brown at the University of Southampton and Dr. A. Yoshida at Nagoya University for their cooperation and encouragement. One of the authors (H.K.) expresses special thanks to The British Council for financially supporting his study in Britain.

References

TSUCHIYA, K. 1977. Gait Analysis (1). Sogo Rehabilitation 5(1):57-63. (in Japanese)

TSUCHIYA, K. 1977. Gait Analysis (2). Sogo Rehabilitation 5(2):56-62. (in Japanese)

Factors Determining Measured Values of Muscular Force under Static Conditions

Kazimierz Fidelus and Andrzej Wit
Institute of Sport, Warsaw, Poland

The purpose of this investigation was to identify the influence of basic and training factors on the values of muscular torques. The basic factors considered were body mass, age, and biorhythms over days, weeks, and years. On the other hand, the training factors were the years of training and the values of training loads of particular exercises as well as of training units.

Muscular force is necessary to perform physical exercises, to execute movements of our own bodies, or to move other bodies. Force, which creates the potential possibility for movement of the human body, is utilized in the skills or techniques of movements (Fidelus et al., 1975; Fidelus, 1979). Both muscular torques and skill must be developed by means of optimal training methods. The trainer must estimate the muscular torques of main parts of the body, how they are utilized in a given exercise, and based on this recognition, apply the applicable loads for training sportsmen (Fidelus and Wit, 1981). This procedure is treated as a regulatory process, with a general scheme as presented in Figure 1. The trainer (as a regulator) requires the athlete (regulative object) to perform proper exercises. The magnitude of muscular torques is dependent on the the value of training loads, which includes the amounts of internal work and power and optimal time of rest between exercises. The purposes of this study were to investigate the changes of muscular torques as a function of eight factors, to describe this function, and to compare the calculated and measured values (Wit, 1980).

Methods

The main method of investigation was the measurement of muscular torques (M_m) of flexor and extensor muscles under static conditions. The

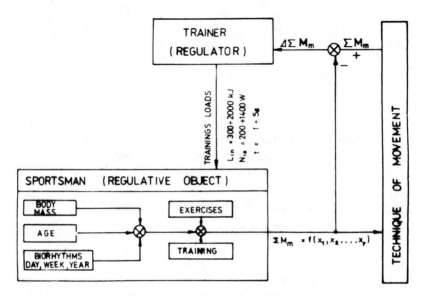

Figure 1—Scheme of muscle torques (M_m) training as a regulative process.

groups of muscles of the following joints were chosen: shoulder, elbow, hip, knee, and trunk. Athletes were informed of their maximal values of torques for each muscular group. Static conditions eliminated the influences of skill and strategies of movements. An attempt was made to account for psychological factors, for instance motivation, by introducing elements of competition during the measurement.

Changes of M_m in relation to body mass, age, years of training, annual biorhythm, and training loads were investigated for 24 top class weightlifters. Effects of diurnal and weekly biorhythms were investigated for 30 students of the Academy of Physical Education in Warsaw. Fifteen of these students were middle class weightlifters and the others were not trained for any sporting event.

Analysis of the measuring system showed that measurement error was less than 10%. The normal distribution of data was checked using Kolmogoroff and Helwig tests. The relationships between parameters were approximated by algebraic description and measured data were tested by means of the least squares method. Calculations were performed using PDP 11/34 and CYBER 73 computers.

Results

Values of muscular torques of athletes increased with greater body mass (m) and additional years of training (t_1). During the first two years, M_m increased about 40% and after that the percentage was smaller. For ex-

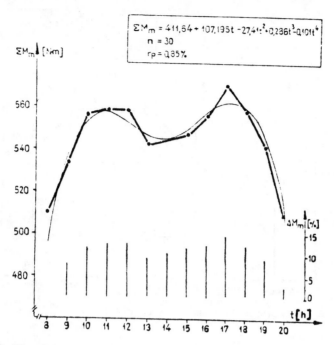

$$\Sigma M_m = 411,64 + 107,195t - 27,44t^2 + 0,236t^3 - 0,101t^4$$
$$n = 30$$
$$rp = 0,85\%$$

Figure 2 — Diurnal changes of the sum of muscular torques.

ample between the 5th and 9th year, the mean increase was about 12%. The relationship between age (t_2) and M_m increased up to 27 or 28 years. Between 20 and 27 years, an increase in M_m of 20% was found. In general, the relation between these parameters and the sum of muscular torques (ΣM_m) can be described as follows:

$$\Sigma M_m = 0.6 \ m^{0.841} \cdot t_1^{0.131} \cdot t_2^{0.683} \qquad (1)$$

The coefficient of correlation between measured data and calculated values was 0.856. Nevertheless, it must be emphasized that this relationship was found for ages 20 to 30 years. Diurnal changes of M_m were investigated every hour between 8 a.m. and 8 p.m. (see Figure 2). This curve has two maxima at 10 to 12 a.m. and 4 to 6 p.m. The mean difference between the maximal and minimal values of M_m was 15%. The influence of diurnal biorhythm on the value of M_m can be described by a fourth degree polynomial.

Weekly changes of ΣM_m are shown in Figure 3. The highest value of muscular torques occurred on Thursday, Friday, and Saturday. The lowest values were obtained on Monday. The weekly differences were up to 10% and the corresponding relationship can be estimated by a second degree polynomial.

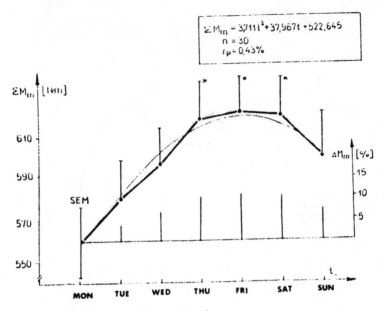

Figure 3—Weekly changes of the sum of muscular torques.

Changes of ΣM_m in annual biorhythm were mostly individual and they were primarily dependent on years of training. These changes were smaller for the beginners (about 10%). For older subjects, for example, weightlifters who had been training for 10 years, changes in the annual cycle were greater and reached 15%. This relationship could be described by a polynomial of the fifth degree.

The greatest increase in muscular torques was accounted for by the training loads. Every physical effort leads to the depletion of energy sources (ATP, CP, G) and after this depletion the resynthesis is accomplished from excess sources. This phenomenon is called supercompensation (Wit et al., 1978). The changes in energy source causes an increase or decrease in muscular force. The fluctuations of muscular torque after a training unit are shown in Figure 4. The form of these fluctuations is characteristic of the process of damp oscillations. Therefore, measured data can be approximated by the following equation:

$$\Sigma M_m = \frac{A}{\omega} e^{-\delta t} \sin(\omega t + p) + at + b \tag{2}$$

where A = amplitude coefficient; ω = own oscillations; t = time; p = shift of phase; a = direction coefficient; and b = constant value.

Using Equation 2 it is possible to describe the supercompensation phenomenon after particular exercises as well as after training, during

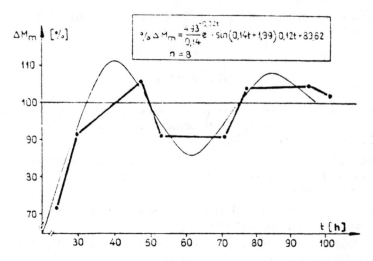

Figure 4 — Percentage changes of the muscular torques after a heavy training of a weight-lifter.

which internal work was from 200 up to 2000 kJ and the mean value of internal power was from 200 to 1400 W. It is then evident that the phenomenon of changes due to training is very general, but the coefficients in Equation 2 are always specific to an individual. It confirms the old saying in the practice of sport: general principles apply but individual treatment is required in training athletes.

Discussion

The eight main factors influencing the value of muscular torques and their relationships with muscle force were investigated. These relationships were estimated by mathematical functions. The completeness of the factors which were taken into account were verified and the differences between measured and approximated data were calculated. The mean error was about 8.5% if the measured values of muscular torques were taken as 100%. The extreme individual differences were between 31.4 and 22.6%. For 10 investigated athletes the mean error was no greater than 15%. Only in the case of one athlete were all differences about 60%. This may have been due to some auxiliary pharmaceutical means used by this competitor.

The differences between measured and calculated data were probably due to three sources of errors: 1) errors of measurement and approximation of data — these errors can be reduced; 2) some unknown factors influencing the muscular torques — these errors should be eliminated by further basic investigations; and 3) psychological states, environmental influences, nutritional states, etc. — these are very important but it was not possible to measure these factors in this study.

References

FIDELUS, K. 1979. Physical effort and its value in different sports. In: F. Landry and W.A.R. Orban (Eds), Biomechanics of Sports and Kinanthropometry, pp. 245-251. Symposia Specialists Inc.

FIDELUS, K., Skorupski, L., & Wit, A. 1975. Optimization of sports effort during training. In: P.V. Komi (Ed.), Biomechanics VB, pp. 351-356. University Park Press, Baltimore.

FIDELUS, K., and Wit, A. 1981. Hauptfaktoren des Muskelon Trainings (Principle factors in muscle training). (in press)

HETTINGER, T. 1964. Isometrisches Muskeltrainings (Isometric Muscle Training). G. Thieme, Stuttgart.

WIT, A. 1980. Zagadnienia regulacji w procesie rozwoju sily miesniowej (Regulation problems in the process of muscular force development). Edited AWF in Warsaw.

WIT, A., Jusiak, R., Wit, B., and Zielinski, J.R. 1978. Relationships between muscular strength and the level of energy sources in the muscle. Acta Physiologica Polonica 29:139-151.

A Quantitative Comparison Method of Display Scheme in Mobility Aids for the Blind

Susumu Tachi
Mechanical Engineering Laboratory
Tsukuba Science City, Japan

Robert W. Mann and Derek Rowell
Massachusetts Institute of Technology
Cambridge, Massachusetts, U.S.A.

The idea of the technological assist of the mobility of the blind person is of a rather recent origin (since World War II). With the advent of electronics and computer sciences, several mobility devices for the blind have been developed. These include, i.e., the Sonic Guide (Kay, 1966) which transmits the ultrasonic sound and receives the reflected sound by the obstacle, and therefore informs the blind user about the environment by means of the audible sound code. Another example is a laser cane.

All of these, however, represent ad hoc solutions and offer no practical suggestions for a generalized design procedure to define optimal displays for specific applications.

Suppose that a device which directs or guides a blind individual has somehow acquired information about the direction of, and width of, the path along which it should lead the blind individual (Tachi et al., 1981). The problem is the choice of sensory display of the path and its safe margins appropriate for presentation to the remaining exterio-receptive senses of the blind individual.

The central features of such a design capability must include a method to quantify the motion or movement of an unhampered blind individual walking in a real or mock-up physical environment, and the means of feeding back to the individual, in real-time, the path information and/or error from the path by means of a very flexible and potentially rich psychophysical sensory display code.

The importance of this mobility environment simulation approach was first proposed in 1965 (Mann, 1965). The realization of the system had to await Conati's fundamental work in 1977 on a system for Telemetered Real-Time Acquisition and Computation of Kinematics (Conati, 1977), a general purpose laboratory system dubbed TRACK, which coupled the high-speed, yet low-cost, laboratory mini-computer with the high-performance, multi-channel, point-monitoring, and data transferring device called Selspot. In this paper the newest version of the TRACK system operating under RSX11-M on a DEC computer PDP 11/60, which extends the ability of the previous version, is first reported.

Using this new system the performance of human subjects employing different auditory display schemes communicating the course they should follow are quantitatively compared. A method for the quantitative comparison is also proposed. An optimal auditory display scheme is sought, by measuring the movement of a human subject to a random course (generated by the computer) which displays to the subject the course error from the desired course in real-time. The transfer function of the subject employing each of several different display schemes is estimated. The effective gains and the effective time delays of the transfer functions for the several display devices are calculated according to the crossover model. Using the sum of the effective gain and the reciprocal of the time delay as the criterion of optimality, the optimal display scheme is sought and the effect of difference between the alternative display schemes is quantitatively evaluated.

Experimental Apparatus

Figure 1 shows the experimental arrangement. The movement of the human subject was measured by the newly revised TRACK system. The system consisted of the raw-data acquisition and handling device, Selspot, marketed by Selcom AB of Sweden, a PDP 11/60 mini-computer, and an auditory display device which was linked to the computer through a Laboratory Peripheral Accelerator (LPA).

The Selspot system used cameras with the lateral photo-electronic plates in their image planes which were sensitive to infrared illumination. Each plate detected the position of the image of a light emitting diode (LED) and thereby provided two-dimensional position data from each camera for up to 30 LEDs which were serially pulsed to at a rate of 315 Hz.

The two cameras were positioned accurately in laboratory coordinates and their two-dimensional image position data were manipulated trigonometrically by the computer to yield three-dimensional data of the LEDs. By arranging three or more LEDs on a plane attached to a segment of a moving human (in this case the abdomen) the location and orientation of the human subject could be tracked in real-time.

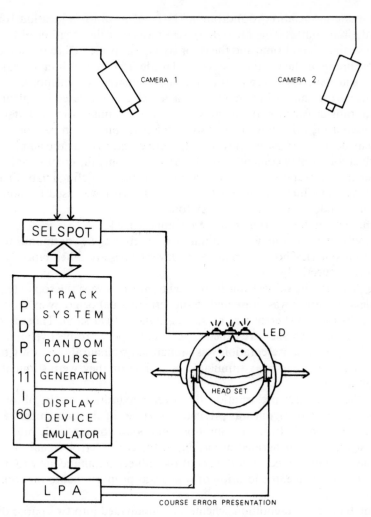

Figure 1 — Experimental arrangement for real-time evaluation of the display devices for the blind.

Band limited random noise was generated by the computer and was used to define the course which the subject should follow. Error in the human's location relative to the indicated course was fed back to the subject via auditory signals through a headset (Electrostat-Dynamic Systems k-340) through the LPA's D to A converters.

Experimental Goal and Method

The computer system emulated several display devices which used amplitude modulation to indicate an error of a subject's location from the desired course. The TRACK system measured the location of a human subject in real-time and the error signal was presented to the subject through one of the emulated devices. The desired course, human trajectory, and the error, were recorded in the computer disk memory.

The performance of the human in each task was evaluated by calculating a transfer function of the movement of the human with each device and then using this transfer function as the criterion for comparison. The display devices compared in this study were categorized as the amplitude modulation display controlled by the error, i.e., only the error signal was presented, corresponding to compensatory display as defined in ordinary tracking experiments in manual control. The error was used to modify the amplitude of a fixed frequency tone.

Three attributes of this tone were compared:

1. Whether the tone was continuous or discrete (i.e., continuous tone and tone burst). Those alternative display schemes are called type-C and type-D, respectively;

2. Whether the subject was instructed to move toward the sound (e.g., the left-side error was presented to the left ear-side) or go away from it (e.g., the right-side error was presented to the left-ear side). Those alternative display schemes are called type-T and type-A, respectively;

3. Whether the presentation was monaural or binaural (i.e., only one side of the two ears was stimulated at a time or they were stimulated simultaneously).

In the monaural presentation the inverse logarithm of the absolute value of the error signal was presented to either of the human ears and are called type-M. In the binaural presentation two amplitude modulation signals were generated according to the course error signal. These signals were presented to both ears of the subject simultaneously to produce a fused image, the location of which was proportional to the course error.

The binaural presentation scheme was subdivided into two parts: one used only the position cue as an indication of the course error (type-B1) and the other used the positional and the loudness cues at the same time (type-B2).

All combinations of these attributes were compared using an experimental procedure as follows:

A subject was asked to side step (right or left) within the TRACK's viewing field according to the auditorily presented error of his location from the indicated random course generated by the computer.

Figure 2 shows the block diagram of this compensatory system and display. The i(t) is the input Gaussian white noise with cutoff frequency (fc) of 0.32 Hz with a zero mean. The y(t) is the observed output of the system. It consists of the response of the human subject with each of the emulated device o(t) and the additive noise n(t).

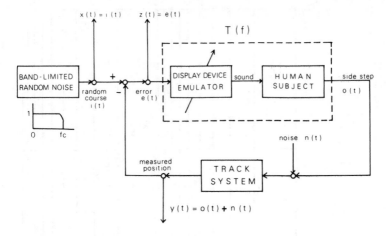

Figure 2—Block diagram of the compensatory system and display used.

The z(t) is the error e(t), i.e., the difference between i(t) and y(t). The transfer function T(f) of each of the emulated devices as interpreted by the human subject is calculated by the following formula:

$$T(f) = \frac{\Phi xy}{\Phi xz} = \frac{\Phi iy}{\Phi iz} = \frac{E[I(f)^* (O(f) + N(f))]}{E[I(f)^* E(f)]}$$

$$= \frac{E[I(f)^* O(f)]}{E[I(f)^* E(f)]} = \frac{O(f)}{E(f)} \tag{1}$$

where Φ is the cross spectrum between signals x(t), y(t), i(t) and z(t) which are indicated as subscripts. The signals x(t), y(t), and z(t) are measured during a finite time so that they have their Fourier transforms.

Experimental Results

All the display schemes, i.e., monaural (M), binaural 1 (B1), or binaural 2 (B2); toward the sound (T) or away from the sound (A); and continuous (C) or discrete (D) were combined. Thus twelve display devices were emulated, which were called MTC-, MTD-, B1TC-, B1TD-, B2TC-, B2TD-, MAC-, MAD-, B1AC-, B1AD-, B2AC-, and B2AD-type, respectively. Two subjects (TS: student, 26, and NA: researcher, 41) used these 12 emulated display devices to follow the random course generated by the computer. An open-loop transfer function of the subject with each of the twelve emulated devices was calculated using formula 1.

Figure 3 shows the amplitudes and the phases of the transfer functions of the subject TS with emulated device MTC-type. The same kind of the

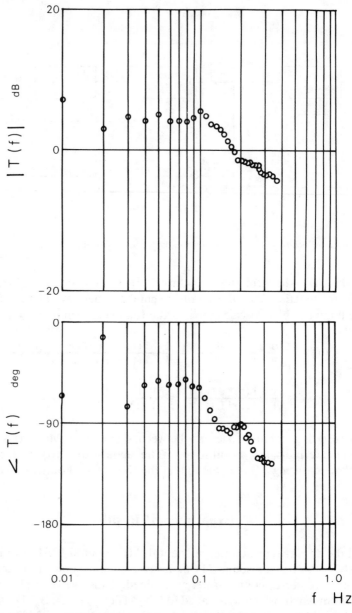

Figure 3 — Examples of the open-loop transfer function (MTC-type for subject TS).

transfer functions were estimated for all twelve emulated devices used by
the two subjects. These results obtained for the side-step task are very
similar to the results obtained for the manual compensatory tracking experiments with controlled elements of K_c (position vehicle) obtained by

McRuer and Jex (1967), although the crossover frequency is extremely lower because of the larger inertia of the body compared with the hand.

Thus as the first order approximation, the crossover model can be applied to the results obtained by the side-step version of the compensatory tracking experiments. According to the crossover model of the Macruer the transfer function T(f) can be described as follows in the region of the crossover frequency:

$$T(f) = \frac{\omega_c \exp(-j\omega T_e)}{j\omega}. \tag{2}$$

where ω_c is the crossover frequency and is equivalent to the human side-stepper's gain compensation K_e with the emulated device and T_e is the effective time delay due to both reaction time and high frequency neuro-muscular dynamics.

In the model the total performance is better if the equivalent gain is higher and the equivalent time delay is smaller. The two parameters K_e and T_e describe the total characteristics of the human side-stepper with the emulated devices. Thus the quantity:

$$EV = K_e + 1/T_e \tag{3}$$

can be considered to be one of the most effective measures or criteria to determine or evaluate quantitatively the effectiveness of the emulated devices.

The effective gain K_e and the effective time delay T_e were calculated using the following formula based on the fitting method usually done for the position vehicle (Hill, 1970):

$$K_e = \omega_c,$$
$$T_e = 1/K_e \left(\pi - \phi_m \cdot \frac{\pi}{180}\right) \tag{4}$$

where ϕ_m is the phase margin measured.

Table 1 shows the results for two subjects with each of the twelve display schemes. In the table it is clear that the B1TD-type is the most effective for subject NA and MTD-type for subject TS.

Conclusion

It was found to be feasible to design the display scheme of the mobility aid using the procedure reported in this paper, and the optimal choices can be made before committing a particular design to a lengthy development process.

Moreover, it will be possible to find out the best display scheme for the particular subject, i.e., a custom-made mobility aid for the blind, by using the method reported in this paper.

Table 1

Equivalent Gain Ke, Equivalent Time Delay Te, and Evaluation Value EV = Ke + 1/Te of 12 Display Schemes for Two Subjects

		MTC	MTD	B1TC	B1TD	B2TC	B2TD	MAC	MAD	B1AC	B1AD	B2AC	B2AD
T	Ke	1.26	1.38	0.63	0.82	0.82	0.82	0.94	1.13	0.50	0.44	0.75	0.82
S	Te	1.26	1.17	3.06	2.29	1.82	1.77	1.78	1.51	3.06	3.41	1.67	1.32
	EV	2.05	2.23	0.96	1.25	1.37	1.38	1.50	1.79	0.83	0.73	1.35	1.57
N	Ke	1.51	1.32	0.57	0.94	1.70	2.26	1.63	1.63	0.88	0.88	0.94	1.19
A	Te	0.78	0.85	2.56	1.98	1.18	0.99	1.14	1.19	1.79	1.61	1.56	1.33
	EV	2.80	2.50	0.96	1.45	2.54	3.27	2.51	2.48	1.44	1.50	1.58	1.95

Acknowledgment

This work was supported by the Japanese Government and the National Eye Institute of the U.S. Government. The research reported herein was conducted while ST was a guest of the Department of Mechanical Engineering, Massachusetts Institute of Technology.

References

ANTONSSON, E.K., and Mann, R.W. 1979. Automatic 3-D analysis using a Selspot centered system. 1979 Advances in Bioengineering (Winter Annual Meeting of the ASME), New York: 51.

HILL, J.W. 1970. A describing function analysis of tracking performance using two tactile displays. IEEE Trans. MMS-11(1):92-101.

KAY, L. 1966. Ultrasonic spectacles for the blind. Proc. Int. Conf. on Sensory Devices for the Blind, St. Dunstans, London.

MANN, R.W. 1965. The evaluation and simulation of mobility aids for the blind. Am. Found. for the Blind Research Bulletin 11:93-98.

McRUER, D.T., and Jex, H.R. 1967. A review of quasi-linear pilot models. IEEE Trans. HFE-8:231-249.

TACHI, S., Tanie, K., Komoriya, K., Hosoda, Y., and Abe, M. 1981. Guide Dog Robot—its basic plan and some experiments with MELDOG MARK I. Mechanism and Machine Theory 16(1):21-29.

TETEWSKY, A.K. 1978. Implementing a real time computation and display algorithm for the Selspot system. M.S. Thesis, Dept. of Mech. Eng., M.I.T.

Method for Measuring the Curve of the Spine by Electronic Spherosomatograph

Czeslaw Wielki
Universite Catholique De Louvain
Louvain-la-Neuve, Belgique

In spite of the development of methods for measuring the curves of the vertebral column, their precision and the interpretation of the results of many techniques leave much to be desired. Technical and scientific progress has allowed principles of electronics to be applied, for the first time, to measure the curves of the spine.

'Wielki's Electronic Spherosomatograph,' created in 1974, allows one to record at the same time the curves of the spine or vertebral column in two perpendicular planes in the 'stand-easy' position or another position of choice. The outline of the apparatus has already been presented to the Medical Congress of Sport at Marseilles in 1978 (Wielki, 1979). It comprises three parts: a detector (D), an electronic control and commands (E), and a recorder (R).

Detection

A contact is made by a little rubber wheel which is attached to a sliding indicator in a guide, which is free to turn about a vertical axis. The possibilities for movement are in the horizontal, vertical, and rotational directions. The extremity of the indicator wheel allows it to follow the spines of the vertebrae and to record the characteristics of the curves. All the contact displacements are converted into electronic signals.

Recording

The signals activate the recorder. Displacements of the contact are converted into electric signals through collectors which break down the curve

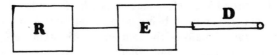

Figure 1—The components of the electronic spherosomatograph.

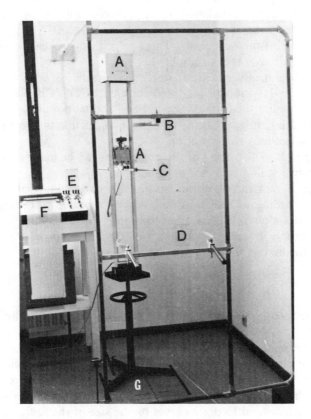

Figure 2—Electronic spherosomatograph.

of the vertebral column into two perpendicular planes. They are recorded on paper tape in blue in the frontal plane (lordosis-kyphosis) and in red in the sagittal plane (scoliosis).

Measuring Method

Following pilot studies in order to keep the measurements of the curves of the vertebral column constant, a standard methodology has been established.

1. Stabilization of the subject in the 'stand easy' position, which is

determined by three fixed references: a) the heels are level, together, or apart, the angle between them being immaterial; b) the hips are stabilized in a level plane with the aid of the semiautomatic stabilizers; c) the head is held erect with the eyes straight ahead and a stabilizer is placed above the brows.

2. Breathing. The detection and recording of the curve of the spine is made during a slow exhalation after a generous inhalation. This procedure involves 'control breathing'.

3. Number of measurements. Three recordings are usually sufficient, from which an average is computed.

4. Points of measurement: a) Starting point: the 7th cervical vertebrae (Cureton, 1941; Hunnebelle, 1969; Minski, 1972). It is illusory to attempt to stabilize the cervical region in a habitual position. b) Finishing point: the 5th lumbar + 4 cm, in such a manner that the position of the pelvis is taken into consideration.

5. The scale. The tracing of the projections can be made on any scale, but usually a scale of 1:2 is employed.

6. Tape movement. Movement is possible in two directions, thus permitting recording of numerous subjects or repeated recordings of the same subject alongside each other for comparative results.

7. Accuracy. In spite of accuracy of the apparatus of 0.01 mm, in practice the limit is 0.5 mm.

Methodology of the Analysis of the Curves of the Spine

The former method of analysis for this recording was by three angles (α, β, γ) but currently this is not sufficient, for it does not allow one to characterize all the curves of the spinal column in their entirety. Therefore, the 'radius method' is used which allows proper expression of the kyphotic curves and lordotic curves by the size of the radius of a circle that is the closest to the curves registered by the electronic spherosomatograph.

The former method of measuring the spine using Wolanski's (1956) simple apparatus (the 'Kyfolordozometer') measured only the distance between the four extreme points of the curves of the spine, and the angle obtained relative to the vertical. Taking into consideration the accuracy of the measurements of the characteristics of the spinal curves using the apparatus described herein, it should be noted that the lordotic and kyphotic curves may be better expressed by the radius of a circle. This has been referred to as the 'Radius Method'. Some results have been presented as examples to the Congrès de Mèdecine du Sport in Lisbon, in May, 1979. In order to establish the value of the r of the kyphotic and lordotic curves of a person, one needs only to apply the theorem of

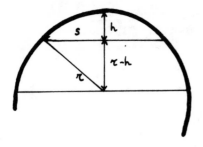

$$r^2 = (r-h)^2 + s^2$$
$$r^2 = r^2 + h^2 - 2rh + s^2$$
$$0 = h^2 - 2rh + s^2$$
$$r = \frac{h^2 + s^2}{2h}$$

Figure 3—Demonstration.

Pythagoras. In practice, this yields the following equation:

$$r = \frac{h^2 + s^2}{2h} \tag{1}$$

where r = the radius of the kyphotic curves, h = the height of the kyphosis, and s = the semi-cord of the kyphosis. (By simply introducing the values of h and s into a computer, the value for r can be immediately obtained.) Therefore, the radius method is employed for the analysis of the curves of the spine, taking into account two variations.

Variation A: Radius Method Using an Intersection Point

So that results from the analysis of the spine can be compared by researchers from different countries, it is proposed to standardize the procedures concerning the reference points of the curves of the spine, using: 1) upper point: C_7 = A, and 2) lower point: L_5 + 4cm = B. A and B are joined by a straight line which cuts the curves at point C, giving two curves: 1) the kyphosis curve with the cord A-C, and 2) the lordosis curve with the cord C-B.

At the midpoint of these cords, a perpendicular is drawn, obtaining h_1, the height of the kyphosis, and h_2, the height of the lordosis. By drawing the vertical tangent at the top of each curve, the reference points D (the top of the kyphosis curve), and E (the top of the lordosis curve) are obtained. The vertical from the upper point A meeting the horizontal from the lower point B yields the point I, allowing one to characterize the forward incline of the entire vertebral column.

This radius method by intersection can be applied to the spinal curves of the normal subjects in a free-standing position. If 'special cases' arise, however, or if subjects are examined in a sitting position, one can note that the straight line between points A and B does not divide the curve of the spine into the two functional curves, the lordotic and kyphotic.

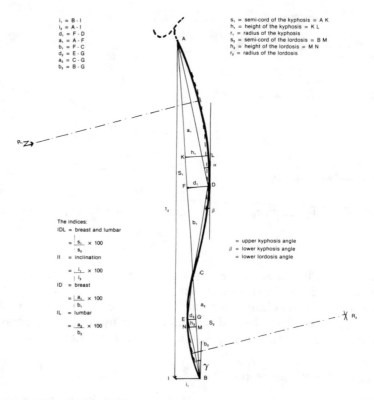

Figure 4—Scheme of the analysis of the physiological curves of the spine by the intersection point.

Variation B: Radius Method Using an Inflexion Point

To complete the analysis of these curves, the radius method by the inflexion method is proposed which is presented according to the same principles.

Following the same procedure for analysis, divide the curves of the spine into the kyphotic and the lordotic curves by establishing the point C at the middle of the straight line between the kyphosis and lordosis. This point C is the inflexion point of the two curves.

Then, join by a straight line, the inflexion point C to the upper point A which gives the cord of the kyphosis A-C. By joining the inflexion point C to the lower point B, the value of the cord of the lordosis C-B is obtained. Measure the inside angle formed between these two cords, referring to it as ω (in the previous method, the angle of intersection was always 180°).

In order to obtain the actual top limits of the kyphosis and lordosis, taking into account the inclination of the cords, draw the tangent not vertically, but parallel to the cords. That is, the top limit of kyphosis D is

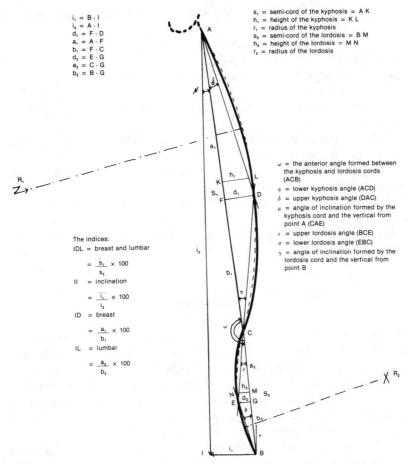

$i_1 = B \cdot I$
$i_2 = A \cdot I$
$d_1 = F \cdot D$
$a_1 = A \cdot F$
$b_1 = F \cdot C$
$d_2 = E \cdot G$
$a_2 = C \cdot G$
$b_2 = B \cdot G$

s_1 = semi-cord of the kyphosis = A K
h_1 = height of the kyphosis = K L
r_1 = radius of the kyphosis
s_2 = semi-cord of the lordosis = B M
h_2 = height of the lordosis = M N
r_2 = radius of the lordosis

ω = the anterior angle formed between the kyphosis and lordosis cords (ACB)
η = lower kyphosis angle (ACD)
δ = upper kyphosis angle (DAC)
μ = angle of inclination formed by the kyphosis cord and the vertical from point A (CAE)
ϵ = upper lordosis angle (BCE)
σ = lower lordosis angle (EBC)
γ = angle of inclination formed by the lordosis cord and the vertical from point B

The indices:
IDL = breast and lumbar

$= \dfrac{s_1}{s_2} \times 100$

II = inclination

$= \dfrac{i_1}{i_2} \times 100$

ID = breast

$= \dfrac{a_1}{b_1} \times 100$

IL = lumbar

$= \dfrac{a_2}{b_2} \times 100$

Figure 5—Scheme of the analysis of the physiological curves of the spine by the inflexion point.

found by the tangent parallel to the cord A-C. The top of the lordosis (E) is obtained by the tangent parallel to the cord of the lordosis C-B.

All the other operations are similar to the preceding method, including the calculation of the radii.

In order to examine in more detail the kyphotic column, draw a line from A to D, and then from D to C. By analogy, proceed in the same pattern by drawing a line from B to E and E to C. By using these six angles, one can further analyze the curves of the spine. This analysis can be applied in a similar fashion in the previous method:

ω = the anterior angle formed between the kyphotic and lordosis cords (ACB).

ν = lower kyphosis angle (ACD).

δ = upper kyphosis angle (DAC).

μ = angle of inclination by the kyphosis cord and the vertical from point A (CAE).

ϵ = upper lordosis angle (BCE).

θ = lower lordosis angle (EBC).

γ = angle of inclination formed by the lordosis cord and the vertical from point B.

In order to further study the characteristics of the spine, take into account the following indices:

1. Breast and lumbar: IDL $= \frac{A-C}{C-B} \times 100$ giving the relationship of the size of the curvatures.

2. Inclination: II $= \frac{B-I}{A-I} \times 100$ giving the degree of inclination of the anatomical curvatures of the subject.

3. Breast: ID $= \frac{A-F}{F-C} \times 100$ indicating the position of the top of the kyphosis.

4. Lumbar: IL $= \frac{C-G}{G-B} \times 100$ indicating the position of the top of the lordosis.

Summary

The analysis of the spinal curves recorded by the electronic spherosomatograph and interpreted by the method of radii (intersection-inflexion) allows one to perform the following functions:

1. Determine the characteristics of the curves of the spine. a) blue pen: the curves of the lordosis and kyphosis. b) red pen: the tendency toward scoliosis (children) or the curves of scoliosis.

2. Follow the evolution of the curves of a child's growth.

3. Discover any kind of deformation like spondylolisthesis, that may be the result of over-specialization in an activity during the early developmental stage of the spine.

4. Diagnose early so that preventive treatment of any deformation which may occur during the formative years may be initiated.

5. Monitor the progress of orthopedic treatments without endangering the subjects by continued use of X-rays.

6. Detect deformations due to postures of work in industrial settings. This method allows one to diagnose problems in time to prevent pathologies of the spinal column.

References

CURETON, T.K. 1941. Bodily posture as an indicator of fitness. Research Quarterly 12(2):348-367.

HUNEBELLE, G., and Damoiseau, J. 1969. Mise au point d'une mèthode rapide d'évaluation des courbures sagitales de la colonne vertébrale. Mouvement 4(3):235-237.

MINSKI, I.A. 1972. Kyfoscoliométre, Hygiéne et Santé, pp. 73-75.

WIELKI, C. 1979. Vers une méthode électronique de mesure des courbures de la colonne vertébrale. Lyon Méditerranée Médical Médicine du Sud Est, Paris. 2:1223-1227, T. XV, Nz. 14.

WOLANSKI, N. 1956. Kyfolordosometre. Culture Physique 12:947-953.

Analysis of the Resultant Floor Reaction Force
in Normal Gait

S. Okumura, S. Hayashi, S. Kawachi,
T. Takeuchi, and K. Furuya
Tokyo Medical and Dental University, Tokyo, Japan

The normal gait is a rhythmic forward movement of the center of gravity and a successive loss and recovery of equilibrium. The human bipedal gait is defined as a rhythmic alternation of periods of swing and support. The loss of equilibrium is produced by propulsion; recovery of equilibrium is due to the so-called restraint which prevents the subject from falling forward.

There are many methods of gait analysis. The floor reaction force is produced as a vector. A force plate divides the floor reaction force into three components (vertical, fore-aft, and lateral). In the ordinary method of analysis of the floor reaction force it is standard to analyze these three components. It is difficult to study the floor reaction force itself.

In this research the scalar quantity of the floor reaction force, i.e., the resultant force during the stance phase in normal gait was calculated.

Procedure

The subjects were 14 men and 16 women of the third and fourth decades. They walked 10 times on a Kistler Force Plate at 80, 100, 120, and 140 steps/min, using a metronome as a pacemaker. The scalar quantity of the resultant vector $FR = \sqrt{FX^2 + FY^2 + FZ^2}$ was calculated every 10 msec from heel strike to toe off from the data of the three components by a microcomputer and the sum of the scalar quantity of the resultant vertical and fore-aft forces was gained during the stance phase (see Figure 1). As the value depends on body weight and the period of the stance phase, it was divided by the body weight and the period of the stance phase. The stance phase can be divided into the restrictive and the propulsive periods at the instant that the fore-aft vector changes its direction

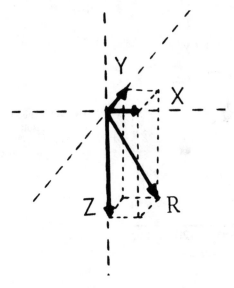

Figure 1—R = resultant, X = lateral, Y = fore-aft, Z = vertical. FR was given as $\sqrt{FX^2+FY^2+FZ^2}$.

from forward to backward in the sagittal plane. Therefore, each value was divided into the restrictive period and the propulsive period and the periods were compared with each other. The sagittal direction of the resultant vector was obtained at the peaks of the fore-aft forces in the restrictive period and the propulsive period, as follows: $\tan\Theta = FY/FZ$. The scalar quantity of the lateral force was extremely small compared with the vertical force, so it was not analyzed.

Results

The Resultant Force

The scalar quantity per kg of body weight of the resultant vector during the stance phase ($\Sigma FR/W$) decreased with the increase of cadence (see Figure 2). $\Sigma FR_1/W$ of the restrictive period and $\Sigma FR_2/W$ of the propulsive period decreased with the increase of cadence and the former was always greater than the latter (see Figure 3). $\Sigma FR/W$ per 10 msec of the stance phase ($\Sigma FR/W \cdot t$) changed little with the increase of cadence (see Figure 4). Comparing the values, however, of the restrictive period and the propulsive period of the stance phase, $\Sigma FR_1/W \cdot t$ increased and $\Sigma FR_2/W \cdot t$ decreased linearly with the increase of cadence (see Figure 5).

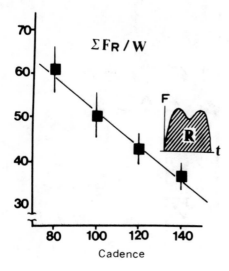

Figure 2 — The sum of the scalar quantity of the resultant force was divided by the body weight. It decreased with the increase of cadence.

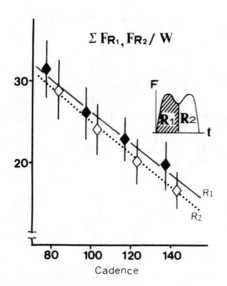

Figure 3 — The sum of the scalar quantity of the resultant force was divided into the restrictive period (R_1) and the propulsive period (R_2). R_2 was always greater than R_1 in spite of the change of cadence.

The Vertical Force

The value of the vertical force was nearly the same with the resultant force. The change with the increase of cadence was also in the same pattern.

The Fore-aft Force

The scalar quantity per kg of body weight of the fore-aft force during the stance phase ($\Sigma FY/W$) also decreased with the increase of cadence (see

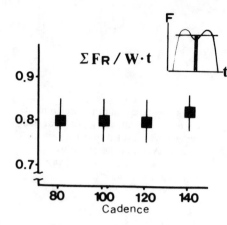

Figure 4 — The sum of the scalar quantity of the resultant force was divided by the body weight and the period of the stance phase. It was nearly constant in spite of the change of cadence.

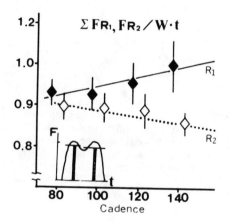

Figure 5 — $\Sigma FR/W$ per 10 msec of the stance phase was divided into the restrictive period (R_1) and the propulsive period (R_2). R_1 increased and R_2 decreased with the increase of cadence.

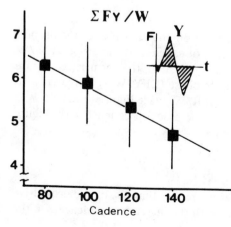

Figure 6 — The sum of the scalar quantity of the fore-aft force was divided by the body weight. It decreased with the increase of cadence as the resultant force.

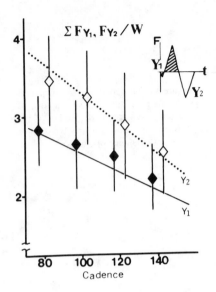

Figure 7 — The sum of the scalar quantity of the fore-aft force per kg of body weight was divided into the restrictive period (Y_1) and the propulsive period (Y_2). Y_2 was always greater than Y_1 contrary to the resultant force.

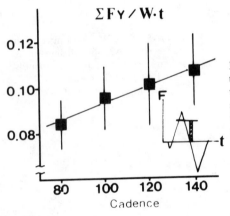

Figure 8 — The sum of the scalar quantity of the fore-aft force was divided by the body weight and the period of the stance phase. It became greater with the increase of cadence contrary to the resultant force.

Figure 6). $\Sigma FY_1/W$ of the restrictive period and $\Sigma FY_2/W$ of the propulsive period decreased with the increase of cadence. Comparing $\Sigma FY_1/W$ and $\Sigma FY_2/W$, the latter was always greater than the former (see Figure 7). $\Sigma FY/W$ per 10 msec of the stance phase ($\Sigma FY/W \cdot t$) increased with the increase of cadence contrary to $\Sigma FR/W \cdot t$ (see Figure 8). $\Sigma FY_1/W \cdot t$ and $\Sigma FY_2/W \cdot t$ increased with the increase of cadence (see Figure 9).

The Direction of the Resultant Vector (Θ)

Θ_1, the restrictive period, was always smaller than Θ_2 of the propulsive period and became slightly greater with the increase of cadence. On the

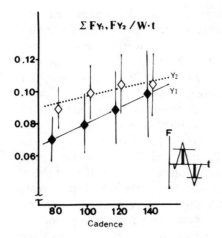

Figure 9—$\Sigma FY/W$ per 10 msec of the stance phase was divided into the restrictive period (Y_1) and the propulsive period (Y_2). Y_1 and Y_2 became greater with the increase of cadence and Y_2 was greater than Y_1.

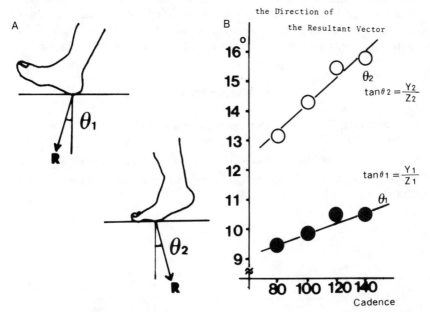

Figures 10A and 10B—The direction of the resultant force was gained as tan. $\Theta = \dfrac{Fy}{Fz} \cdot \Theta_1$ was the direction of the resultant force in the restrictive period and Θ_2 was in the propulsive period. Θ_2 was always greater than Θ_1 and became greater with the increase of cadence.

other hand, Θ_2 increased extremely with the increase of cadence (see Figure 10A and 10B).

Discussion

The increase of cadence alters the maximal magnitudes of the three components and the period of the stance phase. It was found that the sum of the scalar quantity of the resultant force per kg of body weight decreased with the increase of cadence and $\Sigma FR/\bar{w} \cdot t$ was nearly constant.

Comparing the values of the restrictive and the propulsive periods, $\Sigma FR_1/W$ was greater than $\Sigma FR_2/W$, and $\Sigma FR_1/W \cdot t$ was greater than $\Sigma FR_2/W \cdot t$. It was considered that force is needed much more in the restrictive period than in the propulsive period because the center of gravity is lowered in the propulsive period but must be lifted in the restrictive period. $\Sigma FY/W \cdot t$ became greater with the increase of cadence contrary to the result of the resultant force to transfer quickly the center of gravity forward and to stop the foot abruptly. Comparing the values of the restrictive period and the propulsive period, the latter was always greater than the former. This is closely related to the direction of the resultant vector. These results are explained in that in the restrictive period, the direction of the resultant vector (Θ_1) approaches a right angle to gain safe floor contact by flexion of the knee joint, and in the propulsive period, Θ_2 becomes greater to provide a quick forward movement of the center of gravity and to gain increased force of the fore-aft component.

References

ELFTMAN, H. 1939. Forces and energy changes in the leg during walking. Am. J. Phys. 125:339-356.

INMAN, V.T. 1966. Human locomotion. Canad. Med. Assoc. J. 94:1047-1054.

STEINDLER, A. 1935. Mechanics of Normal and Pathological Locomotion in Man. T. Springfield.

C.
Data
Processing

Digital Differentiation Filters
for Biological Signal Processing

Shiro Usui and Itzhak Amidror

Toyohashi University of Technology, Toyohashi, Japan

Low-pass differentiation is often required in processing various biological or biomechanical data. However, both the nature of the biological signals and the use of micro or minicomputers for such applications imply the need for simple, low-order, and fast differentiation methods rather than sophisticated high-order algorithms. This need prompted us to investigate the low-pass first- and second-order digital differentiation from both theoretical and practical points of view. Summarizing the results of this study, presented here is a short description of new simple low-order low-pass differentiation filters which are not only very convenient but are also almost optimum for use.

Background

In processing biological or biomechanical data it is often required to find the derivatives of data signals. This is the case, for example, in the research of velocity and acceleration in human locomotion, as well as in various other fields (Andrews and Jones, 1976; Lesh et al., 1979). Usually, biological and biomechanical signals have low frequency components, which are contaminated by intrinsic biological noise as well as wide band noise due to the use of measuring devices. Additional wide band noise is caused also by the quantization and AD conversion of the signals which are done in order to process the data by digital computers. Such noises are amplified by the operation of differentiation, especially at higher frequencies, and in some cases they may obscure the nature of the results.

These considerations often lead to the choice of low-pass differentiation algorithms (Pezzack et al., 1977; Usui and Ikegaya, 1978) rather than full band differentiation. Yet, due to the conditions in sampling and the noises mentioned, most empirical data of a biological nature have a precision of about 60 dB, which is equivalent to 3 decimal digits (10 to 12 bits), and therefore simple integer algorithms can be used rather than floating point arithmetic. Moreover, since biomechanical data are usually sampled at low rates, the number of sampling points is comparatively small. Thus large spans of data points cannot be used in processing, nor are high-order and very accurate algorithms required. These considerations are especially important when utilizing micro or minicomputers. All these facts prompted this investigation of low order algorithms for low-pass first- and second-order differentiation, from both theoretical and practical points of view, in order to achieve good and useful filters according to the following two criteria:

1. Good approximation to the ideal frequency characteristics in the low frequency range.

2. Simple filters having a small number of coefficients, and furthermore, convenient ones — thus requiring minimum computational efforts.

Low-pass differentiation is often achieved by using a low-pass filter and a differentiating filter in cascade (Usui and Ikegaya, 1978). In this research, however, another approach was used which combines the two stages into one low-pass differentiation filter, thus performing both smoothing and differentiation in one operation. By finding the optimum filters within this class, one can assure that not only will the differentiation filter or the low-pass filter be optimal in themselves, but *the combination of the two operations*, which is indeed the actual goal, will have the best characteristics. It should be noted, however, that our interest was not in finding the theoretically best approximation filters, but rather in finding useful, convenient, and fast low-pass differentiation algorithms.

Another important consideration was the phase distortion. In many cases it is essential to use filters which cause no phase distortion; this is particularly important when some features of the waveform (such as the time of occurrence or the amplitude of a particular peak) are to be subsequently examined. Keeping this consideration in mind, the study was restricted to only symmetric Finite Impulse Response (FIR) filters causing no time delay and no phase distortion within the pass band.

In the following sections, both first- and second-order differentiations are discussed. First, the theoretical optimum low-pass differentiation filters are discussed; these are used thereafter as a base for comparing the properties of various filters. Then the new low-pass differentiation algorithms developed are presented, showing their convenience for practical use due to their simple coefficients, as well as their near-optimal characteristics.

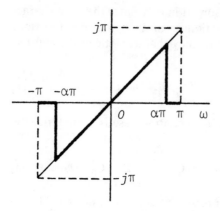

Figure 1 — The graph of $H_\alpha^{(1)}(\omega)$ for an arbitrary value of ω $(0 < \alpha \leq 1)$.

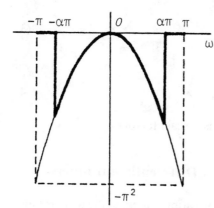

Figure 2 — The graph of $H_\alpha^{(2)}(\omega)$ for an arbitrary value of ω $(0 < \alpha \leq 1)$.

The Ideal Frequency Characteristics
and the Approximation Filters

The ideal low-pass frequency characteristics for first- and second-order differentiation are given by:

$$H_\alpha^{(1)}(\omega) = \begin{bmatrix} j\omega & |\omega| \leq \alpha\pi \\ 0 & \alpha\pi < |\omega| \leq \pi \end{bmatrix} \tag{1}$$

$$H_\alpha^{(2)}(\omega) = \begin{bmatrix} -\omega^2 & |\omega| \leq \alpha\pi \\ 0 & \alpha\pi < |\omega| \leq \pi \end{bmatrix} \tag{2}$$

where $\alpha\pi$ $(0 < \alpha \leq 1)$ denotes the upper limit of the differentiation band (i.e., the cut-off frequency), and j is the complex imaginary unit (see Figures 1 and 2).

The aim is now to find practical algorithms which approximate the ideal frequency characteristic. As we restricted ourselves here to only the case of FIR (nonrecursive) digital filters, the frequency response of the approximation filter will be a polynomial in $e^{-j\omega}$. Since the ideal characteristic of the first-order differentiation is an antisymmetric function, however, the general form of the frequency characteristic of the approximation filters reduces to:

$$F^{(1)}(\omega) = j \sum_{n=1}^{N} C_n \, \text{SIN} \, n\omega. \tag{3}$$

Similarly, since the second-order differentiation case is symmetric, the general form for this case becomes:

$$F^{(2)}(\omega) = \sum_{n=0}^{N} C_n' \, \text{COS} \, n\omega = C_0' + \sum_{n=1}^{N} C_n' \, \text{COS} \, n\omega. \tag{4}$$

The corresponding time-domain expressions (received by inverse Z transform) are given by:

$$Y_k^{(1)} = \frac{d}{2} \sum_{n=1}^{N} C_n \, (X_{k+n} - X_{k-n}) \tag{3'}$$

for first-order differentiation, and:

$$Y_k^{(2)} = \frac{d}{2} \sum_{n=0}^{N} C_n' \, (X_{k+n} + X_{k-n}) \tag{4'}$$

for second-order differentiation (d is a scale factor).

The Optimum Low-pass Differentiation Filters

In order to compare the 'closeness' of various filters to the ideal frequency characteristics, the mean square error measure which is defined as follows was used:

$$E(\alpha) = \int_{-\pi}^{\pi} |H_\alpha(\omega) - F(\omega)|^2 \, d\omega \tag{5}$$

Clearly, for any α, the smaller $E(\alpha)$, the better $F(\omega)$ approximates the ideal $H_\alpha(\omega)$. If Equations (1) to (4) are substituted into $E(\alpha)$ one receives the general error functions for first- and second-order differentiation. Now, by calculating for each α the coefficients $\{C_n\}$ which minimize these error functions, one can find the best approximation filters for any cut-off frequency $\alpha\pi$. This can be done simply by differentiating $E(\alpha)$ with respect to each of the coefficients, C_n, and equating to zero. However, minimum error over the whole range does not imply that the error for ω is minimal. But since the main interest here is in low-pass differentiation filters, this part of the range of ω is the most important for the purpose, and therefore the following restrictions were introduced, which were derived from the behavior of the ideal characteristics at $\omega = 0$:

1. For first-order differentiation:

$$\frac{dF^{(1)}(\omega)}{d\omega}\Big|_{\omega=0} = j, \text{ or according to Equation (3): } \sum_{n=1}^{N} nC_n = 1 \quad (6)$$

2. For second-order differentiation:

$$F^{(2)}(\omega)|_{\omega=0}, \text{ or according to Equation (4): } \sum_{n=0}^{N} C_n' = 0 \quad (7\text{-a})$$

and

$$\frac{d^2 F^{(2)}(\omega)}{d\omega^2}\Big|_{\omega=0} = -2, \text{ or according to Equation (4): } \sum_{n=0}^{N} n^2 C_n' = 2. \quad (7\text{-b})$$

By differentiating the error function under the respective restrictions, the optimum coefficients $\{C_n\}$ (i.e., the optimum filters) for any value of α were found. The filters thus received have the minimum possible mean square error over the whole range, $0 \leq \omega \leq \pi$.

Simple Near-optimal Low-pass Differentiation Filters for Practical Use

The main disadvantage of the optimal filters received above as well as most of the other well-known algorithms, is in their inconvenient coefficients. This is especially important when utilizing micro or minicomputers, when the simplicity of the calculation is an indispensable consideration in the choice of an algorithm. Therefore, presented here are some new low-order low-pass differentiation filters (first suggested by Usui and Ikegaya, 1978). These filters have mean square errors which are quite close to the theoretical minimum; the big advantage, however, is in the simplicity of their coefficients, all of which (except for one scaling factor) are very convenient integer numbers, such as 1 or -1, rather than troublesome floating point fractions. This keeps the algorithm so simple that the essential operations are addition, subtraction, or bit shift operations.

The basic formula of these filters is:

$$_Nf_k^{2L+1} = \frac{1}{2NT(2L+1)} \left[\sum_{n=-L}^{L} X_{k+n+N} - \sum_{n=-L}^{L} X_{k+n-L} \right]. \quad (8)$$

The interpretation of this formula is given in Figure 3. This algorithm is based on the difference between the symmetric weight moving averages taken around the points X_{k+N} and X_{k-N} with a span of $2L + 1$; it is denoted by $_Nf^{2L+1}$. The frequency characteristics of some interesting cases are plotted in Figure 4 along with those of some other algorithms, such as the Lanczos differentiation formula (Lanczos, 1956). Comparing the mean

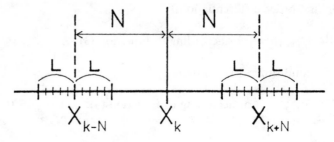

Figure 3 – The interpretation of the new differentiation algorithm.

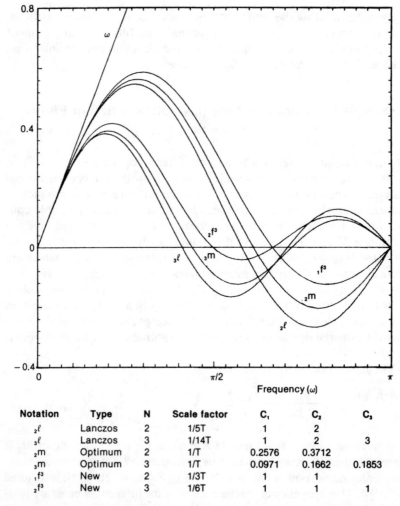

Notation	Type	N	Scale factor	C_1	C_2	C_3
$_2\ell$	Lanczos	2	1/5T	1	2	
$_3\ell$	Lanczos	3	1/14T	1	2	3
$_2m$	Optimum	2	1/T	0.2576	0.3712	
$_3m$	Optimum	3	1/T	0.0971	0.1662	0.1853
$_1f^3$	New	2	1/3T	1	1	
$_2f^3$	New	3	1/6T	1	1	1

Figure 4 – The frequency characteristics of some near-optimum first-order differentiation filters as compared to other filters.

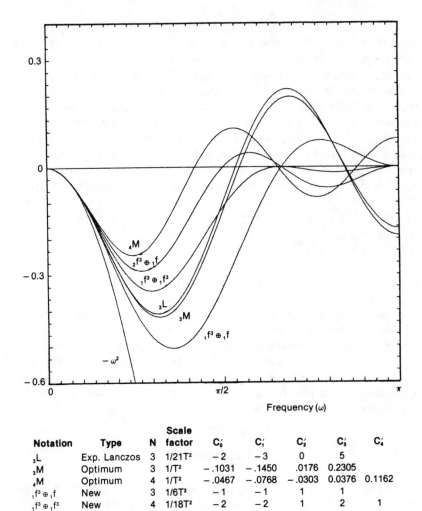

Notation	Type	N	Scale factor	C_0'	C_1'	C_2'	C_3'	C_4'
$_3L$	Exp. Lanczos	3	$1/21T^2$	-2	-3	0	5	
$_3M$	Optimum	3	$1/T^2$	$-.1031$	$-.1450$	$.0176$	0.2305	
$_4M$	Optimum	4	$1/T^2$	$-.0467$	$-.0768$	$-.0303$	0.0376	0.1162
$_1f^3 \oplus _1f$	New	3	$1/6T^2$	-1	-1	1	1	
$_1f^3 \oplus _1f^3$	New	4	$1/18T^2$	-2	-2	1	2	1
$_2f^3 \oplus _1f$	New	4	$1/12T^2$	-1	-1	0	1	1

Figure 5 — The frequency characteristics of some near-optimum second-order differentiation filters as compared to other filters.

square errors of those filters with their coefficients, one can see indeed the advantage of our near-optimal filters: While keeping the mean square error almost the same as in the best filters, these new filters have very simple coefficients, thus enabling very easy and fast calculation.

By applying these filters twice in cascade, corresponding simple near-optimum filters for second-order differentiation are received (Usui et al., 1979). This algorithm can be generalized even further, by temporarily also considering points between the actual data points. If one applies in cascade two first-order differentiation filters which make use of imaginary data points $X_{n+1/2}$ in between the actual points X_n, new second-order differentiation filters are received which finally make use of only the real data points, X_n. These new filters close the gap between the ordinary filters received in the usual way, also in terms of the frequency characteristics. Referring again to Figure 3, this generalization permits one to change N or L not only in integer steps, but also in half steps. When applying two such half-step point algorithms in cascade, the result is a full-step point filter for second-order differentiation. By investigating this new method some interesting low-pass second-order differentiation filters were received, which had an almost minimum error but still used very simple coefficients. The frequency characteristics for some of our second-order differentiation filters are plotted in Figure 5. Note that in both Figures 4 and 5, some of the new filters have been better characteristics than the corresponding optimum filters in the high frequency range.

Conclusion

Having in mind the special need of data processing in various biological or biomechanical applications using micro or minicomputers, some simple low-order low-pass differentiation filters are proposed. These filters are not only very practical for use, but are also almost optimum. The results of this study may be applied to any kind of empirical function differentiation in various fields of research, including biomechanical data processing.

References

ANDREWS, B., and Jones, D. 1976. A note on the differentiation of human kinematic data. Digest 11th Int. Conf. Med. & Biol. Eng. (Ottawa), pp. 88-89.

LANCZOS, C. 1956. Applied Analysis. pp. 321-331. Prentice Hall, London.

LESH, M.D., Mansour, J.M., and Simon, S.R. 1979. A gait analysis subsystem for smoothing and differentiation of human motion data. J. of Biomech. Eng. 101:205-212.

PEZZACK, J.C., Norman, R.W., and Winter, D.A. 1977. An assessment of derivative determining techniques used for motion analysis. J. of Biomech. 10:377-382.

USUI, S., and Ikegaya, K. 1978. Low order low-pass differentiation algorithm for data processing and its evaluation. Trans. IECE of Japan. E61(11):850-857. (in Japanese)

USUI, S., Ohzawa, I., and Ikegaya, K. 1979. Low order low-pass second order differentiation algorithm for data processing. Trans. IECE of Japan. E62(8):552-553. (in Japanese)

A Data Smoothing Method Using Spline Functions and Its Application in Motion Analysis

Shigeru Niinomi, Yoshitaka Suzuki, and Kazuo Tsuchiya
Labor Accidents, Prosthetics and Orthotics Center
Nagoya, Japan

Recently the importance of smoothing data has been widely recognized among the investigators in the field of motion analysis. Especially in order to derive useful physical quantities such as velocity and acceleration from the displacement data of human motion, it is necessary to smooth the data because of the existence of inherent noise and the lack of data points in the measurements which have an important effect on the accuracy of the calculations of the physical quantities.

The purpose of this study was to establish an effective data smoothing method using spline functions (S-functions). Previously, many investigators have reported on methods of smoothing data with S-functions and the validity in applications of analysis of human motion (Reinsch, 1967; Soudan & Dierckx, 1978).

S-functions are defined as the polynomials of a degree m connected smoothly at the points, called knots, so that m-1 order derivatives of them are continuous. S-functions have generally two fine properties. First, S-functions are simple polynomials, so they can be dealt with quite easily and their derivatives or integrals can be calculated simultaneously. The derivatives of S-functions are also S-functions. Second, a local behavior of S-functions does not severely influence the behavior of its S-function in other regions. That is, its behavior in different regions is almost unrelated. This property is contrary to that of usual polynomials.

There are, however, difficult problems in using S-functions to smooth data. Namely, one must determine the suitable number and locations of knots of S-functions with respect to a proper criterion. Various criteria and procedures to derive the (sub)optimal S-functions which smooth any given data have been considered (Reinsch, 1967; Ichida and Yoshimoto, 1979).

Described herein are trivial but effective propositions about the properties of S-functions, and a procedure to obtain the S-function which smooths data is proposed. This procedure does not necessarily give an optimal S-function but gives an effective and definite one. The necessary time to compute the smoothed data is also decreased by using this procedure.

The criterion adopted here is the squared variance for the computation of the coefficients of the S-function where the number and locations of knots are specified. For the determination of the number and locations of knots, the restricted squared variance is adopted as the criterion; the restriction is that squared variance should not be lessened under the specified value.

Definition and Basic Properties of Spline Functions

In this section, the definition and the basic properties of S-functions which are needed to obtain an effective data smoothing procedure are described.

Definition of S-function

Let R be a set of real numbers and let a set of real numbers $(x_1, x_2, ---, x_N)$ be a strictly increasing finite sequence. A function S: $R \to Y$ is called a spline function (S-function) of degree m with the knots $x_1, x_2, ---, x_N$, if S satisfies the following conditions where Y is the set of output values:

1. In each interval (x_i, x_{i+1}), for $i = 0, 1, 2, ..., N$ (where $x_0 = -\infty$ and $x_{N+1} = \infty$), S coincides with some polynomial of degree m or less.

2. S belongs to the class C^{m-1}, that is, its derivatives of order 1, 2, ..., $m - 1$ are continuous everywhere.

Definition of B-spline Function

Let $M_m(x; x_i) = (x_i - x)_+^m$, where $(x_i - x)_+^m = (x_i - x)^m$ for $x_i > x$, and $(x_i - x) = 0$ for $x_i \leq x$. The S-function S_i^m defined with the equation $S_i^m(x) = M_m(x; x_i, x_{i+1} ..., x_{i+m+1})$, $M_m(x; x_i, x_{i+1} ..., x_{i+m+1}) = (M_m(x; x_{i+1}, x_{i+2} ..., x_{i+m+1}) - M_m(x; x_i, x_{i+1} ..., x_{i+m})) / (x_{i+m+1} - x_i)$, is called a B-spline function of degree m. Note: B-spline function S_i^m has a local support, that is, S_i^m equals to 0 out of the interval (x_i, x_{i+m+1}).

Lemma

Any S-function, S, of degree, m, which has knots $x_1, x_2, ..., x_N$, is expressed uniquely as the linear combination of B-spline functions S_0^m, $S_1^m, ..., S_{N-m}^m$; $S(x) = \sum_{i=0}^{N-m} c_i S_i^m(x)$, $c_i \epsilon R$.

Corollary of the Theorem (Schoenberg-Whitney)

Let the values $y(t_1)$, $y(t_2)$,---, $y(t_n)$ ϵ Y be any given data where Y is the set of output values and $t_i(t_i < t_{i+1})$, i = 1, 2, ..., n, are sampling times and let x_1, x_2, ..., x_N be knots. Assume that closed intervals $[t_i, t_{i+1}]$, i = 1, 2, ..., n − 1, include more than m + 1 knots and N − m + 1 < n, then there exists a unique S-function, S, of degree, m, which minimizes the variance, σ^2, where $\sigma^2 = \sum_{i=1}^{n} (S(t_i) - y(t_i))^2/n$, and this S is expressed with the following equation: $S(x) = \sum_{i=0}^{N-m} c_i S_i^m(x)$, c = $(A^t A)^{-1} A^t y$, where c = $(c_0\ c_1\ ...\ c_{N-m})^t$, y = $(y(t_1)\ y(t_2)\ ...\ (t_n))^t$

$$A = \begin{bmatrix} S_0^m\ (t_1)\ S_1^m\ (t_1)\ ...\ S_{N-m}^m\ (t_1) \\ S_0^m\ (t_2)\ S_1^m\ (t_2)\ ...\ S_{N-m}^m\ (t_2) \\ ...... \\ S_0^m\ (t_n)\ S_1^m\ (t_n)\ ...\ S_{N-m}^m\ (t_n) \end{bmatrix}$$

In addition, the following relations are satisfied: $x_1 < t_{m+1}$, $t_1 < x_2 < t_{m+2}$,---, $t_{n-m-1} < x_N$, then the variance, $\sigma^{-2} = 0$. Note: On account of the corollary, the smoothed data $\bar{y}(t_1)$, $\bar{y}(t_2)$,..., $\bar{y}(t_n)$ are given as the linear combinations of the given data $y(t_1)$, $y(t_2)$,···, $y(t_n)$; $\bar{y} = A(A^tA)^{-1} A^t y$, where $\bar{y} = (\bar{y}(t_1)\ \bar{y}(t_2)$,···, $\bar{y}(t_n))^t$.

For specified knots, one can obtain an optimal S-function which fits any given data in the sense of least squared variance.

As for the problem of determining the number and locations of knots with respect to a reasonable criterion, there has been no analytical solution until now. Ichida and Yoshimoto (1979) have described procedures to obtain a suboptimal solution with respect to the criterion A.I.C.

Proposition 1

Let $y(t_1)$, $y(t_2)$, ..., $y(t_n)$ be any given data, and let $\{x_1, x_2, ..., x_N\}$, $\{x'_1, x'_2, ..., x'_M\}$ be two sets of knots which satisfy the following relations and the assumption in the preceding corollary:

$$x_1 = x'_1 = t_1, x_N = x'_M = t_n, N < M$$

If $x'_{k+1} - x'_k < x_{i+1} - x_i$ whenever $(x'_k, x'_{k+1}) \cap (x_i, x_{i+1}) = \varnothing$, for any k = 1, 2,···, M − 1 and i = 1, 2,···, N − 1, then the relation $\sigma'^2 < \sigma^2$ holds, where $\sigma^2 = \sum_{i=1}^{n} (S(t_i) - y(t_i))^2/n$ and $\sigma'^2 = \sum_{i=1}^{n} (S'(t_i) - y(t_i))^2/n$, S and S' are the S-functions for the knots $\{x_1, x_2,---, x_N\}$, $\{x'_1, x'_2, ..., x'_M\}$, respectively, which minimize the variances.

Proposition 2

Let $y(t_1)$, $y(t_2)$, ..., $y(t_n)$, x_1, x_2, ..., x_N and S be the same as in Proposition 1. If the number of knots is increased by adding a knot x' between x_i and x_{i+1} that is, the increased set of knots is $\{x_1, x_2, ..., x_i, x', x_{i+1}, ..., x_N\}$, under the condition that the increased set of knots satisfies the assumption of the corollary, then we have the following relation for any pair of intervals $[x_j, x_{j+1_j}]$, $[x_k, x_{k+1_k}]$, $x_j < x_{j+1_j} < x' < x_k < x_{k+1_k}$;

$$\left|\, e_{[x_j,\, x_{j+1_j}]} - e_{[x_k,\, x_{k+1_k}]} \,\right| \leq$$

$$\left|\, \bar{e}_{[x_j,\, x_{j+1_j}]} - \bar{e}_{[x_k,\, x_{k+1_k}]} \,\right|,$$

where $e_{[x_h,\, x_{h+1_h}]} = \sum_{t_h \in [x_h, x_{h+1_h}]} (S(t_h) - y(t_h))/a$, a is the number of t_h in the closed interval $[x_h, x_{h+1_h}]$ and $\bar{e}$ is defined similarly, except S is replaced with S'.

Lemma

Let x_1, x_2, ..., x_i, $x_{i+\Delta}$, ..., x_N be knots and S be any S-function of degree m which has these knots. If a limit $\Delta \rightarrow 0$ is taken then S belongs to class C^{m-2} not c^{m-1} in the neighborhood of the point x_i. In this case, the knot x_i is called the second multiple knot. The multiplicity of the knot is extended to more than 2, in such a case, the continuously differentiable order is lessened along with the multiplicity.

A Method to Obtain a Smoothing Spline Function

In this section, a simple but effective method to obtain the S-function is described which smooths given data based on the result of the previous section.

In practice, one can observe and measure human motion only for a finite time interval, and obtain a finite number of measurements. So the domain of S-functions is restricted to $[t_1, t_n]$ from here on. This can be done by choosing proper multiplicities of knots $x_1 = t_1$, $x_N = t_n$, so that the S-function is discontinuous at t_1, t_n.

The degree of S-functions can also be fixed at 2 or 3. The second or third degree are reasonable because one only computes first and second derivatives and the difficult problem of determining the degree m with respect to a proper criterion can be avoided.

For the specified knots, the variance is used as the criterion which should be minimized for the computation of the coefficients of the S-function. For the determination of the number and locations of knots, the variance is also adopted but the following restriction is added: min σ_i^2 > δ_1, $\sigma_i^2 = \sum_{t_j \in [t_i, t_{i+1}]} (S(t_j) - y(t_j))^2/a_i$, where a_i is the number of t_j in $[t_i, t_{i+1}]$.

In order to normalize the sampling time, let t_1, t_2, ..., t_n be the sampling times at which measurements are obtained. We correspond integers to these sampling times by the following relation: $t_i \rightarrow i$. We use these integers as a normalized time sequence and express the measurement at time t_i by the symbol $y(i)$ instead of $y(t_i)$. Note: By normalization of time one can compute and stock the matrix $A(A^t A)^{-1} A^t$ for the various numbers and locations of knots in advance. There is no additional restriction on sampling times with this normalization.

Procedure

1. Compute the $\bar{y}$ by the following equation for various numbers of knots which are placed uniformly on the interval $[t_1, t_n]$. Pick up the number of knots, N, such that for this set of knots the variance, σ^2, is minimized under the condition: min $\sigma_i^2 > \delta_1$.

2. If there are intervals $(x_i, x_{i+1}]$ on which $\sigma_i^2 - \min_i \sigma_i^2 > \delta_2$ where δ_2 should be determined empirically in advance, then increase the number of knots in the following manner: let $[x_i, x_{i+1}]$, $[x_{i+1}, x_{i+2}]$,..., $[x_{i+k}, x_{i+k+1}]$ be the succeeded intervals which satisfy the assumption. Fix the knots x_i, x_{i+k+1}, and place $k+1$ knots between the knots x_i, x_{i+k+1} uniformly after deleting the k knots x_{i+1}, x_{i+2},..., x_{i+k}.

3. If $\max_i e_i - \min_i e_i > \delta_3$, then place an additional knot between the intervals which correspond to these max e_i and min e_i under the condition that increased knots satisfy the assumption of the corollary previously given where δ_3 should be determined empirically in advance.

Application

In Figure 1 the data of trajectory of LEDs (light emitting diodes) attached at the anatomical points (trochanter, femoral condyles, and malleolus of the fibula) of the human body during gait (swing phase) are shown. These data are obtained as 3-dimensional coordinates by the instrumentation which consists of two cameras and a digital computer.

In Figure 2 the smoothed data obtained by using this method are shown. The derivatives of these displacement data are computed quite easily by the following equations: $dy(j)/dt = \sum_{i=0}^{N-m} (c_i dS_i^m(j)/dt) k_i$, $d^2y(j)/dt^2 = \sum_{i=0}^{N-m} (c_i d^2 S_i^m(j)/dt^2) k_i^2$, where $k_i = 1/(t_{i+1} - t_i)$. The velocity curves derived from these data are shown in Figure 3. In Figure 4, velocity curves computed directly from the unsmoothed data are shown. Comparison of these two results illustrates clearly the validity of smoothing.

The trajectory of the second LED is the curve of the y-coordinate

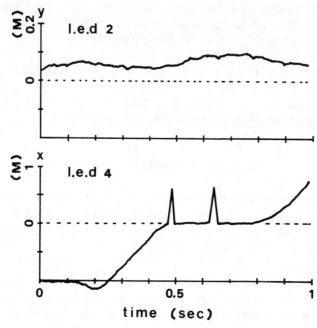

Figure 1—Trajectories of LEDs 2 and 4 which are attached at the trochanter and malleolus of fibulae, respectively.

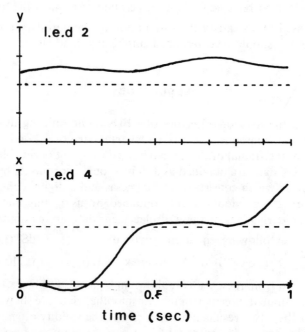

Figure 2—Smoothed curves of the trajectories of Figure 1.

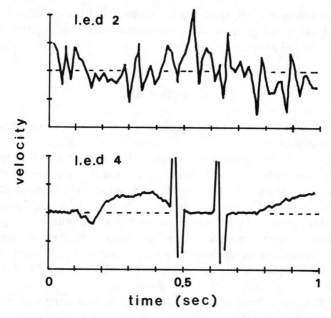

Figure 3 — Velocity curves derived from the trajectories of Figure 1.

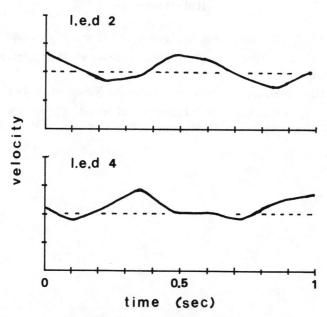

Figure 4 — Velocity curves derived from the smoothed curves of Figure 2.

which expresses the height of the trochanter, and the trajectory of the fourth LED is the curve of the x-coordinate which corresponds to the direction or the progress of the gait.

Conclusion

A data smoothing method with S-functions based on specific properties has been proposed.

The criterion for the determination of the number and locations of knots is reasonable and proper for calculation of an effective S-function quickly, but one cannot obtain analytically an optimal solution for this criterion. If little information about the noise is available and data points are lacking, then the A.I.C. is recommended as the criterion.

By normalizing the sampling times, one can obtain the smoothed data quite easily because all one has to do is calculate the linear equation $\bar{y} = A(A^t A)^{-t} A^t y$, where the matrix $A(A^t A)^{-1} A^t$ is computed and stocked in advance.

Note that in the field of motion analysis the smoothing method must be developed independently because it is difficult to obtain mathematically correct models of human motion as opposed to other fields such as system or control theory.

References

GREVILLE, T. 1969. Theory and Applications of S-functions. Academic Press.

ICHIDA, K., & Yoshimoto, F. 1979. Spline Functions and their Applications. Kyoiku-Shuppan.

REINSCH, C. 1967. Smoothing by spline functions. Numer. Math. 10:177-183.

SOUDAN, K., & Dierckx, P. 1978. Calculation of derivatives and fourie coefficients of human motion data while using spline functions. J. Biomechanics 12:21-26.

The Use of Cluster Analysis in Movement Description and Classification of the Backstroke Swim Start

B.D. Wilson and A. Howard
University of Queensland, St. Lucia, Australia

A number of studies have been published which investigate the technique of performing a skill, by measuring characteristics associated with predefined phases of the movement. These phases have been selected on the basis of divisions of the movement such as pre-flight, take-off, and flight, or on a rather more arbitrary division such as percentages of prepatory and propulsive phases of a motion.

Cluster analysis provides an objective procedure for classifying sets of parameters into groupings. The procedure is widely used in behavioral research. Davies (1978) used the cluster procedure to classify the postures adopted by the Flamingo. Davies interpreted the clustered postures, each cluster being represented by a modal action pattern (MAP), as being indicative of animal behavior. Even though he was interested in the frequency of occurrence of MAP's Davies did not investigate the temporal pattern of movement. We have used cluster analysis previously to describe differences between static postures adopted by minimal brain dysfunction children when horizontally supported.

The purpose of this study was to investigate the use of cluster analysis in describing the movement pattern in a dynamic movement, a backstroke swim start. The objective of the cluster procedure used was to describe the movement with a minimum number of MAP's which are different from each other and can be meaningfully related to the movement.

Table 1

Subject Details

Subject Number	Age (yrs)	Backstroke Swimming Ability	Height (cm)	Mass (kg)	Best 100 m Time (sec)
1	22	Commonwealth Games Champion 1978 (100 m)	193	86	57.9
2	21	State Age Sprint Finalist	188	95	74.0
3	18	State Sprint Champion	186	73	68.0
4	18	State Age Champion	186	78	61.5
5	17	Australian Age Champion	184	70	60.5
6	22	Ranked No. 10 in the World *(100 m)	185	80	57.1
7	17	State Finalist	189	82	71.2
8	18	State Finalist	175	66	72.3
9	22	Ranked No. 7 in the World *(200 m)	194	84	60.1
10	17	State Age Champion	185	72	63.0

*as of June, 1979 (Swimming World Vol. 20 No. 7 July, 1979).

Methods

Subjects

Ten skilled male backstroke swimmers aged between 17 and 22 years were used as subjects. They ranged in backstroke swimming ability from an Australian State finalist to a world ranked No. 7 200 m swimmer (see Table 1 for further detailed descriptions of the subjects). The study was conducted during the Australian swimming team's training camp preceding the 1980 Olympic Games.

Testing Procedure

Following a five-minute swimming warm-up and two practice starts, the subjects performed three trials of their usual variation of the FINA backstroke start and a 12 m sprint. All trials were filmed at a nominal 100 frames/sec using a Photosonics IPL camera fitted with a 100 Hz timing light pulse generator. Subjects were started by an experienced starter under simulated racing conditions. The complete testing procedure is described by Hooper (1981).

Analysis of Data

For each subject, film of two trials was digitized using a PCD motion analyzer connected to a Hewlett Packard 2648 Graphics terminal and PDP 11/34 minicomputer. The coordinates of segment end points for a 14-segment body model were obtained from the frames showing gun firing (smoke emitted) and for every third frame through to head entry. The digitized data were smoothed using a double pass moving three-point averaging method. Vectors, in 2D space, were then computed in terms of segment length and angles between adjacent segments for each body segment for each frame analyzed. These vectors are shown in Figure 1. In subsequent cluster analysis the body position adopted in each film frame was described as a location in 14-dimensional quasi-Euclidean space. The term 'quasi' is used since some of the component angles defining the body position were treated as circular rather than linear measures.

The cluster analysis procedure involves a number of inter-active programs written in UNIX FORTRAN. These programs may be obtained on request from the authors.

The CLUSTER 1 program takes data consisting of the 14 components of each vector (NP) for the 10 subjects and two trials (NS) at each of the frames analyzed (NT) and clusters the NS × NT events.

An agglomerative hierarchical method was used in the cluster programs (Mardia et al., 1979). In successive steps in the cluster program the two closest items or centroids were placed in a cluster. The criterion of condensation of a pair of clusters to form a new cluster was based on the minimum distance in the 14-dimensional space. DENDOGRAM and STICK programs were used to identify and display the clusters or MAP's.

Chi-square and t-tests at the $p \leq 0.01$ level were used to assess the significance of cluster separation. Chi-square tests were also applied to the output of CLUSTER 1 to test the significance of variation between and within subjects as shown by contingency tables of subject-trial by clusters.

Results and Discussion

With CLUSTER 1, large increases in cluster separation were observed in grouping the data into a small number of clusters. Figure 2a shows an exponential trend to decreasing cluster separation with increasing cluster number. This trend is not shown in Figure 2b for cluster separations

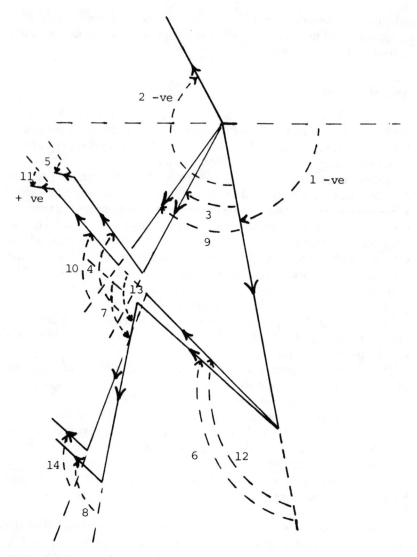

Figure 1—Definition of body segment vectors.

based on random number data, which would not be expected to form clusters. Large cluster separations were noted in the transitions to form 9, 16, 21, and 41 clusters (see Figure 2a). The DENDOGRAM program revealed that a number of clusters were clusters containing one or two items and represented outliers in the data. For example, the leftmost item of data in the dendogram shown in Figure 3 is not incorporated into any other MAP until four groups have been formed. The vertical separation is an indication of the extent of cluster separation (1 to 9 is a scaling of

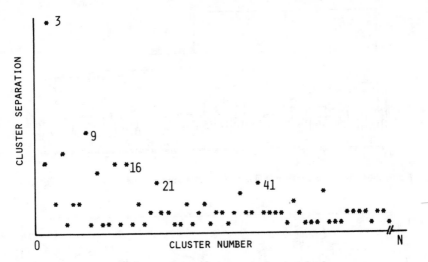

Figure 2a—Cluster separation vs number of clusters (swim start data).

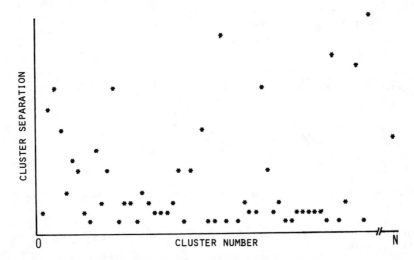

Figure 2b—Cluster separation vs number of clusters (random data).

the significance of a t-test of no difference between clusters). The horizontal bars correspond to events or film frames × subjects. The broken line indicates the level of separation for the six clusters shown.

It should be reemphasized that the objective of the analysis was to describe the movement with a minimum number of MAP's which are different from each other and can meaningfully be related to the movement. Twenty-one MAP's (see Figure 4) were required to distinguish differences between the starting techniques used. The movement of all but one

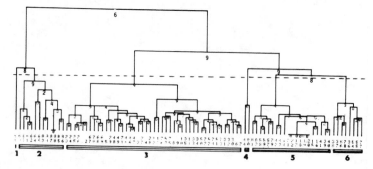

Figure 3—Sample cluster DENDOGRAM for 22 frames/4 trials.

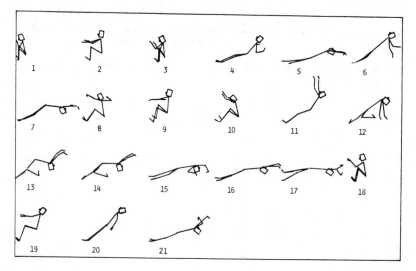

Figure 4—Stick figure output of the 21 modal action patterns produced by CLUSTER 1.

of the subjects, however, could be described with six MAP's (see Figure 5).

The movement pattern for each of the subjects is shown in Figure 5. The numbers refer to the STICK diagram numbers of Figure 4. Eight of the ten subjects used the sequence 18, 19, 16 with the other three subjects adopting an intermediate position, 20, indicative of a greater range of movements. Significant differences between trials and between subjects were noted. Three different starting positions were adopted by the subjects represented by MAP's 1, 3, and 18 in Figure 4. All subjects, however, used starting position 18 in at least one trial. Two different arm movement patterns, over and around arm, as well as asymmetry in arm movements, were noted. Marked inconsistencies in movement pattern between trials were noted in subjects 6, 8, and 9 ($p < 0.01$).

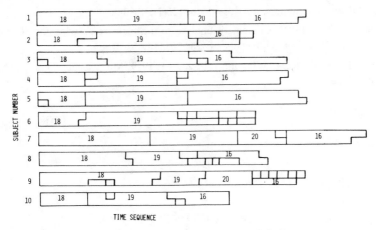

Figure 5 — Movement pattern cluster number, sequence, and timing.

In a correlational study of 21 kinematic variables and anthropometric measures of these 10 subjects, no correlation with time to 8 m was observed for any parameter (Hooper, 1981). Significant differences between subjects were, however, noted for almost all the variables. Description of these differences between subjects proved to be voluminous and the significance of such differences difficult to interpret.

Cluster analysis provides an objective procedure for providing a condensed description of the movement and of describing differences between subjects and trials. The procedure is not limited to clustering postural data. The parameters describing items or events to be clustered do not need to be of the same form or of equal weighting. Analysis with the vectors for the feet given a zero weighting reduced the number of one- and two-item clusters from 14 to 5 and did not change the description of the movement pattern.

A disadvantage of the cluster analysis is that it is purely topological; however, the modal action patterns represented by the stick diagrams can be recovered from the data. A major limitation may be one of computing cost or capacity of minicomputers. In the present problem, clustering 560 events in 14D space required approximately 2 hours computing time on a PDP10 and 30 hours on a PDP 11/34 computer system. In spite of this limitation, the cluster procedure has been demonstrated to be a practical method for describing phases of a movement and differences between subjects. An extension of cluster analysis is to group the subjects according to similarity of movement pattern.

References

DAVIES, W.G. 1978. Cluster analysis applied to the classification of positives in the Chilean Flamingo. Animal Behavior, 26(2):381-388.

HOOPER, S.L. 1981. A kinematic analysis of the backstroke swim start. Unpublished MHMS Thesis, University of Queensland, Australia.

MARDIA, K.V., Kent, J.T., and Bibby, J.M. 1979. Multivariate Analysis. Academic Press.

Determination of Parameters from Intensity Time Plots

K. Nicol

Westfälische Wilhelms-Universität Münster, FRG

There are two main groups of methods used in biomechanics: the measuring and statistical method, and the method of mathematical modelling. When using a method from the first category, physical quantities such as velocities, accelerations, force, etc. vs time are normally recorded. The graphic displays of these measurements are called intensity time plots. From these plots a few parameters are derived which undergo statistical treatment. Though it is uncommon to call this first category of methods 'modelling', it is modelling in certain aspects. The complex reality is described by means of a simpler picture, a so-called model. First the complex event, for instance a sport movement is described by means of variables of external biomechanics, that is by the movement of all body surface points and by all external forces acting on the body. Secondly, this general description of movement and forces is reduced to some points and some forces represented by intensity time plots. Third, the contents of information of these time plots containing (in principle, indefinite) in reality several hundred measurements of intensity are reduced to several parameters. So the whole complex event is described by these few numbers, the 'model' of the event.

In this approach, the selection of the variables is very important. It is primarily carried out using two different strategies. First, the particular variables are evaluated, which correspond to visible characteristics and are therefore used by therapists and coaches, characteristics such as extreme rates of joint angles and rates of angles at the beginning and the end of movement phases. The second group includes variables which might be important according to considerations from the fields of mathematical modelling, such as the integral over the force which is proportional to the impulse or the decline of the velocity time plot and therefore is a measure of the load exerted on the movement system. A third group of variables may be taken if the considerations leading to the variables in the first and the second group cannot be applied. The criteria are often taken from the mathematical discussion of plots so zeros, extreme values, ascent, etc. are evaluated. Moreover, from these 'primary parameters' 'secondary parameters' are derived by mathematical combinations of primary parameters, such as the maximum force divided by the average force.

The parameters just described may be taken from the plot as a total, but often the movement is divided into phases and the variables are taken for each phase separately. In this way, a great number of variables can be evaluated which hopefully include the most important information from the intensity time plots as a means of describing the movement. But as there is no criterion concerning which variables should be evaluated, the tendency is to evaluate as many variables as possible from an economical point of view, and to subject them all to statistical treatment. For instance, when an investigation of the shot put was conducted (Ballreich et al., 1980) 46 variables were derived from one force-time plot. Moreover, as it was not reasonably certain that every important parameter was evaluated, it was necessary to store all the intensity time plots for several years, thus wasting a great deal of computer storage space for a long time.

Method

The new method that shall be described here is directed at these two points. Variables shall be derived that contain, in principle, all information that is contained in the original intensity time plot so that there is no need for storing the intensity time plot itself. Moreover, from these few variables every other variable can be derived later on when needed.

Normally, one intensity-time plot does not have to be evaluated alone. As there are a great many plots which are similar, they only have to be compared with each other. So one 'reference plot' s(r) is taken out of the number of plots x(y) to be compared and mathematical variables are derived for every plot x(y) forming an image of the actual plot to the reference plot. By doing this, one single reference plot and one set of variables for every plot to be compared contain the same information as all plots together.

There are several possibilities for obtaining the variables according to the principle outlined. Our method was selected under the point of view that: 1) variables should be used which in turn coincide with conventional variables so that they may be used for interpretation without transformation, 2) a method should be followed which can be easily understood by amateurs in the fields of physics and mathematics (which, for instance, is not the case with the Fourier and La Place transformations which are somtimes used), and 3) a method should be used which requires only a little time on the computer. This is important because the method should be used on inexpensive microcomputers which normally use 'BASIC,' a slow language.

For better understanding the method is described in detail in relation to intensity time plots which are similar to the vertical force while walking. The force time plot has two maxima and one minimum and is zero before and afterwards. The origin of the time axis is given by an external signal at any time before the intensity time plot becomes different from zero.

First, the time plot measured is divided into phases which in this case (1) begin at:

x_b — where the plot becomes different from zero for the first time,
x_1 — the first maximum, x_2 — the minimum, x_3 the second maximum, and which (2) end at $x_1 - 1$, $x_2 - 1$, $x_3 - 1$ and x_e, where the plot becomes zero again, respectively.

Secondly, the following mathematical operations are applied to every phase of the measured plot separately. This will be illustrated for Phase 1.

1. Horizontal shift and linear stretch. Every x-rate of Phase 1 is imaged to the homologous r-rate by means of the transformation equation

$$x = x_b + \frac{x_1 - x_b}{r_1 - r_b} r = x_b + f \cdot r \tag{1}$$

So q(r) is obtained (see Figure 1).

2. Vertical, linear, and square stretch. Every q-rate of Phase 1 is transformed by

$$u(r) = a \cdot q(r) + b \cdot q(r)^2 \tag{2}$$

in such a way that

$$u(r_b) = s(r_b) = 0; \quad u(r_1) = s(r_1) \quad \text{and} \tag{3}$$

$$\int_{r_b}^{r_1} u \, dr = \int_{r_b}^{r_1} s \, dr \tag{4}$$

In this way: 1) all abscissae of the measuring plots are imaged to the abscissa of the reference plot, 2) the ordinates of both plots are equal at the beginning and at the end of the phase, and 3) both plots have the same surface. With these requirements the coefficients of Equation 2 can be calculated:

$$a = (s_1 - b \cdot y_1^2) / y_1 \tag{5}$$

$$b = \frac{s_1 / y_1 - I_s / I_y \cdot f}{y_1 - S / I_y} \tag{6}$$

where $s_1 = s(r_1)$, $y_1 = y(x_1)$, $I_s = \int_{r_b}^{r_1} s \, dr$, $I_y = \int_{x_b}^{x_1} y \, dx$, and $S = \int_{x_b}^{x_1} y^2 \, dx$.
Then the transformations are applied to the other phases where y and s are not related to zero in general, but to the minimum of the phase. The u-plot that is finally yielded equals the reference plot at least for nine points and has an integral equal to the integral of the reference plot which means that both are basically the same. In this way every plot can be represented with a certain amount of error by one single reference plot and by the following 16 parameters:

2 zero rates	x_b; x_e
6 rates of extremes	x_1; x_2; x_3; y_1; y_2; y_3
4 rates of integrals $I_{y, i, k} = \int_{x_i}^{x_k} y \, dx$	$I_{y, b, 1}$; $I_{y, 1, 2}$; $I_{y, 2, 3}$; $I_{y, 3, e}$
4 rates of square integrals $S_{i, k} = \int_{x_i}^{x_k} y^2 \, dx$	$S_{b, 1}$; $S_{1, 2}$; $S_{2, 3}$; $S_{3, e}$

All parameters are independent of the reference plot so no decision has to be made when the parameters are calculated. The most suitable plot, i.e., a very typical one, can be selected later as a reference plot under the condition of minimizing errors of approximation.

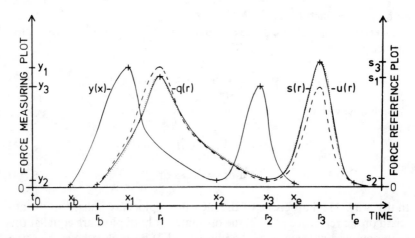

Figure 1—Typical force time plots of a pilot study. x(y) = one out of 100 measured plots (left ordinate), q(r) and u(r) = x(y) after horizontal and vertical transformation, respectively (right ordinate), and s(r) = the reference plot.

Accuracy of the Method

In order to assure the accuracy of the method, the following checks should be carried out:

1.

$$y(x) = c \cdot x \quad x = 0 \ldots x_1$$
$$s(r) = d \cdot r^2 \quad r = 0 \ldots r_1$$

The coefficients are calculated by

$$a = 0$$
$$b = s_1 / y_1^2$$

So the approximating plot is

$$u(r) = d \cdot r^2$$

In other words, u is identical with the reference plot so that there is no error at all.

2.

$$y(x) = c \cdot x^2 \quad x = 0 \ldots x_1$$
$$s(r) = d \cdot r^3 \quad r = 0 \ldots r_1$$

The coefficients obtained are

$$a = \frac{3}{8} \frac{s_1}{y_1}$$

$$b = \frac{5}{8} \frac{s_1}{y_1^2}$$

The approximating plot is

$$u(r) = d \cdot r_1^3 \left[\frac{3}{8} \left(\frac{r}{r_1} \right)^2 + \frac{5}{8} \left(\frac{r}{r_1} \right)^4 \right]$$

The comparison with the reference plot

$$s(r) = d \cdot r_1^3 \left(\frac{r}{r_1} \right)^3$$

gives a maximum error rate of 0.017%.

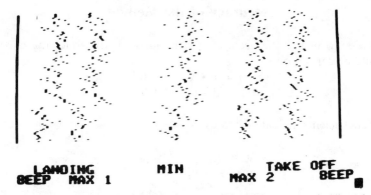

LANDING BEEP MAX 1   MIN   MAX 2 TAKE OFF BEEP

Figure 2 — Time parameters (horizontal axis, one cycle length between vertical lines) for 100 cycles, beginning at the top. In terms of Figure 1 from left to right: t_o, x_b, x_1, x_2, x_3, x_e, t_o'.

Application

For the purpose of demonstration, a pilot study was carried out using a measuring system for biomechanics described in Nicol and Liebscher (1983) for measurement and the method just described for evaluation. Besides other instrumentation the biomechanical system consists of a capacitance force platform for vertical force, a capacity/voltage transducer, and a microcomputer. A program was developed that: 1) activates the loudspeaker of the computer with a frequency of 1 Hz to indicate cycles, 2) measures the force 256 times per cycle, and 3) makes an extra sound when 110 cycles have been carried out to indicate the end of measurement. The subjects were asked to perform one two-phase vertical jump per cycle. Two examples of the force time plots obtained are shown in Figure 1 [plots y(x) and s(r)]. The 16 parameters described above were evaluated for the cycles 5 to 104. The evaluation of one single plot by means of a program that was written in the 'BASIC' language took about 30 sec; using 'FORTRAN' or 'ASSEMBLER' programs the evaluation time can be reduced to about 5 sec. After one typical plot has been selected as a reference plot, all other plots can be removed from the computer, as the parameters and the reference plot contain the same information as the raw data. All other variables which might be needed in the future such as ascent rates, impulse in arbitrary time intervals, half time rates, etc. can be derived from the reconstructed plots.

A distinction has already been made between primary variables which are derived from data and secondary variables which are derived from primary variables mathematically. Using the technique described, the number of primary variables can be limited to 16. Their prime task is to maintain the information of measurement, whereas the secondary variables are tailored to the purpose of interpretation. Nevertheless, the primary variables can be used for interpretation too, because they were selected from this standpoint.

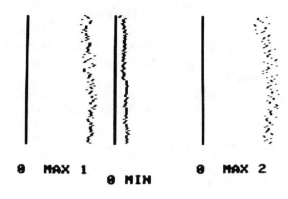

Figure 3 — Extremes y_1, y_2, y_3. The line left of each plot indicates zero.

Figure 2 shows on the horizontal time axis from left to right the beginning of the sound (zero point of the axis), the time of landing, first maximum of the force-time plot, the minimum, the second maximum, the time of take-off, and the beginning of the sound of the next cycle. The 100 measures are plotted one below the other, and the first measure is the top most line. It can be seen that the parameters of the plot show a comparatively high degree of accuracy, but they vary greatly in relation to the time of the sound. This variation is in part periodical. The period increases regularly from 7 to 15 cycles.

Figure 3 shows in the same arrangement the extreme rates of the plots. The minimum is relatively constant. The maxima show the same tendency: an increase at the end of the measurement.

Outlook

The primary and the secondary variables can be subjected to statistical treatment as usual. As the primary variables already contain all the information related to any single plot, it is very possible that in some cases the statistical treatment of the primary variables already contain information concerning the results of the treatment of the secondary variables although this treatment has not yet been carried out. This is, for instance, the case for the statistical mean value if the secondary variable is a linear combination of N primary parameters, as

$$M_s = \sum_{n=1}^{N} a_n \cdot M_{p,\,n}$$

(M_s = mean value of the secondary variable, and $M_{p,\,n}$ = mean value of the primary variable n). This question will be considered in the future.

References

BALLREICH, R., Nicol, K., & Ernst, H. 1980. Kugelstoss Maenner (men's shot put). In: R. Ballreich and A. Kuhlow (Eds.), Beitraege zur Biomechanik des Sports. Schorndorf.

NICOL, K., and Liebscher, F.F. 1983. An integrated system for biomechanics designed for small working groups and for teaching. Journal of Human Movement Science. (in press)

Day-to-Day Reproducibility of Selected Biomechanical Variables Calculated from Film Data

J. Grainger, R. Norman, D. Winter, and J. Bobet
University of Waterloo, Ontario, Canada

One of the most serious and justified criticisms of much of the bio-mechanical literature has been that generalizations concerning movement phenomena have often been based on case studies or from measurements of a small number of trials on too few subjects. One of the reasons for this is that cinematography has been heavily employed as a data acquisition method and, notwithstanding the relatively recent development of on-line or semi-automated data acquisition systems, it still takes a relatively long time to reduce film data. This use of a small number of trials and subjects can only be justified if the reliability of the cinematographic variables is high. Despite this, there are very few cinematography studies in the literature in which trial-to-trial, day-to-day or subject-to-subject variability has been reported on kinetic variables. Such reports have occasionally appeared on temporal and kinematic variables (Winter et al., 1974; Bates et al., 1979; Foley et al., 1979). The purpose of this study was to examine the trial-to-trial and day-to-day variability of selected kinematic and kinetic biomechanical variables obtained from film analysis. This work was part of a larger project in which mechanical energy, metabolic cost, and EMG activity were studied during prolonged load carrying.

Methodology

Four physically active males (age 20.8 ± 1 years; weight 72 ± 3.4 kg) each carried loads of 20, 25, and 32 kg placed, alternately, high and low (center of mass at the level of the ear lobe and xiphoid process, respectively) on their backs. The back pack consisted of a modified pack frame with fully adjustable shoulder and belt straps. To change the magnitude and position of the load, lead weights were fastened to the frame.

All subjects walked over a level, indoor, 90 m route for 110 min. They maintained a target velocity of 1.56 m/sec (5.6 km/hr) by using their lap time as feedback. Cadence and step length were allowed to vary. At least three days were allowed between testing sessions for any one subject.

Two non-consecutive strides were filmed in the sagittal plane by means of a stationary 16 mm camera (50 frames/sec) after 15 and again after 107 min of walking. An eight-member linked segment model [feet, shanks, thighs, head, arms and trunk combined (HAT), pack] was digitized, scaled to life size, converted to absolute coordinates and smoothed by means of a Butterworth, fourth order, recursive, low-pass digital filter cutting off at 2.5, 3.0, or 3.6 Hz, depending upon the anatomical marker. A complete kinematic analysis of the model was followed by the calculation of the instantaneous potential and kinetic energy levels for each segment.

Work rates (energy change per stride divided by stride time) were then calculated in two ways. The mechanical work, assuming passive energy transfers both within and between segments (Wwb), was determined by first summing the potential and kinetic energy curves within each segment, then across all 'segments' to produce a single total body energy curve. The absolute changes in this curve were then summed to produce the work output for the stride. The second method summed the absolute energy changes during the stride separately for each energy component of each 'segment', then added these sums to give a total body work output. The implicit assumption in this calculation was that there were no energy transfers (Wn). For a more detailed discussion of the above techniques and necessary equations the reader is referred to Pierrynowski et al. (1980). In all, 104 different strides were studied in the above way including a repeat, for each subject, of one of the load magnitude/load placement combinations. Only the data for these repeated trials are treated here.

Results and Discussion

The values of selected variables are reported in Table 1.

Day 1 to Day 2 Differences

The two trials for each subject were pooled and the Day 1 to Day 2 differences averaged across subjects were analyzed using a matched pairs t-test in which the significance level was set at 0.10 to improve the power of the test. None of the dependent measures proved significantly different from Day 1 to Day 2. Scanning of the Day 1 to Day 2 means in Table 1 show virtually no difference (less than 1%) in the average veloci-

Table 1

Subject × Trial × Day Means and Standard Deviations of Selected Variables

		A Day 1	A Day 2	B Day 1	B Day 2	C Day 1	C Day 2	D Day 1	D Day 2	X Day 1	X Day 2
Average velocity (m/sec)	T_1	1.49	1.55	1.58	1.55	1.55	1.56	1.56	1.53		
	T_2	1.52	1.56	1.59	1.62	1.55	1.53	1.59	1.63		
	X	1.51	1.56	1.59	1.59	1.55	1.55	1.58	1.58	1.56 (.04)[1]	1.57 (.02)
Stride length (m)	T_1	1.46	1.49	1.58	1.58	1.68	1.66	1.66	1.65		
	T_2	1.49	1.5	1.59	1.59	1.67	1.65	1.59	1.66		
	X	1.48	1.5	1.59	1.59	1.68	1.66	1.63	1.66	1.6 (.09)	1.6 (.08)
Shank angle at heel strike (deg)	T_1	79	81	78	84	76	72	81	82		
	T_2	83	80	74	80	85	76	86	83		
	X	81	81	76	82	81	74	84	83	81 (3)	80 (4)
Shank peak angular velocity during swing (rad/sec)	T_1	7	7.10	7.35	6.64	7.1	7.79	7.31	7.07		
	T_2	6.72	6.99	7.06	6.99	7.09	7.28	7.3	7.28		
	X	6.86	7.05	7.21	6.82	7.1	7.54	7.31	7.18	7.33 (.46)	7.15 (.3)
Mechanical work rate (W_n) (watts)	T_1	514	511	544	500	499	547	448	464		
	T_2	502	546	492	496	551	532	451	490		
	X	508	529	518	498	526	540	450	477	501 (34)	511 (29)
Mechanical work rate (W_{nb}) (watts)	T_1	86	103	150	95	86	70	77	58		
	T_2	92	124	107	124	71	61	81	62		
	X	89	114	129	110	79	66	79	60	94 (24)	88 (28)

[1](Standard Deviation)

Table 2

Single Trial and Averages of 2, 3, and 4 Trials of Selected Variables

Variable	Subject	T_1	T_1+T_2	$T_1+T_2+T_3$	$T_1+T_2+T_3+T_4$
Shank angle	A	79	81	81	81
at heel strike	B	78	76	79	79
(deg)	C	76	81	78	78
	D	81	84	83	84
	X	79	81	80	81
	SD	2.1	3.3	2.2	2.6
Shank peak	A	7	6.86	6.94	6.96
velocity	B	7.35	7.21	7.02	7.02
(rad/sec)	C	7.1	7.1	7.33	7.32
	D	7.31	7.31	7.23	7.25
	X	7.19	7.12	7.13	7.14
	SD	.7	.19	.18	.17
Mechanical	A	514	508	509	519
work rate;	B	544	518	512	508
W_n (watts)	C	499	526	532	533
	D	448	450	454	464
	X	501	501	502	506
	SD	40.1	34.5	33.4	29.8
Mechanical	A	86	89	94	102
work rate;	B	150	129	117	120
W_{wb} (watts)	C	86	79	76	73
	D	77	79	72	70
	X	100	94	90	92
	SD	33.8	23.8	20.5	24

ty and stride length. This indicates that the self-pacing technique worked extremely well in enabling the subjects to maintain the target velocity of 1.56 m/sec.

The individual angular velocity and work rate values show a day-to-day variation of only 2.5 to 6.8%. Even the individual subject Day 1 to Day 2 values, averaged over two trials per subject, showed maximum differences of less than 10% for all of the variables except Wwb which showed a difference of 27.9% (Subj. A).

Trial-to-Trial Differences

As would be expected the Trial 1 to Trial 2 differences were higher than the Day 1 to Day 2 differences averaged across the two trials, but were still 11% or less, again except for Wwb which now showed a maximum difference of 40% (Subj. B, Day 1). Since no statistically significant differences between the Day 1 and Day 2 mean values were detected for any variable, as noted previously, the four trials for each subject were pooled and the means of 2, 3, and 4 trials for some of the variables were calculated. These and the single trial scores appear in Table 2.

Inspection of the data shows that the mean of three trials is not appreciably different from the mean of four trials for most subjects on all of the variables. Indeed, the mean of two trials is similar to the mean of three on most variables. The single trial values are much less stable. This is particularly true of the Wn and Wwb values where the segment velocities are squared and in the latter case the individual segment energy curves were summed. Slight phase differences in these energy curves among the segments, from trial to trial have an appreciable effect, therefore, on the Wwb values even though the amplitudes of the segment kinematic time histories may be nearly identical from trial to trial.

Examples of two trials on two days of the shank angular velocity and total body energy time histories for one complete stride are shown in Figures 1 and 2.

It should be noted that the two trials are not consecutive walking strides but rather are separated by about 50 strides. This notwithstanding, the four angular velocity curves are nearly overlays. On the other hand, the total body energy curves are somewhat less consistent from trial to trial, presumably due to effects of the phase angle differences noted above.

Other kinetic variable time histories such as joint torque curves show more trial-to-trial variability than the energy curves (Winter, 1981). Figure 3 shows plots of the moments of force about the ankle, knee and hip for a normal subject walking at a natural cadence. The first trial (WN35A) was assessed a month before the latter four trials (WN35K,L,M,N) each done minutes apart and which replicated the cadence and stride length of the earlier trial. In spite of the fact that the kinematics appeared identical there were drastic changes in the moment of force patterns at the hip and knee. The knee showed a resultant extensor torque during all of the stance for the earlier trial, but primarily a flexor torque during the latter trials. The hip had corresponding compensatory changes to prevent the knee from collapsing in the same trials. The torque curves, of course, are based on second derivative data while the energy curves are based, to a large extent, on squared first derivatives of displacement data. Consequently, slight slope differences in the associated displacement curves, often not even noticeable to the eye, become amplified in the calculations.

Conclusions

These conclusions are tentative until more subjects are evaluated.

1. Even in a well-learned motor task such as normal walking, single

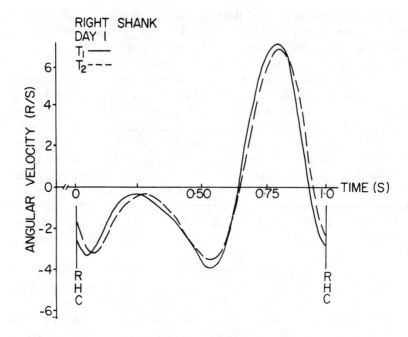

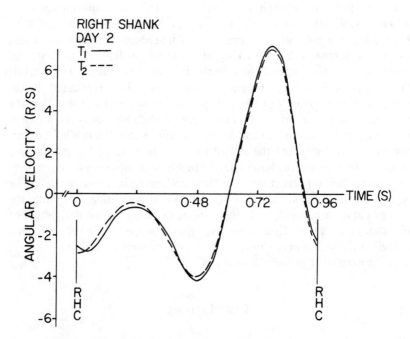

Figure 1 – Trial-to-trial and day-to-day shank angular velocity time histories for one walking stride at 5.6 km/hr. RHC = right heel contact.

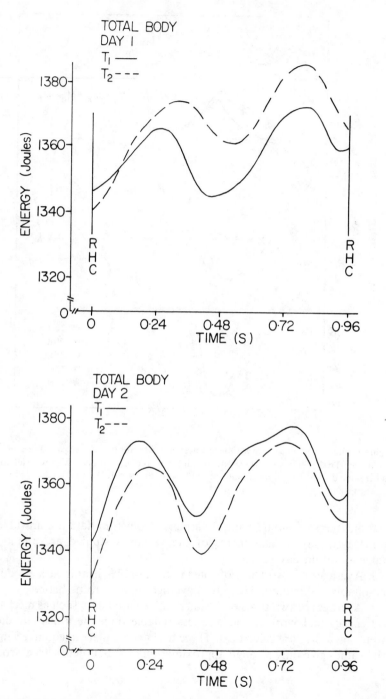

Figure 2 – Trial-to-trial and day-to-day total body energy time histories for the same stride as in Figure 1.

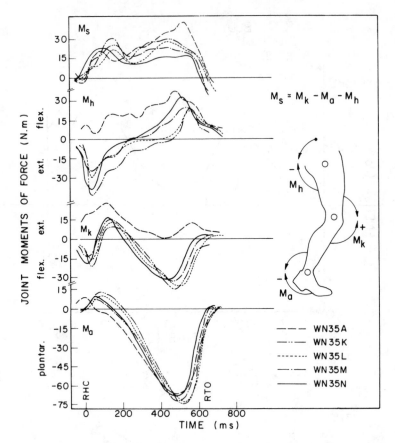

Figure 3 — Joint moments of force time histories of 5 trials for one subject walking at about 4.8 km/hr. Trial WN35A was done one month before the other trials, which were all done on the same day. MS = support moment; RTO = right toe-off. With permission, Physiotherapy Canada.

trial measures of biomechanical variables from film data are unstable.

2. Day-to-day measurements of the variables studied are about as stable as within day values.

3. Averages of two trials of kinematic variables in this task produce reasonably stable data. Three trial averages appear to be better.

4. Averages of two trials of first derivative based data such as mechanical energy and work output are less reliable than the kinematic data from which they are calculated. Three trial averages are adequate. Single trial values give only a rough approximation of the subject's 'true' score.

Acknowledgment

This work was funded by the Canadian Department of Supply and Services Contract No. 85U79-00059 and coordinated through the Defense and Civil Institute of Environmental Medicine.

References

BATES, B.T., Osternig, L.R., Mason, B.R., and James, S.L. 1979. Functional variability of the lower extremity during the support phase of running. Medicine and Science in Sports 11(4):328-331.

FOLEY, C.D., Quanbury, A.O., and Steinke, T. 1979. Kinematics of normal child locomotion — A statistical study based on TV data. J. Biomechan. 12(1):1-8.

PIERRYNOWSKI, M.R., Winter, D.A., and Norman, R.W. 1980. Transfers of mechanical energy within the total body and mechanical efficiency during treadmill walking. Ergonomics 23(2):147-156.

WINTER, D.A. 1981. Use of kinetic analyses in the diagnostics of pathological gait. Physiotherapy Canada 33(4):209-214.

WINTER, D.A., et al. 1974. Kinematics of normal locomotion — A statistical study based on TV data. J. Biomechan. 7:479-486.